The Cricket as a Model Organism

Hadley Wilson Horch • Taro Mito
Aleksandar Popadić • Hideyo Ohuchi
Sumihare Noji
Editors

The Cricket as a Model Organism

Development, Regeneration, and Behavior

Springer

Editors
Hadley Wilson Horch
Departments of Biology and Neuroscience
Bowdoin College
Brunswick, ME, USA

Aleksandar Popadić
Biological Sciences Department
Wayne State University
Detroit, MI, USA

Sumihare Noji
Graduate school of Bioscience and Bioindustry
Tokushima University
Tokushima, Japan

Taro Mito
Graduate school of Bioscience and Bioindustry
Tokushima University
Tokushima, Japan

Hideyo Ohuchi
Department of Cytology and Histology
Okayama University
Okayama, Japan

Dentistry and Pharmaceutical Sciences
Okayama University Graduate School of Medicine
Okayama, Japan

ISBN 978-4-431-56476-8 ISBN 978-4-431-56478-2 (eBook)
DOI 10.1007/978-4-431-56478-2

Library of Congress Control Number: 2016960036

Printed on acid-free paper

This Springer imprint is published by Springer Nature
The registered company is Springer Japan KK
The registered company address is: Chiyoda First Bldg. East, 3-8-1 Nishi-Kanda, Chiyoda-ku, Tokyo 101-0065, Japan

Preface

Crickets inhabit all areas of the world with the exception of subarctic and arctic regions. Encompassing about 2,400 species, they are the most diverse lineage of the "leaping" insects. Their defining characteristic is the chirping sound made by males during mating. For the past 100 years, detailed studies at the behavioral, acoustic, and neurophysiological level have revealed fundamental aspects of mating behavior and the complexity of the aggressive interactions among males. Around the world, crickets are also considered an important food source. They are frequently reared on an industrial scale to satisfy demands from zoos and pet stores as well as from food processing plants. Crickets are served as a common street snack throughout Southeast Asia and can frequently be found as an ingredient in commercially produced protein bars and baked goods. Remarkably, the food conversion efficiency of house crickets (*Acheta domesticus*) is five times higher than beef, and if their fecundity is taken into account, this efficiency increases 15–20-fold (Nakagaki and Defoliart 1991). If one considers the ever-increasing human population growth, our survival on Earth may depend on altering our eating habits and consuming new sources of food, such as insects (O. Deroy 2015).

The present volume aims to provide recent scientific updates on research on crickets in general, with the emphasis on *Gryllus bimaculatus*. We believe that this species can serve as a representative model for basal, hemimetabolous insect lineages. In this mode of development, an embryo develops into a miniature adult (first nymph), which in turn undergoes a number of successive molts before turning into an adult. In comparison, the development and overall biology of the premier insect genetic model system, *Drosophila melanogaster*, are highly derived and representative of only one insect group (Diptera – flies). As this book demonstrates, it is rather the cricket, as exemplified by *Gryllus*, which should be considered to represent a typical insect. Until recently, though, only very limited functional and genetic manipulation studies were feasible in non-drosophilid species. This all changed dramatically in the past 10 years, as the cricket community made rapid progress adapting existing and new experimental techniques in *Gryllus*. The main impetus behind all these advancements can be traced to the 2006 Nobel Prize in

Physiology or Medicine to Andrew Z. Fire and Craig C. Mello for their discovery of RNA interference (RNAi). The advent of the RNAi methodology provided a powerful tool to study almost any insect species. As shown in this book, many researchers have performed functional analyses of a variety of cricket genes, yielding important information about the biology and development of this organism.

In 2015, the first available rough draft assembly of the whole genome of *Gryllus bimaculatus* was completed at Tokushima University. This was a milestone event, enabling researchers to study hundreds of new genes (public access is planned for 2017). In addition to making transgenic crickets, it is now possible to use site-specific approaches such as TALENs and zinc-finger nucleases to alter the *Gryllus* genome at a targeted region. Furthermore, the CRISPR/Cas-based genome-editing system has been adapted for use in the cricket. These newly available genome-editing techniques can spearhead the detailed examination of gene function and the production of gene-edited crickets that can serve as models for human diseases. In theory, such genetically engineered crickets can be used to screen various chemicals to find drug candidates for genetic disorders and to produce human therapeutic proteins or metabolites.

In 2012, we organized the 2nd International Conference on Cricket Research in Tokushima, Japan. (In fact, the first conference was canceled because of the Great East Japan Earthquake of March 11, 2011.) At that second meeting, we proposed the publication of this book and invited participants to contribute chapters representing their fields and their work. We want to thank all the authors for their contributions and support throughout the development of this book. We hope that this volume will inspire scientists in various disciplines to use the cricket model system to ask interesting and innovative questions.

Brunswick, ME, USA — Hadley Wilson Horch
Tokushima, Japan — Taro Mito
Tokushima, Japan — Sumihare Noji
Okayama, Japan — Hideyo Ohuchi
Detroit, MI, USA — Aleksandar Popadić
September, 2015

References

Nakagaki BJ, DeFoliart GR (1991) Comparison of diets for mass-rearing Acheta domesticus (Orthoptera: Gryllidae) as a novelty food, and comparison for food conversion efficiency with values reported for livestock. J Econ Entomol 84:891–896

O'Deroy O (2015) Eat insects for fun, not to help the environment. Nature 521:395

Contents

Part I
Development and Regeneration

Chapter 1
History of Cricket Biology

Gerald S. Pollack and Sumihare Noji

Keywords Neuroethology • Communication • Wind-evoked escape • Circadian rhythms • Learning • Evolution • Endocrinology • Development • Regeneration

1.1 Introduction

"Cricket Behavior and Neurobiology" (Huber et al. 1989) summarized what were, at that time, the main areas of research on crickets. In his preface to the book, Franz Huber wrote "We hope this book stimulates cross-fertilization between different biological fields and encourages scientific progress." One purpose of the present book is to document the considerable progress and cross-field interaction in studying crickets that has taken place in the years since "Cricket Behavior and Neurobiology" was published. Another goal is to make the case that crickets are excellent model organisms for studying problems in a broad range of biology extending beyond behavior and neurobiology. Although mankind's millennia-old interest in crickets can be ascribed to their fascinating and easily observed behaviors, such as singing and fighting (Laufer 1927), recent advances have made it possible to address questions in areas including molecular, developmental, and evolutionary biology. Bentley and Hoy (1974), summarizing genetic, behavioral, and neurobiological studies on the mechanisms underlying production and responding to cricket songs, likened crickets to decathlon athletes: "We view the cricket as a kind of decathlon performer in neurobiology: it may not excel at any one thing, but it can be counted on for a sound

G.S. Pollack (✉)
Department of Biological Sciences, University of Toronto Scarborough, Toronto, ON, Canada
e-mail: gerald.pollack@utoronto.ca

S. Noji
Tokushima University, Tokushima City 770-8501, Japan
e-mail: noji@tokushima-u.ac.jp

H.W. Horch et al. (eds.), *The Cricket as a Model Organism*,
DOI 10.1007/978-4-431-56478-2_1

performance in every event." The technical and conceptual advances covered in the present volume show that the range of events in which crickets can be viewed as serious contenders has now broadened considerably.

In this chapter, we highlight some of the most exciting advances in cricket biology since 1989, some of which are discussed in greater detail in the chapters that follow.

1.2 Neuroethology

As its title suggests, "Cricket Behavior and Neurobiology" focused on neuroethology. Most of the topics it considered remain active research areas today. Here we summarize recent progress in a few of these areas.

1.2.1 Acoustics

The most conspicuous aspect of cricket behavior for humans is their singing, and it is not surprising that studies on acoustic communication have been in the forefront of cricket research for more than a century (Regen 1913). Approximately half of "Cricket Behavior and Neurobiology" was devoted to acoustic signaling and hearing, and these remain important areas of research today.

The biomechanics underlying sound production was understood by Bennet-Clark (1989), but the underlying neuronal mechanisms were not. Crickets sing by rubbing their front wings together in repeated scissor-like motions. Because these periodic wing movements resemble those during flight, a common expectation was that singing and flying would turn out to be closely related, both in terms of neuronal circuitry and in evolution. The notion of shared circuitry holds at the level of the motor neurons that control movements of the forewings, but a paper that appeared shortly after publication of "Cricket Behavior and Neurobiology" showed that the motor patterns are generated by separate networks of interneurons (Hennig 1990). Interestingly, although the motor neurons for flight and singing are in the thoracic ganglia, some of the neurons comprising the underlying pattern-generating networks are found in abdominal neuromeres of the metathoracic ganglion (Robertson et al. 1982) or even, as has recently been shown for singing, in free abdominal ganglia (Schöneich and Hedwig 2012; Chap. 10). A dominant hypothesis for the evolution of insect wings is that they are derived from appendages that occurred on all body segments of primitive arthropods, including the abdomen (Kukalova-Peck 1987; Dudley and Yanoviak 2011). The occurrence of flight interneurons in abdominal neuromeres has been offered as support of this hypothesis (Robertson et al. 1982), and the same idea might apply to the evolution of wing movements underlying song. If so, then flight and singing might indeed be related evolutionarily, if not mechanistically.

1.2.1.1 Hearing and Acoustic Signal Processing

The dual function of cricket hearing, for intraspecific communication and for predator detection, was well established in 1989 (Pollack and Hoy 1989; Weber and Thorson 1989). Since then, however, new insights about the neural processing of behaviorally relevant acoustical signals have come to light. These are due in part to the application of new techniques and approaches, such as Ca imaging to study neuronal processing at the subcellular level (Baden and Hedwig 2006), and to the use of analytical methods borrowed from information theory and decision theory (Marsat and Pollack 2004, 2006). Important recent advances, e.g., in understanding how brain circuits analyze communication signals (see Chap. 11), have also resulted from the continued application of more traditional methods such as intracellular recording and staining (albeit with exceptional skill and patience and an abundance of hard work!).

Flying crickets steer away from ultrasound stimuli, a behavior believed to be a defense against predation by echolocating bats. This response is driven by the identified ultrasound-tuned auditory interneuron, AN2 (reviewed by Pollack and Hoy 1989). Recent work has shown that the fine structure of AN2 responses is important for evoking steering responses. Specifically, as discussed in Chap. 11, steering responses are driven by bursts of closely spaced action potentials in AN2, which in turn are driven by bursts in ultrasound-tuned receptor neurons (Marsat and Pollack 2006; Sabourin and Pollack 2009). Thus, the neural "code" for bat-avoidance behavior is established at the earliest possible level of sensory processing, in sensory receptor neurons themselves. AN2 terminates in the brain, where a number of ultrasound-tuned interneurons, both local and descending, are found (Brodfuhrer and Hoy 1990). The detailed circuitry intervening between AN2 and output to motor neurons, however, awaits further investigation.

Female crickets respond to conspecific song by walking or flying toward its source. This behavior is pattern selective; that is, the probability and strength of response depend on how closely the stimulus temporal pattern matches the species-typical song pattern, for example, the rate at which sound pulses occur. Both receptor neurons and a first-order interneuron called AN1 respond robustly to stimuli having a wide range of temporal patterns (although, as detailed in Chap. 11, signs of species specificity are evident even at these early levels of processing). Sharp specificity for song-like rhythms first appears in the brain (Schildberger et al. 1989). A network of brain neurons receives input from AN1, and neurons within the network display differing levels of specificity, with the most selective neurons showing a relationship between firing rate and stimulus pattern that precisely matches behavioral selectivity (Kostarakos and Hedwig 2013). The structure and physiology of this filter network are discussed further in Chap. 11.

The increasingly detailed understanding of the neural mechanisms underlying both sound production and signal analysis opens the door to one of the most intriguing, but least understood, questions in neuroethology: how do neural circuits evolve so as to support species-specific behaviors? It should soon be possible to know, at the level of single neurons and their synaptic interactions, the sorts of adaptations in neuronal circuits that result, for example, in production of chirps instead of trills in a cricket's song and that allow selective recognition of species-specific song patterns.

1.2.2 Wind-Evoked Escape

Like cockroaches and other orthopteroid insects, crickets bear conspicuous "rear-end antennae," or cerci, on the abdomen. The role of the cerci and their associated neural structures in wind-evoked escape has been appreciated for nearly 70 years (Roeder 1948). Recent advances, using Ca imaging to reveal the spatial patterns of neuronal activity and computational approaches to understand the neuronal algorithms underlying information processing, have established the cricket cercal system as an important model for understanding sensory processing.

A properly oriented escape response, i.e., away from the wind (and the approaching predator that might have produced it), requires information about stimulus direction. Directional selectivity is encoded first by the mechanical properties of the several hundred filiform hairs on each cercus, which are constrained to move in a restricted plane by the shape and orientation of the sockets in which they sit (Jacobs et al. 2008). The single sensory neuron associated with each hair inherits the hair's directionality and passes this on to the central nervous system through its spiking activity. Anatomical studies (Jacobs and Theunissen 1996) show that the pattern of central projections of individual sensory neurons forms a map of stimulus direction, which is sampled by the dendrites of specific directionally tuned interneurons. Ca-imaging studies have made it possible to observe this stimulus-dependent activation pattern in real time, along with the activation of specific dendritic fields of individual interneurons as well as the outputs of the interneurons as reflected in their spiking activity (Ogawa et al. 2008; Chap. 14). This wealth of data has led to an exceptionally comprehensive view of how the properties of sensory neurons, the patterns of their connectivity with interneurons, and postsynaptic integrative mechanisms cooperate to extract behaviorally relevant information from a sensory stimulus.

1.3 Circadian Rhythms

Another long-studied area, dating back at least to the 1930s (Lutz 1932), is the circadian rhythmicity of cricket behavior. Most cricket species are primarily active at night. Early researchers showed, by employing reversed light/dark cycles, that this activity pattern was set by the daily alteration of light levels caused by the earth's rotation. They also showed, however, that circadian rhythmicity was endogenous, that is, it persisted even when crickets were held under constant conditions that lacked any timing cues. By 1989, endogenous rhythms for behaviors such as locomotion and singing had been demonstrated in a number of cricket species

(Loher 1989), and surgical and other manipulations, such as covering the eyes and/or ocelli with opaque paint, had begun to identify the anatomical loci of clocks, with most evidence pointing to the optic lobes (Tomioka and Andelsalam 2004). It is fair to say, however, that the field of insect circadian rhythms has been dominated over the past few decades by work on *Drosophila*, where genetic and molecular approaches have provided a wealth of information on how systems of interacting molecules may comprise the core of the clockwork mechanism (Peschel and Helfrich-Förster 2011). In *Drosophila*, as well as in a variety of other organisms ranging from fungi to mammals, circadian oscillations result from molecular negative feedback loops in which transcription factors, acting in concert with other molecules, inhibit their own transcription. Genetic approaches have also led to the identification of specific groups of neurons in the *Drosophila* brain that are required to generate rhythmicity and to distribute it to effector circuits.

In the past several years, new methods, including RNA interference (RNAi) (Tomioka et al. 2009), have made it possible to investigate the molecular bases of rhythmicity in genetically less tractable insects, including crickets. These experiments have shown that both the molecular identities of the key players and the locations of the main "clock cells" in the brain vary among taxa. For example, a transcriptional repressor, timeless (TIM), is at the heart of the *Drosophila* clock (Peschel and Helfrich-Förster 2011). Flies with mutated *tim* genes are arrhythmic (Seghal et al. 1994). However suppression of the expression of a cricket homolog of *tim*, *Gbtim*, using RNAi does not eliminate circadian rhythmicity (Danbara et al. 2010), indicating that the molecular details of the cricket's clock differ from that of *Drosophila*. See Chap. 6 for further details.

1.4 Learning

That insects are capable of learning has long been appreciated. Von Frisch (e.g., 1914) exploited the learning capabilities of bees to investigate their perception of shapes, colors, and odors, and Tinbergen (1952) showed that digger wasps used memorized visual cues to locate their cryptic nests. Surprisingly, the learning abilities of crickets received little attention until relatively recently. Indeed, learning was mentioned only briefly in "Cricket Behavior and Neurobiology" (Honneger and Campan 1989). The past several years, however, have led to an appreciation of the considerable abilities of crickets to modify their behavior based on experience. Like bees and other "smart" insects, crickets can learn to associate shapes, colors, and odors with positive or negative reinforcement. Pharmacological and RNAi methods have begun to reveal the biochemical and molecular mechanisms underlying memory formation and retrieval, including the role of NO in forming long-term memory (Takahashi et al. 2009; Mizunami and Matsumoto 2010), and the roles of biogenic amines in aversive and appetitive learning (Matsumoto et al. 2006). Crickets have thus joined honeybees and fruit flies (Giurfa 2013) as important model systems for studying the cellular and molecular basis of learning and memory in insects. Additional information is provided in Chap. 9.

In addition to studies directed at mechanisms for learning, recent work on crickets has studied the effects of experience on behavior from a functional point of view. The attractiveness to females of male songs varies according to the song's temporal structure, but this is not a static situation: females can adjust their selectivity for song patterns based on their recent acoustic experience, becoming less likely to respond to a moderately attractive song after having heard a highly attractive stimulus (Wagner et al. 2001; Bailey and Zuk 2009). Thus, females can calibrate their choosiness for a potential mate according to the quality of those that are available. It is worth noting that "learning" in this case need not be mechanistically related to the associative conditioning discussed earlier. Rather, it is possible that the change in female selectivity results from short-term neurohormonal mechanisms that are brought in to play by exposure to the highly attractive song.

1.5 Genetics

Cricket songs consist of series of brief sound pulses with stereotypic, species-specific rhythms that are generated by networks of neurons in the central nervous system (see Chap. 10). The life histories of some cricket species, which overwinter as eggs, ensure that singing by adults (the only stage capable of song) and development of juveniles do not overlap in time, ruling out the possibility that the song's structure is learned. This suggests that song structure is genetically determined. Bentley and Hoy (1972) showed that this is indeed the case by crossbreeding closely related species with distinct songs. The resulting hybrids sang songs that were intermediate in structure to those of the parent species. The pattern of inheritance was consistent with multigenic control of song structure.

In complementary experiments, Hoy and Paul (1973) measured the song preferences of females in behavioral tests and found that hybrid females preferred the songs of hybrid males, demonstrating that the neural circuits for song recognition, like those for production, are genetically determined.

Although the genetic basis of singing and song recognition is clear, the identities of the relevant genes remain unknown. Bentley (1975), using chemical mutagenesis followed by behavioral screens, showed that genetics could be used profitably to study cricket behavior. He screened early-instar offspring of mutagenized crickets for the loss of a reliable behavior, evasive jumping in response to wind stimulation, and isolated two distinct mutants, both of which affected the development of the wind-sensitive hairs on the cricket's cerci that mediate wind-evoked escape. He went on to investigate the consequences of the loss of sensory input on the anatomy and physiology of interneurons that mediate escape jumping. However, perhaps surprisingly in view of their ease of rearing, high fecundity, reasonably short generation time, experimentally accessible nervous systems, and elaborate behaviors, crickets have not attracted the attentions of behavioral geneticists to the same extent as other invertebrate model systems such as *Drosophila melanogaster* and *Caenorhabditis elegans*.

Bentley (1975) employed chemicals to induce random mutations in the genome. Modern methods can produce targeted mutations using zinc finger nucleases (ZFNs) or transcription activator-like effector nucleases (TALENs; Watanabe et al. 2012). More recently, the type II bacterial CRISPR/Cas9 system has been demonstrated to be an efficient gene-targeting technology with the potential for multiplexed genome editing (Barrangou 2013). This powerful system has recently been adapted for use in crickets (Awata et al. 2015; Chap. 21). When combined with genome sequence data, these methods will allow a more systematic investigation of the roles of specific genes in the physiology and behavior of crickets.

1.5.1 Evolutionary Change in Cricket Acoustical Communication

A problem closely related to the genetics of singing and song preference is the evolution of acoustic communication. Alexander (1962) noted that the more than 2,000 species of stridulating crickets, the songs of which show considerable diversity, are believed to have descended from a single progenitor. For communication to remain effective, both the male's song and the female's song preference must coevolve. Alexander noted that one way to assure that this occurred would be for song production and song recognition to be controlled by the same genes, a hypothesis supported by the crossbreeding experiments described above. He also speculated that one way in which genetic coupling might be implemented would be for the neural circuits controlling song structure and song preference to share neural elements. Only males sing, but females can produce song-like movements of the wings under certain circumstances. These movements are silent because the female wing lacks the cuticular specializations for sound production; nevertheless they suggest that, even though mute, females might possess neural circuits for singing (Huber 1962). The circuits might, for example, be used to produce an internal song template that the female could use to assess the structure of a received song (Hoy 1978). The pattern generators for song are situated in the thoracic and abdominal segments, whereas circuits for song recognition are in the brain (see Chaps. 10 and 11), but communication between these, e.g., by efference copy, could provide a way for them to share neural elements despite their anatomical separation.

Pollack and Hoy (1979) tested the shared-element idea by randomizing the normally highly stereotyped sequence of brief intervals separating the sound pulses of the song. They reasoned that if females compared the song they heard against an internal template, the randomized song would be a poor match and so should be rejected. However, females responded as readily to this song as to the nonrandomized version, arguing against the template-matching hypothesis.

Pires and Hoy (1992) adopted a different approach, exploiting the fact that the tempo of cricket acoustic communication is temperature dependent: the sound-pulse rate increases with temperature, and warm females prefer higher pulse rates. Using implanted heating coils, Pires and Hoy raised the thoracic, but not brain, temperature of females, reasoning that if song preference was influenced by thoracic circuits for song production, females with warm thoraces should prefer songs with higher pulse rates. Heating the thorax, however, did not affect song preference. Similarly, heating the brain alone was also without effect. Female preference was shifted only when both brain and thorax were heated. Recent experiments, discussed in Chap. 10, have shown that important components of the song-pattern generator reside in the abdominal nervous system. It would thus be interesting to learn whether manipulating both thoracic and abdominal temperature would shift female song preferences.

Although experiments testing the hypothesis of shared neuronal elements are inconclusive, recent genomic studies provide support for genetic coupling between song production and song preference. Shaw and colleagues (Shaw et al. 2007; Shaw and Lesnick 2009; Wiley et al. 2012) used a quantitative trait locus (QTL) approach to study the genetics of singing and song preference in the genus *Laupala*, a Hawaiian group of morphologically cryptic species that are distinguished by their songs. They focused on two species, *L. paranigra* and *L. kohalensis*, the songs of which have low and high pulse rates, respectively (0.7 and 3.7 pulses/s). Interspecific crosses between these species generated hybrids in which the occurrence of genetic material from each parent species could be correlated with song pulse rate. They found eight regions of the genome (QTLs) that were predictive of hybrid pulse rate (Shaw et al. 2007), four of which also predicted the selectivity for pulse rate shown by females (Shaw and Lesnick 2009; Wiley et al. 2012). These results demonstrate a physical linkage among genes controlling song production and song preference, although whether the same genes are involved in both behaviors is not yet known. This genetic architecture is believed to have facilitated the very rapid evolution of acoustic communication in the *Laupala* group.

1.6 Endocrinology

Insect endocrinology is another long-established discipline (e.g., Wigglesworth 1934), but there have been significant advances in recent years (see Chap. 15). In 1994, K.H. Hoffmann and his colleagues found that allostatins or allatostatin-like molecules are produced in neurosecretory cells of the brain and are delivered to the corpora allata through nervous connections and/or via hemolymph (Neuhuser et al. 1994). Allatostatins inhibit the biosynthesis of juvenile hormones in the cricket. Juvenile hormones (JHs) play a crucial role in the regulation of development and reproduction of insects. They are involved in embryogenesis, molting, and metamorphosis during the preimaginal stages but also vitellogenesis and ovarian development in adult females and spermatogenesis and growth of the accessory reproductive glands in adult males (Wang et al 2004). JHs are synthesized and released from the corpora allata (CA). The biosynthesis of JH may be regulated by both peptidergic and aminergic input (Rachinsky et al. 1994). The peptidergic signals were classified as allatotropins (ATR; stimulatory) or allatostatins (AST; inhibitory) on the basis of their effects on JH production by the CA in homologous or heterologous bioassays in vitro (Wang et al. 2004). The peptides can be classified into three different peptide families, the YXFGL-amide allatostatin superfamily (A-type, cockroach allatostatins), the W(X6)W-amide allatostatins (B-type, cricket allatostatins), and the lepidopteran (*Manduca sexta*) allatostatin (C-type). Hoffmann and his colleagues isolated five members of the A-type AST family from methanolic extracts of brains of the Mediterranean field cricket, *Gryllus bimaculatus* (Meyering-Vos et al. 2006). They also isolated cricket- or B-type allatostatins (W(X6)W-amides) and found that they are expressed in the central nervous system and the digestive tract of female adult crickets, indicating the B-type allatostatins of *G. bimaculatus* are brain-gut peptides (Wang et al. 2004). In order to know the functions of the allatostatin A-type gene in the cricket *Gryllus bimaculatus*, they performed RNAi experiments and found that knockdown of allatostatin A-type gene expression results in elevation of JH titers in the hemolymph, a reduced body weight in larval and adult crickets, and incomplete formation of the imaginal molt in addition to reduction of the egg and testes development and the oviposition rate.

Insect Malpighian (renal) tubules generate a flow of primary urine that is subsequently modified by reabsorptive processes in the lower tubule and in the hindgut (ileum and rectum) before being voided. Primary urine production is accelerated by diuretic hormones, which stimulate ion transport into the lumen along with osmotically obliged water. Insect kinins are neuropeptides first identified as myotropins (Nachman et al. 1986) but later shown to stimulate secretion by Malpighian tubules of the mosquito *Aedes aegypti* and to depolarize a lumen-positive transepithelial voltage Vt (Hayes et al. 1989). In addition, *Acheta* diuretic peptide (Coast and Kay 1994) and a diuretic hormone receptor have been isolated (Reagan 1996). However, no systematic study in these substances has been conducted so far.

1.7 Mechanisms of Cricket Development and Regeneration

Mechanisms of insect development have been extensively studied in *Drosophila melanogaster*, a holometabolous insect. Many reviews and books about *Drosophila* have been published. By contrast, the corresponding mechanisms in the hemimetabolous insects have not been studied in as much depth. Thus, many researchers have had the impression that all insects develop similarly to *Drosophila*. In the past decade, however, much has been learned about developmental gene networks in two hemimetabolous insects, the milkweed bug *Oncopeltus fasciatus* and the cricket *Gryllus bimaculatus*. This progress is due mainly to the application of RNA interference (RNAi) to suppress expression of specific genes. In particular, gene functions of *Drosophila* homologs expressed in embryogenesis have been investigated in *G. bimaculatus* as shown in this book (c.f. Chaps. 2 and 4). Recent studies on the cricket have revealed that the developmental mechanisms found in *Drosophila* are very specific and may be an extreme case resulting from evolutionary pressure to adapt to a specific environment. Thus, the early developmental mode found in the cricket, rather than that of *Drosophila*, may serve as a better example of general insect development.

In 1972, D.T. Anderson published a paper entitled "The development of hemimetabolous insects" (Anderson 1972). He described very detailed morphological changes during early development of the cricket. However, further studies on *Gryllus* development had not been performed until 1997. Nakamura et al. (2010) found that dynamic cell movement is important during germ band formation. In the cricket, orthodenticle, a homeodomain-containing transcription factor, and signaling by wingless/Wnt, a family of signal transduction proteins, play crucial roles in the A/P pattern formation. Mito et al. (2010) also discuss recent evidence suggesting that insect developmental modes may have evolved by heterochronic shifts, while retaining certain universal metazoan features.

Late developmental processes of the cricket are also interesting. For example, Niwa and colleagues (2000) found that the expression pattern of decapentaplegic (dpp), which is a key morphogen important in development, is significantly divergent among *Gryllus*, *Schistocerca* (grasshopper), and *Drosophila* embryos, while expression patterns of other morphogens, hedgehog (hh) and wingless (wg), are conserved. Furthermore, since the divergence was found between the pro/mesothoracic and metathoracic *Gryllus* leg buds, they were able to conclude that diversity of leg morphology is closely correlated with divergence in dpp expression pattern during leg development in *Gryllus bimaculatus*.

The mushroom body of the cricket displays continuous neurogenesis throughout the life of the animal. The miniature mushroom body of the newly hatched larva grows throughout the larval stages, and new Kenyon cells are added even during adulthood. Malaterre and colleagues (2002) found that the positions of Kenyon cell somata in the mushroom body cortex and the arborization patterns of axons in the lobes are age related. They concluded that cricket mushroom bodies undergo remodeling throughout life as a result of continuous neurogenesis. It is thus possible to investigate neurogenesis, axonal morphology, and lobe development even during the adult stage. This work lays the foundation for future studies of the functional role of this developmental plasticity.

The cricket has also been used to study mechanisms underlying leg regeneration. Meinhardt (1982) proposed a boundary model to explain regeneration phenomena including insect leg regeneration. At that time, no developmental genes had yet been isolated, even within *Drosophila*. However, Meinhardt predicted that four factors (genes) would be necessary for creating new organizing regions. More than a decade later, Campbell and Tomlinson (1995), relying on *Drosophila* data, predicted that the four factors are dpp, Wnt, hh, and epidermal growth factor (EGF), and these factors may be expressed in blastema cells formed in the most distal region of the amputated leg. Although the nymphal leg system was used extensively by scientists interested in the mechanisms of appendage regeneration until 1986, very few groups are currently working on this system. This was mainly due to the paucity of molecular data and the lack of tools for functional analyses. However, the situation gradually changed as sequencing techniques improved and RNA interference (RNAi) proved to work efficiently in cricket nymphs (nyRNAi). We now know that the four factors predicted by Campbell and Tomlinson are in fact expressed in the blastema of the regenerating cricket leg (Mito et al. 2002). Work from the Noji lab has identified a phenomenon called "regeneration-dependent RNAi." In this phenomenon, nymphal injection of RNAi against a given factor reveals no phenotype in normally developing nymphs but leads to a defect in regeneration after amputation (Nakamura et al. 2008). With this method, they found that large atypical cadherins of Fat (Ft) and Dachsous (Ds) are involved in leg regeneration with the Hippo signaling pathway (Bando et al. 2011). They speculate that this system may be involved in determination of positional information in the leg segment. Details are introduced in other sections of this book. The cricket is a useful experimental system to elucidate molecular mechanisms underlying various biological phenomena.

1.8 Concluding Remarks

Bentley and Hoy's comparison of crickets with decathletes was based on the ability to apply behavioral, neurophysiological, neuroanatomical, and classical (forward) genetic methods to study the same organism. As summarized above and expanded upon in the following chapters, molecular, genomic, reverse-genetic, and computational methods have been successfully developed and adapted, enlarging the already broad toolkit available for cricket research. There is both room and need in biology for versatile model systems other than the current triumvirate of flies, worms, and mice. We hope that this book will convince at least some of its readers that crickets are worthy of serious consideration in this regard.

References

Alexander RD (1962) Evolutionary change in cricket acoustical communication. Evolution 16:443–462

Anderson DT (1972) The development of hemimetabolous insects. Academic, London/New York

Awata H, Watanabe T, Hamanaka Y, Mito T, Noji S, Mizunami M (2015) Knockout crickets for the study of learning and memory: dopamine receptor Dop1 mediates aversive but not appetitive reinforcement in crickets. Sci Rep 5:15885

Baden T, Hedwig B (2006) Neurite-specific Ca^{2+} dynamics underlying sound processing in an auditory interneurone. J Neurobiol 67:68–80

Bailey NW, Zuk M (2009) Field crickets change mating preferences using remembered social information. Biol Lett 5:449–451

Bando T, Mito T, Nakamura T, Ohuchi H, Noji S (2011) Regulation of leg size and shape: involvement of the Dachsous-fat signaling pathway. Dev Dyn 240:1028–1041

Barrangou R (2013) CRISPR-Cas systems and RNA-guided interference. Wiley Interdiscip Rev: RNA 4:267–278

Bennet-Clark H (1989) Songs and the physics of sound production. In: Huber F, Moore TE, Loher W (eds) Cricket behavior and neurobiology. Cornell University Press, Ithaca, pp 227–261

Bentley DR (1975) Single gene cricket mutations: effects on behavior, sensilla, sensory neurons, and identified interneurons. Science 187:760–764

Bentley DR, Hoy RR (1972) Genetic control of the neuronal network generating cricket (*Teleogryllus Gryllus*) song patterns. Anim Behav 20:478–492

Bentley DR, Hoy RR (1974) The neurobiology of cricket song. Sci Am 231:34–44

Brodfuhrer PD, Hoy RR (1990) Ultrasound sensitive neurons in the cricket brain. J Comp Physiol A 166:651–662

Campbell G, Tomlinson A (1995) Initiation of the proximodistal axis in insect legs. Development 121:619–628

Coast G, Kay II (1994) The effects of *Acheta* diuretic peptide on isolated Malpighian tubules from the house cricket *Acheta domesticus*. J Exp Biol 187:225–243

Danbara Y, Sakamoto T, Uryu O, Tomioka K (2010) RNA interference of timeless gene does not disrupt circadian locomotor rhythms in the cricket *Gryllus bimaculatus*. J Insect Physiol 56:1738–1745

Dudley R, Yanoviak SP (2011) Animal aloft: the origins of aerial behavior and flight. Integr Comp Biol 51:926–936

Giurfa M (2013) Cognition with few neurons: higher-order learning in insects. Trends Neurosci 36:285–294

Hayes TK, Pannabecker TL, Hinckley DJ, Holman GM, Nachman RJ, Petzel DH, Beyenbach KW (1989) Leucokinins, a new family of ion transport stimulators and inhibitors in insect Malpighian tubules. Life Sci 44:1259–1266

Hennig RM (1990) Neuronal control of the forewings in two different behaviours: stridulation and flight in the cricket, *Teleogryllus commodus*. J Comp Physiol A 167:617–627

Honneger H-W, Campan R (1989) Vision and visually guided behavior. In: Huber F, Moore TE, Loher W (eds) Cricket behavior and neurobiology. Cornell University Press, Ithaca, pp 147–178

Hoy RR (1978) Acoustic communication in crickets: a model system for the study of feature detection. Fed Proc 37:2316–2323

Hoy RR, Paul RC (1973) Genetic control of song specificity in crickets. Science 180:82-83

Huber F (1962) Central nervous control of sound production in crickets and some speculations on its evolution. Evolution 16:429–442

Huber F, Moore TE, Loher W (eds) (1989) Cricket behavior and neurobiology. Cornell University Press, Ithaca, 565 pp

Jacobs GA, Miller JP, Aldworth Z (2008) Computational mechanisms of mechanosensory processing in the cricket. J Exp Biol 211:1819–1828

Kostarakos K, Hedwig B (2013) Calling song recognition in female crickets: temporal tuning of identified brain neurons matches behavior. J Neurosci 32:9601–9612

Kukalová-Peck J (1987) New carboniferous diplura, monura, and thysanura, the hexapod ground plan, and the role of thoracic side lobes in the origin of wings (insecta). Can J Zool 65:2327–2345

Laufer B (1927) Insect musicians and cricket champions of China. Field Museum of Natural History, Chicago

Loher W (1989) Temporal organization of reproductive behavior. In: Huber F, Moore TE, Loher W (eds) Cricket behavior and neurobiology. Cornell University Press, Ithaca, pp 83–113

Lutz FE (1932) Experiments with Orthoptera concerning diurnal rhythm. Am Mus Novit 550:1–24

Malaterre J, Strambi C, Chiang AS, Aouane A, Strambi A, Cayre M (2002) Development of cricket mushroom bodies. J Comp Neurol 452:215–227

Marsat G, Pollack GS (2004) Differential temporal coding of rhythmically diverse acoustic signals by a single interneuron. J Neurophysiol 92:939–948

Marsat G, Pollack GS (2006) A behavioral role for feature detection by sensory bursts. J Neurosci 26:10542–10547

Matsumoto Y, Unoki S, Aonuma H, Mizunami M (2006) Critical role of nitric oxide-cGMP cascade in the formation of cAMP-dependent long-term memory. Learn Mem 13:35–44

Meinhardt H (1982) Models of biological pattern formation. Academic Press, London

Meyering-Vos M, Merz S, Sertkol M, Hoffmann KH (2006) Functional analysis of the allatostatin-A type gene in the cricket *Gryllus bimaculatus* and the armyworm *Spodoptera frugiperda*. Insect Biochem Mol Biol 36:492–504

Mito T, Inoue Y, Kimura S, Miyawaki K, Niwa N, Shinmyo Y, Ohuchi H, Noji S (2002) Involvement of hedgehog, wingless, and dpp in the initiation of proximodistal axis formation during the regeneration of insect legs, a verification of the modified boundary model. Mech Dev 114:27–35

Mito T, Nakamura T, Noji S (2010) Evolution of insect development: to the hemimetabolous paradigm. Curr Opin Genet Dev 20:355–361

Mizunami M, Matsumoto Y (2010) Roles of aminergic neurons in formation and recall ofassociative memory in crickets. Front Behav Neurosci doi: 10.3389/fnbeh.2010.00172

Nachman RJ, Holman GM, Cook BJ (1986) Active fragments and analogs of the insect neuropeptide leucopyrokinin: structure-function studies. Biochem Biophys Res Commun 137:936–942

Nakamura T, Mito T, Bando T, Ohuchi H, Noji S (2008) Dissecting insect leg regeneration through RNA interference. Cell Mol Life Sci 65:64–72

Nakamura T, Yoshizaki M, Ogawa S, Okamoto H, Shinmyo Y, Bando T, Ohuchi H, Noji S, Mito T (2010) Imaging of transgenic cricket embryos reveals cell movements consistent with a syncytial patterning mechanism. Curr Biol 20:1641–1647

Neuhuser T, Sorge D, Stay B, Hoffmann KH (1994) Responsiveness of the adult cricket (*Gryllus bimaculatus and Acheta domesticus*) retrocerebral complex to allatostatin-1 from a cockroach, *Diploptera punctata*. J Comp Physiol B 164:23–31

Niwa N, Inoue Y, Nozawa A, Saito M, Misumi, Ohuchi H, Yoshioka H, Noji S (2000) Correlation of diversity of leg morphology in *Gryllus bimaculatus* (cricket) with divergence in dpp expression pattern during leg development. Development 127:4373–4381

Ogawa H, Cummins GI, Jacobs GA, Oka K (2008) Dendritic design implements algorithm for synaptic extraction of sensory information. J Neurosci 28:4592–4603

Peschel N, Helfrich-Förster C (2011) Setting the clock – by nature: circadian rhythms in the fruitfly *Drosophila melanogaster*. FEBS Lett 585:1435–1442

Pires A, Hoy RR (1992) Temperature coupling in cricket acoustic communication II. Localization of temperature effects on song production and recognition networks in *Gryllus firmus*. J Comp Physiol A 171:79–92

Pollack GS, Hoy RR (1979) Temporal pattern as a cue for species-specific calling song recognition in crickets. Science 204:429–432

Pollack GS, Hoy RR (1989) Evasive acoustic behavior and its neurobiological basis. In: Huber F, Moore TE, Loher W (eds) Cricket behavior and neurobiology. Cornell University Press, Ithaca, pp 340–363

Rachinsky A, Zhang J, Tobe SS (1994) Signal transduction in the inhibition of juvenile hormone biosynthesis by allatostatins: roles of diacylglycerol and calcium. Mol Cell Endocrinol 105:89–96

Reagan JD (1996) Molecular cloning and function expression of a diuretic hormone receptor.from the house cricket, *Acheta domesticus*. Insect Biochem Mol Biol 26:1–6

Regen J (1913) Über die Anlockung des Weibchens von *Gryllus campestris* L. durch telephonisch übertagene Stridulationslaute des Männchens. Akad Wiss Math Nat Kl Abt I (Wien) 132:81–88

Roberston RM, Pearson KG, Reichert H (1982) Flight interneurons in the locust and the origin of insect wings. Science 217:177–179

Roeder KD (1948) Organization of the ascending giant fiber system in the cockroach (*Periplaneta Americana*). J Exp Zool 108:243–261

Sabourin P, Pollack GS (2009) Behaviorally relevant burst coding in primary sensory neurons. J Neurophysiol 102:1086–1091
Schildberger K, Huber F, Wohlers DW (1989) Central auditory pathway: neuronal correlates of phonotactic behavior. In: Huber F, Moore TE, Loher W (eds) Cricket behavior and neurobiology. Cornell University Press, Ithaca, pp 423–458
Schöneich S, Hedwig B (2012) Cellular basis for singing motor pattern generation in the field cricket (*Gryllus bimaculatus* DeGeer). Brain Behav 2(6):707–725
Sehgal A, Price JL, Man B, Young MW (1994) Loss of circadian behavioral rhythms and per RNA oscillations in the *Drosophila* mutant timeless. Science 263(5153):1603–1606
Shaw KL, Lesnick SC (2009) Genomic linkage of male song and female acoustic preference QTL underlying a rapid species radiation. Proc Natl Acad Sci U S A 106:9737–9742
Shaw KL, Parsons YM, Lesnick SC (2007) QTL analysis of a rapidly evolving speciation phenotype in the Hawaiian cricket *Laupala*. Mol Ecol 16:2879–2892
Takahashi T, Hamada A, Miyawaki K, Matsumoto Y, Mito T, Noji S, Mizunami M (2009) Systemic RNA interference for the study of learning and memory in an insect. J Neurosci Methods 179:9–15
Tinbergen N (1952) The study of instinct. Clarendon Press/Oxford University Press, New York. 237 pp
Tomioka K, Andalsalam S (2004) Circadian organization in hemimetabolous insects. Zool Sci 21:1153–1162
Tomioka K, Sakamot T, Moriyama Y (2009) RNA interference is a powerful tool for chronobiological study in the cricket. Sleep Biol Rhythm 7:144–151
von Frisch K (1914) Der Farbensinn und Formensinn der Biene. Zool Jb Physiol 35:1–188
Wagner WE, Smeds MR, Wiegmann DD (2001) Experience affects female responses to male song in the variable field cricket *Gryllus lineaticeps* (Orthoptera, Gryllidae). Ethology 107:769–776
Wang J, Meyering-Vos M, Hoffmann KH (2004) Cloning and tissue-specific localization of cricket-type allatostatins from *Gryllus bimaculatus*. Mol Cell Endocrinol 227:41–51
Watanabe T, Ochiai H, Sakuma T, Horch HW, Hamaguchi N, Nakamura T, Bando T, Ohuchi H, Yamamoto T, Noji S, Mito T (2012) Non-transgenic genome modifications in a hemimetabolous insect using zinc-finger and TAL effector nucleases. Nat Commun 3:1017
Weber T, Thorson J (1989) Phonotactic behavior of walking crickets. In: Huber F, Moore TE, Loher W (eds) Cricket behavior and neurobiology. Cornell University Press, Ithaca, pp 310–339
Wigglesworth VB (1934) The physiology of ecdysis in *Rhodnius prolixus* (Hemiptera). II Factors controlling moulting and metamorphosis. Q J Microsc Sci 77:191–223
Wiley C, Ellison CK, Shaw KL (2012) Widespread genetic linkage of mating signals and preferences in the Hawaiian cricket *Laupala*. Proc Biol Sci 279:1203–1209

Chapter 2
Early Development and Diversity of *Gryllus* Appendages

Jin Liu and Aleksandar Popadić

Abstract Among insects, orthopterans such as *Gryllus bimaculatus* display an extraordinary diversity regarding the arrangement and morphology of their appendages. In the head region, previous studies have shown that despite the superficial similarities in the morphology of mandibulate mouthparts between holometabolous and hemimetabolous species, the development of these appendages may be regulated in different ways. At present, a comprehensive analysis in any hemimetabolous mandibulate species is lacking; therefore studying the mouthparts in *Gryllus* will significantly improve the current understanding of the evolution of mouthparts in insects. Orthopteran wings are also quite distinct, featuring the hardened, leathery protective forewings (FWs) and the membranous flying hind wings (HWs). Furthermore, the FWs in *Gryllus* are characterized by a complex vein-intervein arrangement, similar to the ancestral hardened wings observed in fossils, providing a unique opportunity to understand the evolution of wing sclerotization in basal insects. Finally, orthopterans feature one of the best-known examples of appendage modification in insects – the presence of the greatly enlarged jumping hind leg. Studies of gene expression and functional analyses suggest that this enlargement is controlled by the Hox gene *Ultrabithorax* (*Ubx*), which acts as a "trigger" for differential leg growth. Furthermore, rather than acting on all genes in the leg development network, *Ubx* seems to selectively upregulate growth factors such as *decapentaplegic* (*dpp*) and *Lowfat* in *Gryllus*. Hence, cricket hind leg can serve as an exceptional model for combined studies of both tissue growth and segmental patterning during embryonic leg development. Overall, this review formulates a general framework that can be used for future studies on the development and diversification of insect appendages.

Keywords *Gryllus bimaculatus* • Embryogenesis • Mandibulate mouthparts • T1–T3 legs • *Decapentaplegic* • *Wingless* • *Extradenticle* • *Dachshund* • *Distal-less* • *Ultrabithorax*

J. Liu • A. Popadić (✉)
Biological Sciences Department, Wayne State University, Detroit, MI 48202, USA
e-mail: apopadic@biology.biosci.wayne.edu

H.W. Horch et al. (eds.), *The Cricket as a Model Organism*,
DOI 10.1007/978-4-431-56478-2_2

2.1 Introduction

The appendages of the two-spotted field cricket, *Gryllus bimaculatus*, display extraordinary morphological diversity in the head and thoracic segments. Despite such individual segment variation, the morphology of mouthparts, wings, and legs in *Gryllus* conforms to a typical orthopteran body plan. In the head region, there are three segments that bear gnathal appendages: mandibles, maxillae, and labia. Crickets have mandibulate mouthparts that are used for chewing and characterized by stubby mandibles, which serve as grinding plates (Fig. 2.1a). The maxillary and labial appendages are similar and have branched morphology, with the latter also being fused into a single structure. Among the three thoracic segments, the prothorax (T1) is wingless, while the meso- (T2) and metathorax (T3) each carry a pair of wings. The forewing (FW) and hind wing (HW) exhibit distinct differences in shape, size, texture, and pigmentation (Fig. 2.1b). The FW is thickened and hardened, mainly black with the main function in protecting the HW. In contrast, the HW is transparent and membranous with a primary function in flight. In addition to wings, each of the three thoracic segments also bears a pair of legs. Whereas two pairs of four legs are generally rather similar, the hind (T3) legs are greatly enlarged and modified for jumping and represent the most distinctive orthopteran feature (Fig. 2.1c).

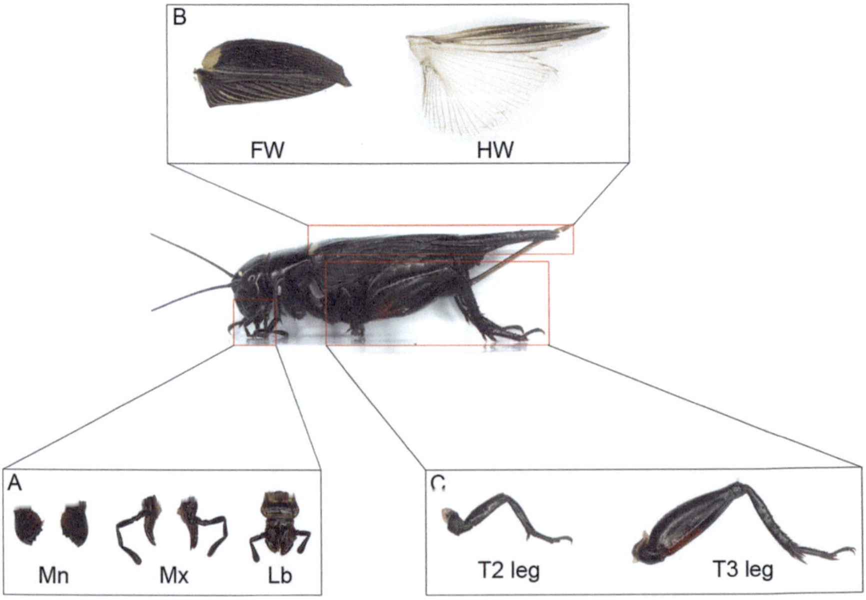

Fig. 2.1 The morphological diversity of *Gryllus* appendages. (**a**) The three gnathal appendages, mandible, maxillae, and labium, establish distinct morphologies. (**b**) The forewing and hind wing of *Gryllus* establish distinct shape and coloration. (**c**) The T2 and T3 legs show significant difference in size. Abbreviations: *Mn* mandible, *Mx* maxilla, *Lb* labium, *FW* forewing, *HW* hind wing

At present, the studies of the molecular mechanism that generate such distinct morphologies of appendages in *Gryllus* had almost exclusively focused on the development of the legs. The data on mouthparts is limited to a single expression study (Zhang et al. 2005), although more information is available in a related house cricket species, *Acheta domesticus* (Rogers et al. 1997, 2002). The development of wings in orthoptera is yet to be analyzed. At the same time, studies in other, mainly holometabolous model species have provided classical insights into the development of these appendages (de Celis et al. 1996; Ng et al. 1996; Kim et al. 1996; Neumann and Cohen 1998). Here we review the available information and discuss whether and to what degree the current developmental models can be applied to *Gryllus*. We also provide a framework for future studies in this and other orthopteran species that can be used to improve the current understanding of the development and differentiation of the head and thoracic appendages in hemimetabolous insects.

2.2 *Gryllus* Mouthparts

Gryllus mouthparts represent the ancestral form of gnathal appendages that are also found in other basal insect lineages such as cockroaches or primitively wingless silverfish. The more derived groups featured a trend toward further specialization, leading to the development of the haustellate (an adaptation to piercing and sucking; Hemiptera) or sponging (Diptera) mouthparts (Snodgrass 1993). Despite their distinct morphological differences, though, the identities of gnathal segments are controlled by the same set of three Hox genes in all insects studied so far (Hughes and Kaufman 2000; Rogers et al. 2002; Martinez-Arias et al. 1987; Merrill et al. 1987; Shippy et al. 2000, 2006; DeCamillis et al. 2001; Curtis et al. 2001; DeCamillis and ffrench-Constant 2003; Brown et al. 2000). Thus, it is now generally accepted that the evolution of mouthparts was governed by the changes in the expression (Rogers et al. 1997, 2002) and/or functions of *Deformed* (*Dfd*), *proboscipedia* (*pb*), and *Sex combs reduced* (*Scr*).

Previous studies of a number of mandibulate species, including the house cricket, *Acheta domesticus*, have shown that the expression patterns of these genes are conserved (Passalacqua et al. 2010; Rogers et al. 1997, 2002; Hrycaj 2010; Curtis et al. 2001; DeCamillis and ffrench-Constant 2003; DeCamillis et al. 2001; Shippy et al. 2000, 2006; Brown et al. 2000). First, *Dfd* is localized throughout the entire mandibular and maxillary segments and their appendages. Second, the expression of *pb* is more posterior and can be observed in the outer branches of the developing maxillary and labial appendages. Note that *pb* is never observed in the proximal portion of either appendage. Third, the *Scr* is primarily expressed in the labial segment. Of these three genes, only *Scr* pattern has been determined in *Gryllus*, and its localization in the labial segment follows the consensus observed in other mandibulate species (Zhang et al. 2005). Hence, it is likely that the expression patterns of *Dfd* and *pb* in *Gryllus* may also follow a mandibulate consensus pattern.

At present, the main functional insight into the genetic mechanisms that control the development of mandibulate mouthparts was generated through the studies of *Tribolium castaneum*, a holometabolous species (Curtis et al. 2001; DeCamillis and ffrench-Constant 2003; DeCamillis et al. 2001; Shippy et al. 2000, 2006; Brown et al. 2000) . Among hemimetabolous groups, the only functional data available is from *Oncopeltus fasciatus*, a hemipteran that has haustellate mouthparts (Hughes and Kaufman 2000). Hence, the potential new insight from a species such as *Gryllus* would be critical for an in depth understanding of evolutionary transition from mandibulate to haustellate insects. In addition, the presence of such information would allow for a direct comparison of genetic mechanisms that control the morphology of mandibular appendages in holo- (*Tribolium*) and hemimetabolous insects (*Gryllus*). In the former, the depletion of a head Hox gene generally causes a distinct identity transformation of the affected gnathal appendage(s). Specifically, *Dfd* RNAi transforms mandibles to legs, without any change in the identity of maxillae (Brown et al. 2000). Similarly, the loss of function of *pb* changes maxillary and labial palps into legs (DeCamillis and ffrench-Constant 2003; DeCamillis et al. 2001; Shippy et al. 2000), while loss of *Scr* results in the complete transformation of labium to antennae (Curtis et al. 2001; DeCamillis et al. 2001; Shippy et al. 2006). It is intriguing that in *Periplaneta americana* (cockroach), which is also a mandibulate insect, the labial appendage displays a mixture of leg and antennae morphology in *Scr* RNAi adults (Hrycaj et al. 2010). This finding suggests that insights from *Tribolium* cannot be directly applied and generalized to other mandibulate lineages. Since *Periplaneta* is also distantly related to orthopterans – it is likely that situation in *Gryllus* would be more similar to it than to *Tribolium*. For example, the loss of function of *Dfd* should result in antenna-like mandibles and maxillae with mixed leg/antenna morphology. The similar mixed identity should also be observed in maxillae in the absences of *pb* or in the labium in the absence of the *Scr*. In contrast, the double depletions of *Dfd/pb* and *Scr/pb* should transform maxillae and labium into antennae, respectively. Future studies confirming these predictions would provide a general framework for detailed understanding of the development of mouthparts in mandibulates and provide a critical insight in the evolution and diversification of gnathal appendages in insects in general.

2.3 Wing Morphology

As illustrated in Fig. 2.1b, the two pairs of wings in *Gryllus* exhibit very different morphologies. At present, the molecular mechanisms underlying such differences have not been studied in this species. However, the now classical studies in *Drosophila, Tribolium, and Precis* have shown that the Hox gene *Ultrabithorax* (*Ubx*) controls the identity of the HW (Weatherbee et al. 1998, 1999; Tomoyasu et al. 2005). This is further corroborated by the recent analysis of *Ubx* in *Acheta domesticus* (Turchyn 2010), in which *Ubx* RNAi transforms HW into FW in adults.

These findings support the notion that *Ubx* controls the hind wing identity by altering the forewing program in a species-specific manner. Hence, divergence of fore- and hind wings can now be understood and examined by analyzing the downstream genes that are up- or downregulated by *Ubx* in each lineage.

In Coleoptera (beetles), the FWs are modified into firm wing cases (elytra) that protect the hind wings underneath and as such can be used to gain an insight into the genetic mechanisms governing the "hardening" wing program. Recent studies in *Tribolium* show that *apterous* (*ap*) and *achete-scute homolog* (*ASH*) are two essential factors creating exoskeleton in elytra (Tomoyasu et al. 2005, 2009). Specifically, the depletion of *ap* can cause loss of exoskeleton of the intervein regions of the elytra, whereas *ASH* RNAi leads to removal of exoskeleton patches that surround the bristles. Based upon these observations, it has been proposed that the combined input from *ap* (functioning as the intervein selector) and *ASH* (functioning as the bristle selector) controls the sclerotization of subsequent elytron regions. Intriguingly, even when both *ap* and *ASH* are knocked down in *Tribolium*, the veins remain sclerotized (Tomoyasu et al. 2009). This suggested that there is another factor that is involved in the hardening of the veins (a putative vein selector). At present, it is not known if, and to what degree, such mechanisms can be generalized to other insects with hardened FW. In *Gryllus*, the vein-intervein arrangement is quite different from the one present in beetles. While the veins in *Tribolium* run parallel along the elytra forming the longitudinal intervein regions, the veins in *Gryllus* are arranged in a complex parallel and perpendicular pattern that divides the FW into checkered intervein territories. These perpendicular crossveins that generate such a meshwork in *Gryllus* do not exist in the *Tribolium* elytra. Hence, the hardening of FW in crickets may have a different genetic underpinning when compared to beetles, further highlighting the significance of determining the molecular basis of wing diversification in *Gryllus*.

Among the general public, crickets are perhaps best known for their chirping (stridulation), which is an integral part of their mating behavior. The cricket song is produced by special structures located in the male FW (Huber et al. 1989; Montealegre et al. 2011): the plectrum (scraper) of the left FW, the stridulatory file (teeth) of the right FW, and a resonator (harp and/or mirror) located on both FWs (Fig. 2.2). During stridulation, the scraper sweeps along the row of teeth to produce the sound, while the harp and mirror serve as the acoustic tuner. In addition to their location on the FW in crickets and katydids, these structures can occasionally be found in other body regions as well. For example, the scraper in grasshoppers is located on the hind legs, whereas the FW contains the file (Snodgrass 1930). Also in rare instances, as observed in sandgropers (*Cylindracheta psammophila*), the scraper and the file are located in the mouthparts (Rentz 1991). Future studies of genetic regulation of chirping in *Gryllus* would provide a greatly needed complement to studies of mating behavior in this species. Consequently, by utilizing the power of functional testing and genome engineering, the crickets have the potential to become one of the premier animal models for studying behavior at the genetic level. More details on chirping and behavior of *Gryllus* are discussed in later chapters.

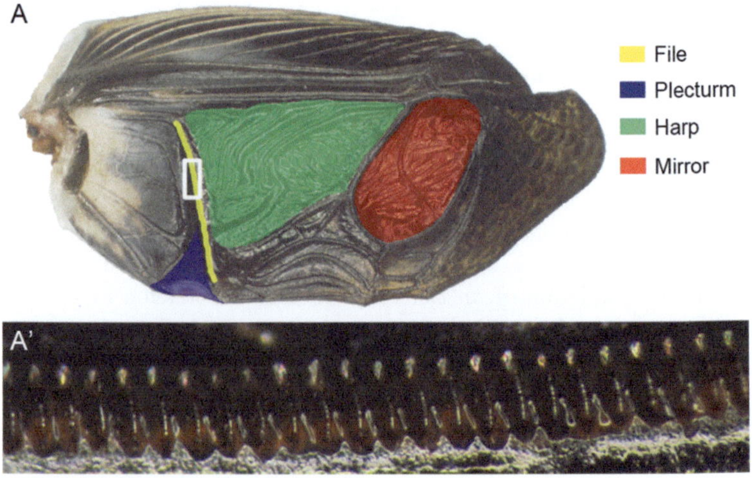

Fig. 2.2 The stridulatory organs on the forewing of male *Gryllus*. (**a**) The file (*yellow*), plectrum (*blue*), harp (*green*), and mirror (*red*) on the forewing. (**a'**) The magnified view of the file (*white rectangular* in (**a**))

2.4 Leg Morphology

In terms of its overall morphology, the *Gryllus*' most distinctive feature is the presence of the greatly enlarged, "jumping" hind legs (Fig. 2.1c). These appendages are based on the common hexapod leg design and represent one of the most recognizable examples of allometric growth in insects (Mahfooz et al. 2007; Turchyn 2010). All three pairs of legs in *Gryllus* share the same modular organization along the proximal/distal axis and are composed of six segments: coxa, trochanter, femur, tibia, tarsus, and claws. The classical functional studies in *Drosophila* (Struhl 1982), as well as more recent results from a variety of species (Chesebro et al. 2009; Hrycaj et al. 2010; Mahfooz et al. 2007; Khila et al. 2009) have established that the "default morphology" of all ventral appendages is the metathoracic (T2) leg. In orthopterans (crickets, katydids, and grasshoppers), the jumping hind legs are generated by the differential growth of the femoral and tibial segments. Here, we first discuss the proximal-distal (P/D) axis formation during early leg development, which establishes the proper positioning of leg segments. Then we focus on the enlargement of the T3 leg, which takes place during mid-late developmental stages.

2.4.1 Leg Patterning

Nearly all of the present understanding of insect leg patterning was inferred from classic experiments of *Drosophila* leg discs (reviewed in Morata 2001), which showed that the elaboration of the leg proximal/distal axis is governed by the activities of *Distal-less* (*Dll*), *dachshund* (*dac*), and *homothorax* (*hth*)/*extradenticle* (*exd*). *Dll* is expressed in the center of the leg disc, where it specifies the distal leg segments; *dac* is expressed in the middle disc region, forming the intermediate leg segments; and *hth*/*exd* are expressed on the disc periphery regulating the formation of the proximal most leg segments. The mechanism driving these patterns was described as a "gradient model" (Fig. 2.3) and was originally proposed by Lecuit and Cohen (1997). The essence of this model is the formation of the central-peripheral gradient of two key morphogens: Wingless (Wg) and Decapentaplegic

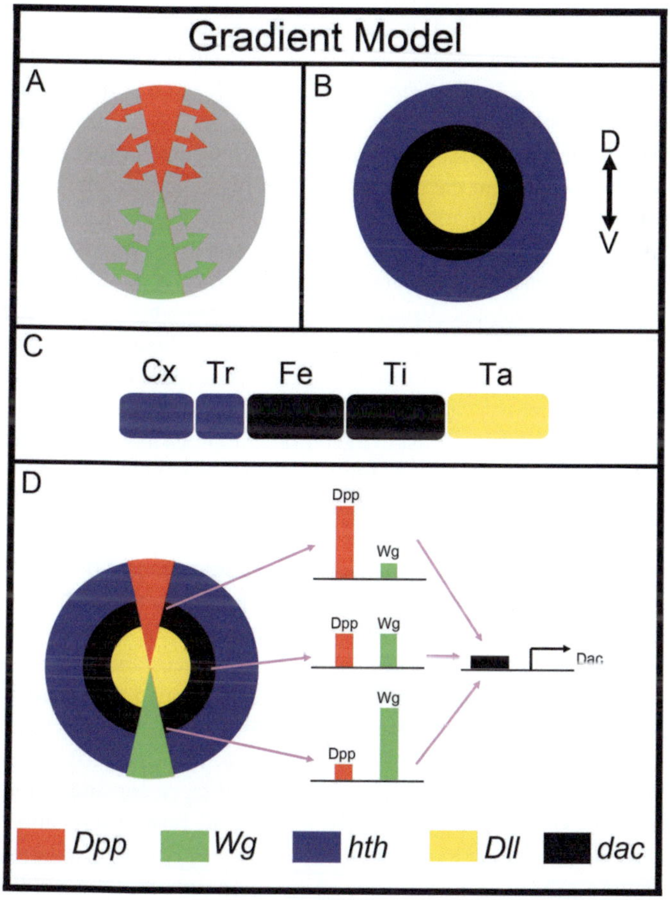

Fig. 2.3 The gradient model of *Drosophila* leg P/D axis formation, redrawn from Morata (2001). (**a**) The formation of *Wg* and *Dpp* gradient in *Drosophila* leg disc. The *arrows* illustrate how *Dpp* and *Wg* signals diffuse from their original (early) expression domains. (**b**) The gradient of *Wg* and *Dpp* concentration causes the activation of *Dll* in the center, *dac* in the middle, and *hth*/*exd* in the periphery of the leg disc, respectively. (**c**) The different subdomains in an adult leg, as illustrated by different segments. Generally, the coxa and trochanter are determined by *hth*/*exd*, femur and tibia by *dac*, and tarsal segments by *Dll*, respectively. (**d**) The main drawback of the gradient model, as illustrated by the difficulty in explaining *dac* expression pattern (Redrawn from Estella et al. 2012). The cis-regulatory module of *dac* must interpret very different ratios of Dpp:Wg signaling depending on the position in the leg disc. *Cx* coxa, *Tr* trochanter, *Fe* femur, *Ti* tibia, *Ta* tarsus

(Dpp). In the leg disc, these molecules are initially expressed as a dorsal (*dpp*) or ventral (*wg*) stripe, respectively. While the highest levels of Wg and Dpp are in the center, they begin to diffuse from their central location causing a drop in their expression levels in regions closer to the periphery of the disc (Fig. 2.3a). These differences in concentrations of Wg and Dpp, in turn, trigger the expression of *Dll* and *dac*, which display dose-dependent responses to the two morphogens (Fig. 2.3b). In the center of the leg disc (featuring the highest concentrations of Wg and Dpp), *Dll* is activated while *dac* is repressed. In the middle region (exhibiting the intermediate levels of Wg and Dpp), it is *dac* that is turned on while *Dll* is no longer expressed. In the periphery of the leg disc (characterized by the low levels of both morphogens), neither *dac* nor *Dll* is activated (Morata 2001; Lecuit and Cohen 1997). Instead, *hth* and *exd* are expressed in this region. Following the formation of such *Dll-dac-hth/exd* patterning, the leg segmentation becomes noticeable (Fig. 2.3c). While the gradient model has been broadly accepted over the last two decades to represent the general mode of leg development in insects, some of its aspects remained difficult to envision at the molecular level. As shown in Fig. 2.3d, in order to form a concentric *dac* expression domain in the middle region of the leg disc, the cis-regulatory module of *dac* must respond to distinct input levels of Wg and Dpp signals (Estella et al. 2012). This, however, is inconsistent with the key assumption that *dac* is activated by intermediate level of both Wg and Dpp. Therefore, the establishment of *Dll-dac-hth/exd* regulatory cascade in *Drosophila* leg imaginal discs cannot be solely explained by the central-peripheral gradient of Wg and Dpp. In contrast to the holometabolous mode, the appendages in hemimetabolous insects originate as limb buds that gradually extend in the distal direction during embryogenesis. In species such as *Gryllus*, *dpp* is initially localized only in the distal tip, whereas *wg* is expressed along the entire ventral margin of the limb bud (Niwa et al. 2000). Under such circumstance, only the Dpp signal would be capable of forming a proximal-distal gradient, since the level of Wg signal remains constant along the P/D axis. And yet, the conserved proximal-middle-distal expression of *hth*, *dac*, and *Dll* is still established, respectively. These observations in *Gryllus* suggest that the gradient model, at least in its strict sense, cannot fully account for the leg patterning in hemimetabolous species either.

To account for the observed inconsistencies of the gradient model, a new explanation was proposed recently (Estella et al. 2012). According to the "cascade model" (Fig. 2.4), Wg and Dpp are only required to turn on *Dll* and *epidermal growth factor receptor* (*EGFR*) in the center of the leg disc at the initialization of P/D axis. Consequently, *Dll* activates *dac* expression in the middle domain, whereas the repression of *dac* in the center region is maintained by the activity of EGFR. Compared to the gradient model, the cascade model is also more applicable to the observed situation of P/D axis patterning in *Gryllus*. At an early embryonic stage, both *wg* and *dpp* are expressed in the distal tip of the limb bud (Fig. 2.4, bottom), allowing them to act together and activate *Dll* to initialize the P/D axis (Inoue et al. 2002; Niwa et al. 2000). As the limb bud starts to elongate, neither of the morphogens are required for the maintenance of P/D axis. Instead, *Dll* activates *dac* expression in the intermediate leg region. At the same time, the repression of

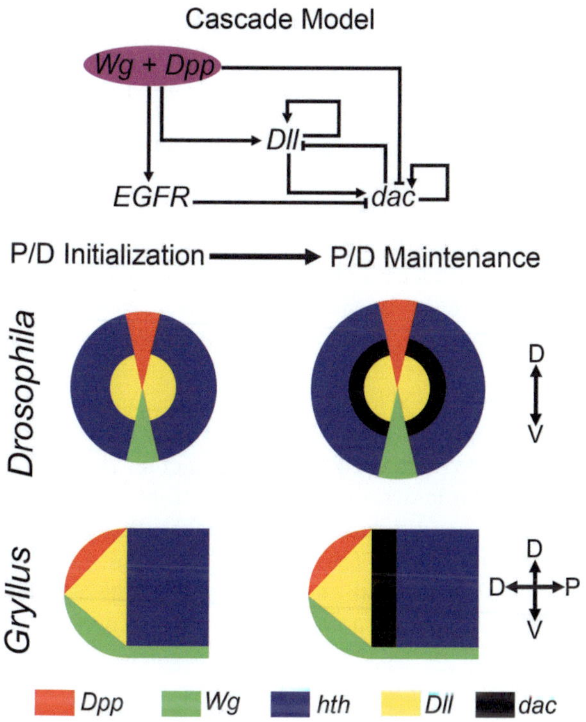

Fig. 2.4 The cascade model of P/D axis formation of an insect leg, redrawn from Estella et al. (2012). This model postulates the presence of two steps: initial phase (*left*) and maintenance phase (*right*). During the initial phase, the coexistence of *Wg* and *Dpp* in the center of *Drosophila* wing disc and distal tip of limb buds in *Gryllus* initiates P/D axis formation by activating *Dll* and *EGFR*. Later in the second phase, *Dll* activates *dac* in the middle concentric ring of *Drosophila* leg disc and middle region of *Gryllus* limb bud, after which both genes maintain their expression in a Wg+Dpp-independent manner. The expression of *hth*, in the periphery region of *Drosophila* leg disc and *Gryllus* limb bud, does not require Wg or Dpp. Note that the absence of *dac* in the distal domain is caused by joint Wg and Dpp repression during the initialization, while its expression at later stages of leg development is maintained by EGFR

dac in the distal region of *Gryllus* legs may be retained by the activity of EGFR, as recently reported by Nakamura et al. (2008b).

2.4.2 *T3 Leg Allometric Growth*

The key feature of insect T3 leg evolution is their lineage-specific differential enlargement. In basal insect lineages, such as thysanurans (firebrats) or archaeognathans (bristletails), all three pairs of legs are very similar in terms of their size and morphology (Mahfooz et al. 2004, 2007). Then, during the radiation of winged species, there was a trend toward lineage-specific enlargement of hind legs. This trend generated the situation that exists today, with differences between lineages encompassing both the involvement of different leg segments as well as varying degrees of enlargement of those segments. Orthopterans feature the largest differential growth of T3 legs, which can be almost twice as large as T1 or T2 legs. This, in turn, makes species such as *Gryllus* excellent models for elucidating the mechanisms of allometric leg growth.

From the conceptual standpoint, differential growth should be associated and coordinated with both leg patterning and joint formation. This is because a particular leg segment territory first has to be defined (i.e., separated by joints from other segments), before it can reach its final size. However, these processes have been traditionally studied separately, and presently very little is known about how they may be coordinated at the molecular level. The observation that *dpp* is differentially expressed in *Gryllus* hind legs at later embryonic stages provides a starting point for a more synergistic insight into the T3 leg enlargement (Niwa et al. 2000). As postulated by the cascade model (Fig. 2.4; (Estella et al. 2012)), the *dpp* plays an essential and conserved role during the initialization of the proximal-distal (P/D) axis in insects. During the mid-developmental stages, though, its pattern in *Drosophila* is transformed into a set of four complete circumferential rings in the tarsus while other leg segments exhibit only a patchy dorsal expression (Fig. 2.5). Recent studies have revealed that the circumferential expression domains are essential to create sharp boundaries of Dpp between leg segments, which induce a Jun N-terminal kinase (JNK)-reaper-dependent apoptosis required for the development of the leg joints (Manjon et al. 2007). Furthermore, this was also proposed to be the "ancestral mechanism" for joint formation in the distal leg regions. As illustrated in Fig. 2.5, while circumferential *dpp* expression is restricted to only tarsal segments in *Drosophila*, it expands to encompass both the femur and tibia in *Gryllus*. Similar observations were reported in another orthopteran, the grasshopper *Schistocerca americana* (Jockusch et al. 2000), suggesting that *dpp* may be involved in the formation of the femur-tibia and tibia-tarsus joints in more basal insect lineages.

The most significant divergence in *dpp* expression patterns in *Gryllus* is observed during later leg differentiation, when divisions between segments become more apparent (Niwa et al. 2000). While the previously complete circumferential rings turn into separate ventral and dorsal patches in T1 and T2 legs, this is not the case in hind legs where the rings are retained and become much wider (Fig. 2.5). As pointed by Niwa et al. 2000, these changes in expression also coincide with the increase in size of the hind legs. In light of the previously documented role of *dpp* signaling in wing tissue growth in flies (Hamaratoglu et al. 2011; Schwank and Basler 2010), it is tempting to postulate the similar causal relationship between the differential expression of *dpp* and differential growth of T3 legs in *Gryllus*. Similar results were observed in grasshoppers (Jockusch et al. 2000), but not in *Drosophila or Tribolium* (Manjon et al. 2007; Niwa et al. 2000), suggesting that this role of *dpp* may be unique to orthopterans.

The previous comparative analyses in several holo- and hemimetabolous insects have shown a tight association between the Hox gene *Ultrabithorax* (*Ubx*) and differential growth of hind legs (Mahfooz et al. 2004). In each instance, the

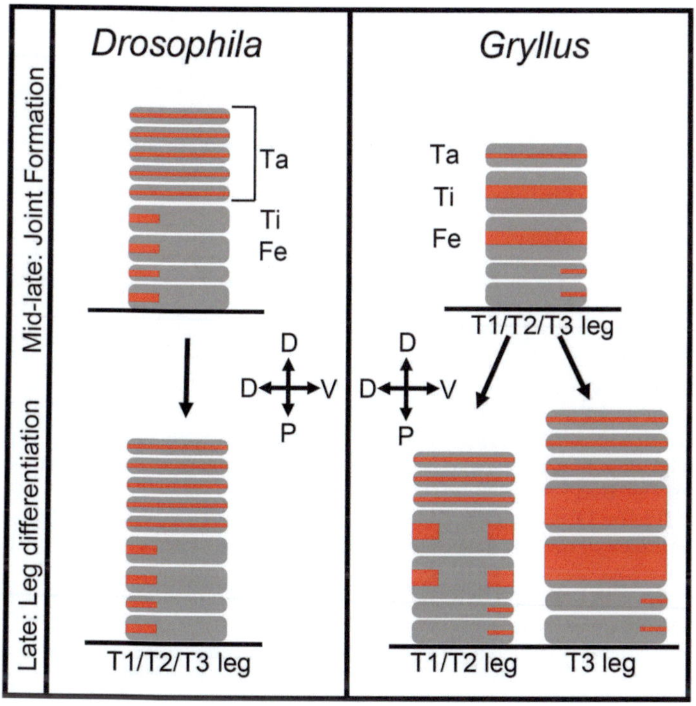

Fig. 2.5 The divergent expression patterns of *dpp* (*red*) between *Drosophila* and *Gryllus* during mid-late leg development, drawn accordingly to Manjon et al. (2007) and Niwa et al. (2000). During mid-late stage, the circumferential ringlike expression patterns of *dpp* are restricted in the tarsal segments in *Drosophila*, whereas they expand into the femur and tibia in *Gryllus*. During late stages, the expression pattern of *dpp* differentiates between T1/T2 leg and T3 leg in *Gryllus*, which is not found in *Drosophila*

expression of *Ubx* in particular leg segments is associated with the disproportionate enlargement of those segments. Furthermore, these enlarged segments display significant shortening and size reduction when *Ubx* is depleted (via RNAi) during embryogenesis (Khila et al. 2009; Mahfooz et al. 2004). These results confirm that *Ubx* can act as a "common trigger" for hind leg growth and diversification. In crickets, *Ubx* expression starts early during limb bud development and precedes *dpp* expression in T3 femur and tibia (Mahfooz et al. 2007; Zhang et al. 2005). In light of the functional studies that show that *Ubx* RNAi causes a reduction in size of these two segments in house crickets (Mahfooz et al. 2007), it is likely that similar mechanism may exist in *Gryllus* as well. Recent studies have suggested another growth factor, *Lowfat*, as a potential *Ubx* target due to its differential expression in T3 legs (Bando et al. 2011). At the same time, though, the majority of genes that were shown to be actually involved in leg growth and patterning (such as *EGFR*, *Fat*, *Dachsous*, and *Four-jointed*) exhibit common patterns in all legs (Nakamura et al. 2008a, b). These results indicate that *Ubx* may trigger the enlargement of T3 leg by selectively acting on a portion of growth regulators instead of upregulating the entire leg growth network. Thus, future studies should focus on determining whether *Ubx* can indeed activate the expression of *dpp* or *Lowfat* and if such activation is orthopteran specific or represents a more general way of generating differential growth of T3 legs.

References

Bando T, Hamada Y, Kurita K, Nakamura T, Mito T, Ohuchi H, Noji S (2011) Lowfat, a mammalian Lix1 homologue, regulates leg size and growth under the Dachsous/Fat signaling pathway during tissue regeneration. Dev Dyn 240(6):1440–1453. doi:10.1002/dvdy.22647

Brown S, DeCamillis M, Gonzalez-Charneco K, Denell M, Beeman R, Nie W, Denell R (2000) Implications of the Tribolium Deformed mutant phenotype for the evolution of Hox gene function. Proc Natl Acad Sci U S A 97(9):4510–4514

Chesebro J, Hrycaj S, Mahfooz N, Popadic A (2009) Diverging functions of Scr between embryonic and post-embryonic development in a hemimetabolous insect, Oncopeltus fasciatus. Dev Biol 329(1):142–151

Curtis CD, Brisson JA, DeCamillis MA, Shippy TD, Brown SJ, Denell RE (2001) Molecular characterization of Cephalothorax, the Tribolium ortholog of Sex combs reduced. Genesis 30 (1):12–20

de Celis JF, Barrio R, Kafatos FC (1996) A gene complex acting downstream of dpp in Drosophila wing morphogenesis. Nature 381(6581):421–424. doi:10.1038/381421a0

DeCamillis M, ffrench-Constant R (2003) Proboscipedia represses distal signaling in the embryonic gnathal limb fields of Tribolium castaneum. Dev Genes Evol 213(2):55–64. doi:10.1007/s00427-002-0291-7

DeCamillis MA, Lewis DL, Brown SJ, Beeman RW, Denell RE (2001) Interactions of the Tribolium Sex combs reduced and proboscipedia orthologs in embryonic labial development. Genetics 159(4):1643–1648

Estella C, Voutev R, Mann RS (2012) A dynamic network of morphogens and transcription factors patterns the fly leg. Curr Top Dev Biol 98:173–198. doi:10.1016/B978-0-12-386499-4.00007-0

Hamaratoglu F, de Lachapelle AM, Pyrowolakis G, Bergmann S, Affolter M (2011) Dpp signaling activity requires Pentagone to scale with tissue size in the growing Drosophila wing imaginal disc. PLoS Biol 9(10):e1001182. doi:10.1371/journal.pbio.1001182

Hrycaj SM (2010) Unraveling the molecular mechanisms of insect diversity. Wayne State University, Detroit

Hrycaj S, Chesebro J, Popadic A (2010) Functional analysis of Scr during embryonic and post-embryonic development in the cockroach, Periplaneta americana. Dev Biol 341(1):324–334. doi:10.1016/j.ydbio.2010.02.018

Huber F, Moore TE, Loher W (1989) Cricket behavior and neurobiology. Comstock Pub. Associates, Ithaca

Hughes CL, Kaufman TC (2000) RNAi analysis of Deformed, proboscipedia and Sex combs reduced in the milkweed bug Oncopeltus fasciatus: novel roles for Hox genes in the hemipteran head. Development 127(17):3683–3694

Inoue Y, Niwa N, Mito T, Ohuchi H, Yoshioka H, Noji S (2002) Expression patterns of hedgehog, wingless, and decapentaplegic during gut formation of Gryllus bimaculatus (cricket). Mech Dev 110(1–2):245–248

Jockusch EL, Nulsen C, Newfeld SJ, Nagy LM (2000) Leg development in flies versus grasshoppers: differences in dpp expression do not lead to differences in the expression of downstream components of the leg patterning pathway. Development 127(8):1617–1626

Khila A, Abouheif E, Rowe L (2009) Evolution of a novel appendage ground plan in water striders is driven by changes in the Hox gene Ultrabithorax. PLoS Genet 5(7):e1000583. doi:10.1371/journal.pgen.1000583

Kim J, Sebring A, Esch JJ, Kraus ME, Vorwerk K, Magee J, Carroll SB (1996) Integration of positional signals and regulation of wing formation and identity by Drosophila vestigial gene. Nature 382(6587):133–138. doi:10.1038/382133a0

Lecuit T, Cohen SM (1997) Proximal-distal axis formation in the Drosophila leg. Nature 388 (6638):139–145. doi:10.1038/40563

Mahfooz NS, Li H, Popadic A (2004) Differential expression patterns of the hox gene are associated with differential growth of insect hind legs. Proc Natl Acad Sci U S A 101 (14):4877–4882. doi:10.1073/pnas.0401216101

Mahfooz N, Turchyn N, Mihajlovic M, Hrycaj S, Popadic A (2007) Ubx regulates differential enlargement and diversification of insect hind legs. PLoS One 2(9):e866. doi:10.1371/journal.pone.0000866

Manjon C, Sanchez-Herrero E, Suzanne M (2007) Sharp boundaries of Dpp signalling trigger local cell death required for Drosophila leg morphogenesis. Nat Cell Biol 9(1):57–63. doi:10.1038/ncb1518

Martinez-Arias A, Ingham PW, Scott MP, Akam ME (1987) The spatial and temporal deployment of Dfd and Scr transcripts throughout development of Drosophila. Development 100 (4):673–683

Merrill VK, Turner FR, Kaufman TC (1987) A genetic and developmental analysis of mutations in the Deformed locus in Drosophila melanogaster. Dev Biol 122(2):379–395

Montealegre ZF, Jonsson T, Robert D (2011) Sound radiation and wing mechanics in stridulating field crickets (Orthoptera: Gryllidae). J Exp Biol 214(Pt 12):2105–2117. doi:10.1242/jeb.056283

Morata G (2001) How Drosophila appendages develop. Nat Rev Mol Cell Biol 2(2):89–97. doi:10.1038/35052047

Nakamura T, Mito T, Bando T, Ohuchi H, Noji S (2008a) Dissecting insect leg regeneration through RNA interference. Cell Mol Life Sci 65(1):64–72. doi:10.1007/s00018-007-7432-0

Nakamura T, Mito T, Miyawaki K, Ohuchi H, Noji S (2008b) EGFR signaling is required for re-establishing the proximodistal axis during distal leg regeneration in the cricket Gryllus bimaculatus nymph. Dev Biol 319(1):46–55. doi:10.1016/j.ydbio.2008.04.002

Neumann CJ, Cohen SM (1998) Boundary formation in Drosophila wing: Notch activity attenuated by the POU protein Nubbin. Science 281(5375):409–413

Ng M, Diaz-Benjumea FJ, Vincent JP, Wu J, Cohen SM (1996) Specification of the wing by localized expression of wingless protein. Nature 381(6580):316–318. doi:10.1038/381316a0

Niwa N, Inoue Y, Nozawa A, Saito M, Misumi Y, Ohuchi H, Yoshioka H, Noji S (2000) Correlation of diversity of leg morphology in Gryllus bimaculatus (cricket) with divergence in dpp expression pattern during leg development. Development 127(20):4373–4381

Passalacqua KD, Hrycaj S, Mahfooz N, Popadic A (2010) Evolving expression patterns of the homeotic gene Scr in insects. Int J Dev Biol 54(5):897–904. doi:10.1387/ijdb.082839kp

Rentz DCF (1991) Orthoptera. In: CSIRO (ed) The insects of Australia, vol 1, 2nd edn. Melbourn University Press, Carlton, pp 369–393

Rogers BT, Peterson MD, Kaufman TC (1997) Evolution of the insect body plan as revealed by the sex combs reduced expression pattern. Development 124(1):149–157

Rogers BT, Peterson MD, Kaufman TC (2002) The development and evolution of insect mouthparts as revealed by the expression patterns of gnathocephalic genes. Evol Dev 4(2):96–110

Schwank G, Basler K (2010) Regulation of organ growth by morphogen gradients. Cold Spring Harb Perspect Biol 2(1):a001669. doi:10.1101/cshperspect.a001669

Shippy TD, Guo J, Brown SJ, Beeman RW, Denell RE (2000) Analysis of maxillopedia expression pattern and larval cuticular phenotype in wild-type and mutant tribolium. Genetics 155 (2):721–731

Shippy TD, Rogers CD, Beeman RW, Brown SJ, Denell RE (2006) The Tribolium castaneum ortholog of Sex combs reduced controls dorsal ridge development. Genetics 174(1):297–307. doi:10.1534/genetics.106.058610

Snodgrass RE (1930) Insects, their ways and means of living, vol 5, Smithsonian scientific series. Smithsonian Institution Series, New York

Snodgrass RE (1993) Principles of insect morphology. Cornell University Press, Ithaca/New York
Struhl G (1982) Genes controlling segmental specification in the Drosophila thorax. Proc Natl Acad Sci U S A 79(23):7380–7384
Tomoyasu Y, Wheeler SR, Denell RE (2005) Ultrabithorax is required for membranous wing identity in the beetle Tribolium castaneum. Nature 433(7026):643–647. doi:10.1038/nature03272
Tomoyasu Y, Arakane Y, Kramer KJ, Denell RE (2009) Repeated co-options of exoskeleton formation during wing-to-elytron evolution in beetles. Curr Biol 19(24):2057–2065. doi:10.1016/j.cub.2009.11.014
Turchyn N (2010) The cellular and genetic mechanisms underlying the morphological diversity of insects. Wayne State University, Detroit
Weatherbee SD, Halder G, Kim J, Hudson A, Carroll S (1998) Ultrabithorax regulates genes at several levels of the wing-patterning hierarchy to shape the development of the Drosophila haltere. Gene Dev 12(10):1474–1482. doi:10.1101/Gad.12.10.1474
Weatherbee SD, Nijhout HF, Grunert LW, Halder G, Galant R, Selegue J, Carroll S (1999) Ultrabithorax function in butterfly wings and the evolution of insect wing patterns. Curr Biol 9(3):109–115
Zhang H, Shinmyo Y, Mito T, Miyawaki K, Sarashina I, Ohuchi H, Noji S (2005) Expression patterns of the homeotic genes Scr, Antp, Ubx, and abd-A during embryogenesis of the cricket Gryllus bimaculatus. Gene Expr Patterns 5(4):491–502. doi:10.1016/j.modgep.2004.12.006

Chapter 3
Leg Formation and Regeneration

Tetsuya Bando, Yoshimasa Hamada, and Sumihare Noji

Abstract In contrast to higher vertebrates, orthopteran nymphs have remarkable regenerative capacity for regrowing complex morphological structures and organs. In this review, we summarize the molecular basis of tissue regeneration in the cricket *Gryllus bimaculatus*. In this species, the lost part of a leg can be regenerated epimorphically from blastema cells, a population of dedifferentiated proliferating cells. Blastema cell proliferation is regulated by JAK/STAT and Salvador/Warts/Hippo signaling pathways. The positional information for leg regrowth, which includes the recognition of amputated position and proper regeneration, is maintained by Dachsous/Fat signaling. The regrowth of lost leg segments is reconstructed through the expressions of genes in the *hedgehog*, *wingless*, *decapentaplegic*, and *Egf* signaling pathways and epigenetic modifiers *E*(*z*) and *Utx*. The insights obtained reveal the high level of conservation between insects and vertebrates, suggesting that *Gryllus* may be a suitable model for human regenerative medicine studies.

Keywords Blastema • Positional information • JAK/STAT signaling • Salvador/Warts/Hippo signaling • Dachsous/Fat • Histone H3K27me3

3.1 Leg Formation

Hemimetabolous insect legs, much like vertebrate limbs, develop from a limb bud (Fig. 3.1) that directly develops into the final adult appendage through successive nymphal stages. This is in contrast to the *Drosophila* leg, which develops from a specific imaginal disc, which is a sheet of cells specified during larval development.

T. Bando (✉) • Y. Hamada
Department of Cytology and Histology, Okayama University Graduate School of Medicine, Dentistry and Pharmaceutical Sciences, 2-5-1, Shikata-cho, Kita-ku, Okayama City, Okayama 700-8530, Japan
e-mail: tbando@cc.okayama-u.ac.jp

S. Noji
Tokushima University, 2-24, Shinkura-cho, Tokushima City 770-8501, Tokushima, Japan

H.W. Horch et al. (eds.), *The Cricket as a Model Organism*,
DOI 10.1007/978-4-431-56478-2_3

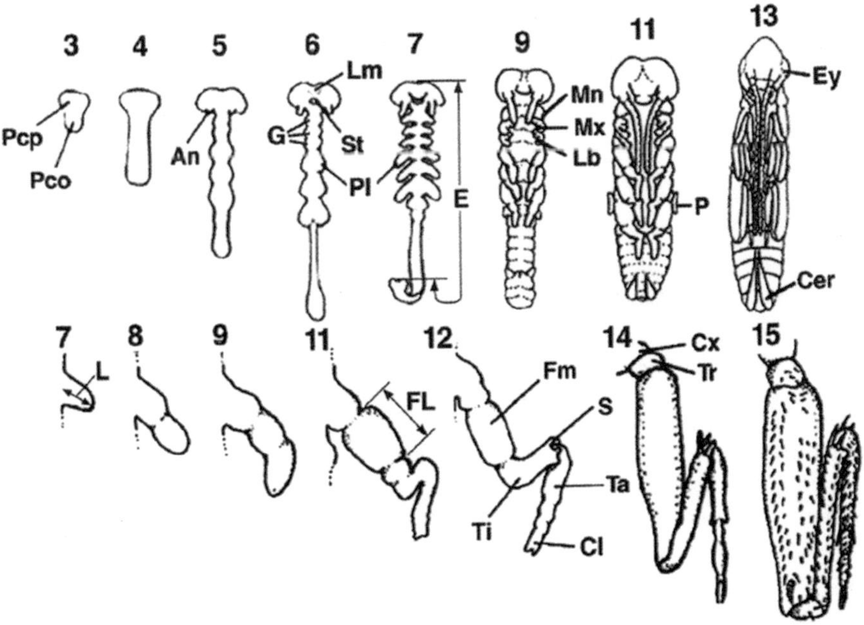

Fig. 3.1 Embryonic development (*upper column*) and limb development (*lower column*) of the cricket *Gryllus bimaculatus* (Niwa et al. 1997)

Although the leg developmental processes differ considerably between crickets and *Drosophila*, the signaling molecules involved in development and regeneration are similar. This suggests that the knowledge of leg development in *Drosophila* can be used as a starting point to understand the molecular mechanisms controlling leg development in hemimetabolous insects.

On the basis of experimental data from studies of *Drosophila*, Meinhardt (1983) proposed that the insect leg is divided into posterior, dorsal/anterior, and ventral/anterior compartments during its growth along proximodistal (P/D) axis. The region where the boundaries between these compartments intersect defines the presumptive distal tip of the leg, and the diffusible morphogens are secreted from this site, further promoting appendage growth and fate specification along the P/D axis (boundary model). Subsequent studies confirmed that *Drosophila* limb is, indeed, divided into three compartments with morphogens secreted from each of them, as predicted by the boundary model (Basler and Struhl 1994; Campbell et al. 1993; Diaz-Benjumea et al. 1994). Hedgehog (Hh), secreted from posterior compartment, induces secretion of Decapentaplegic (Dpp) and Wingless (Wg) from dorsal/anterior and ventral/anterior compartments, respectively (Basler and Struhl 1994; Campbell et al. 1993; Diaz-Benjumea et al. 1994). The P/D axis is initiated at the boundary between *dpp*-expressing dorsal/anterior compartment and *wg*-expressing ventral/anterior compartment (Campbell et al. 1993). The distal region is patterned by a distal-to-proximal gradient of epidermal growth factor receptor (EGFR) tyrosine kinase activity (Campbell 2002) that is established according to the concentration of EGF-related ligands. The source of these ligands is located at the presumptive tip of the limb bud that requires the highest level of EGFR activity, whereas more proximal regions require progressively less EGF (Campbell 2002). This observation indicates that EGF corresponds to the predicted morphogen produced by the distal organizer and that the distal leg segments (such as the tarsus) may be patterned by a distal-to-proximal gradient of EGFR activity (Galindo et al. 2002). These findings thus provided the molecular basis of the boundary model, now termed the "molecular boundary model."

We examined whether this new model could be applied to the development of hemimetabolous insect legs. For this purpose, we observed the expression patterns of the *hh*, *wg*, and *dpp* homologues in *Gryllus bimaculatus* (named *Gb'hh*, *Gb'wg*, and *Gb'dpp*, respectively) and found no striking differences between *Drosophila* and *Gryllus* expression patterns (Niwa et al. 2000). These genes were all expressed in the *Gryllus* limb bud, the only difference being the expression pattern of *dpp*. *Gb'dpp* transcripts were detected as a spot in the dorsal tip of the limb bud, whereas *dpp* transcripts form a stripe along the dorsal side of the anteroposterior (AP) boundary in the *Drosophila* leg imaginal disc. The morphogenetic significance of this difference is currently unknown. However, these results allowed us to conclude that the molecular boundary model is a fundamental and universal mode for forming the P/D axis in the insect leg.

Determination of segment identity in the developing hemimetabolous insect leg is the next step in the leg patterning process. Nymphs of hemimetabolous insects have six leg segments, arranged along the P/D axis in the following order: coxa, trochanter, femur, tibia, tarsus, and claw. These leg segments are not formed simultaneously, but sequentially from the presumptive trochanter/femur, femur/tibia, and tibia/tarsus boundaries, while the initial structure consists only of the presumptive coxa and claw (Miyawaki et al. 2002). During formation of the leg segments, a newly formed segment appears to be intercalated between the proximal and distal structures starting with the trochanter and working its way distally. Segment identity may be determined by the *Gb'homothorax* (*Gb'hth*), *Gb'dachshund* (*Gb'dac*), and *Gb'Distal-less* (*Gb'Dll*) genes in the cricket limb bud (Fig. 3.2) (Inoue et al. 2002), similar to what is seen in the *Drosophila* imaginal disc (Goto and Hayashi 1999; Kojima 2004). Their expression patterns are likely to be regulated by Hh, Wg, and Dpp, which are also expressed in the limb bud (Kojima 2004). Recent findings have shown that in *Drosophila* legs the sharp discontinuity of Dpp signaling is required for forming the leg joint. Based upon expression patterns of *Gb'dpp*, it may also be involved in the formation of the leg joint in the cricket leg bud as well (Manjon et al. 2007).

The Notch signaling pathway has a central role in the segmentation of the arthropod leg. In flies and spiders, *Notch* and its ligands are expressed in developing legs, and depletion of *Notch* causes segment fusion and leg shortening (Tajiri et al. 2011). In *Gryllus*, *Gb'Notch* and its ligand *Gb'Delta* were expressed in developing legs in a pattern of concentric rings. RNAi knockdown of *Gb'Notch*

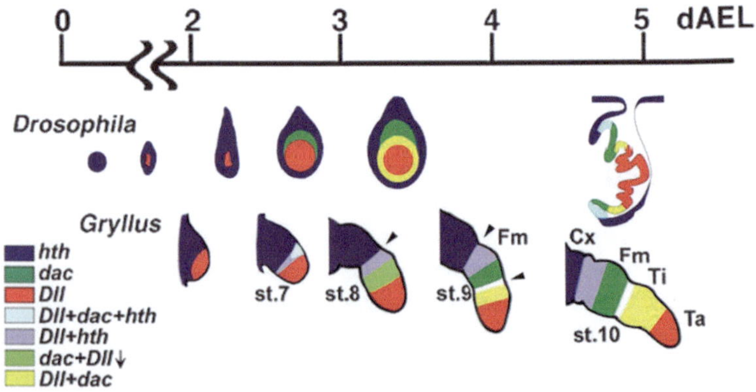

Fig. 3.2 Expression patterns of leg patterning genes *homothorax* (*hth*), *dachshund* (*dac*), and *Distal-less* (*Dll*) in the leg disc of *Drosophila* and developing limb of *Gryllus* (Inoue et al. 2002)

expression markedly reduced leg length with a loss of joints and disrupted expression patterns of leg patterning genes (Mito et al. 2011). In *Drosophila* legs, activated EGF signaling is present in all segments and is necessary for joint formation (Galindo et al. 2005). A similar pattern is observed for *Gb'Egfr* expression during leg development (Nakamura et al. 2008a). Expression of *Gb'Egfr* was severely reduced in *Gb'Notch* RNAi-treated legs, revealing that *Gb'Notch* depletion results not only in the loss of leg joints but also in a reduction of EGF signaling activity. These results suggest that Notch signaling is essential for the correct specification of segment boundaries and normal leg growth in *Gryllus* (Mito et al. 2011).

Intrasegmental pattern formation follows the determination of leg segment identity. This process is characterized by the pattern of spines and hairs specific to each segment. Thus far, we do not know the mechanisms that drive intrasegmental pattern formation. However, information about the regulation of the pattern within a segment has been obtained from intercalary regeneration experiments in hemimetabolous insects (Meinhardt 1983).

3.2 Leg Regeneration

Nymphs of hemimetabolous insects such as cockroaches and crickets have a remarkable regenerative capacity (Fig. 3.3). Cricket nymphs are able to regenerate lost parts of tissues such as antennae, legs, or cerci. *Also, the relatively large size of nymphal legs makes them easy to manipulate surgically*. When cricket nymphs lose a part of their leg, the injured region is immediately covered by a scab, which is a clot of body fluids (Fig. 3.3c–e). The injured region is made up of wound epidermis under the scab and is covered by a newly formed cuticle in the next nymphal stage (Fig. 3.3f). At the distal position of amputated leg, a blastema begins to form, which

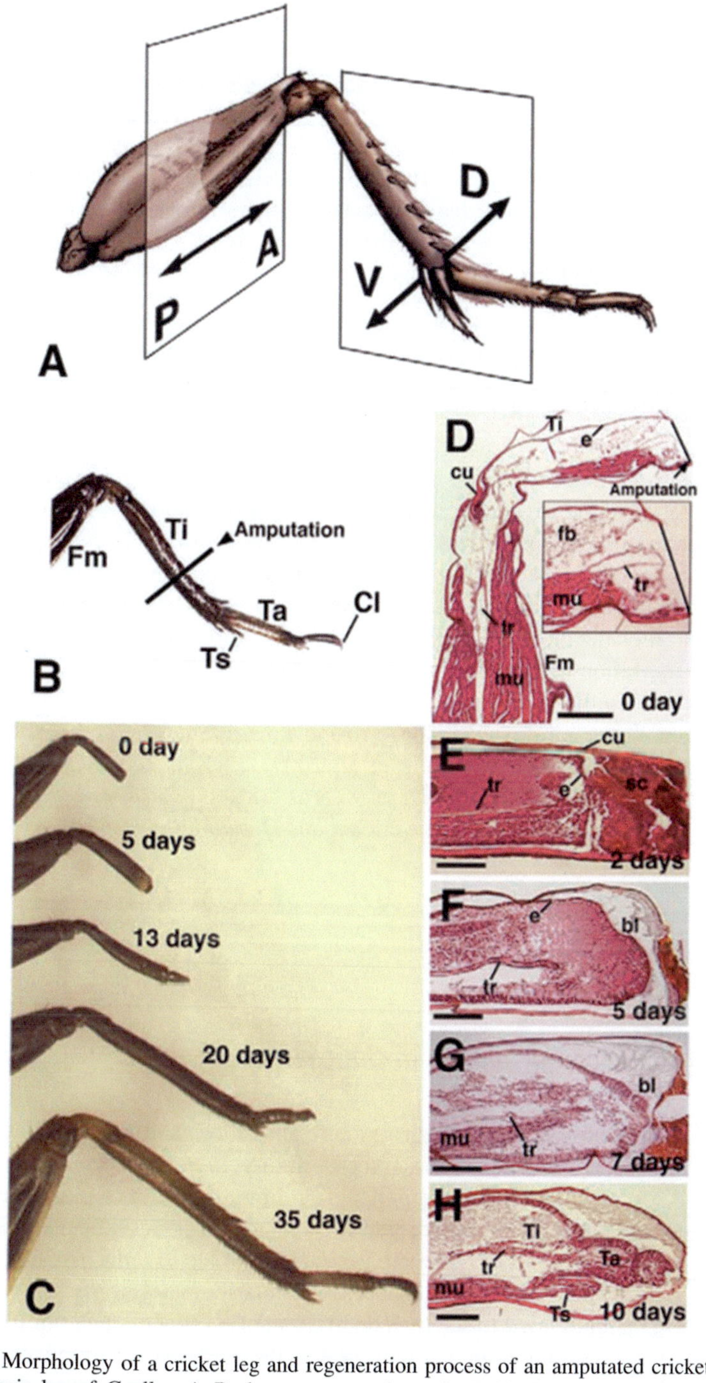

Fig. 3.3 Morphology of a cricket leg and regeneration process of an amputated cricket leg. (**a**) Metathoracic leg of *Gryllus*. A–P, the anteroposterior axis; D–V, the dorsoventral axis. (**b**, **c**) When a leg of a cricket nymph is amputated in the tibia, the lost part is regenerated completely through four molts within about 35 days after amputation. (**d**–**h**) Morphological change in a regenerating leg after amputation (details are described in the text). *Cl* claw, *Fm* femur, *Ta* tarsus, *Ti* tibia, *Ts* tibial spur, *cu* cuticle, *e* epidermis, *bl* blastema, *fb* fat body, *mu* muscle, *sc* scab, *tr* trachea. Scale bars: 200 mm in (**d**); 100 mm in (**e**–**h**) (Mito et al. 2002)

is a population of dedifferentiated proliferating cells. These cells undergo rapid proliferation and differentiate into multiple cell types to restore the complex structure of the lost portion. The amputated position is determined based on positional information along the P/D axis to regenerate the leg epimorphically (Fig. 3.3g–h). The molecular basis of each regeneration step is a long-standing question, and regeneration research in the cricket using recent molecular techniques has answered some of these questions.

3.3 Wound Healing

The injured region is covered by a scab and is re-epithelized by wound epidermis. Wound healing is the trigger for regeneration; however, the molecular basis of the wound healing process is still unclear in the cricket. From our knowledge of other insects, coagulation is the first response to wound healing in insects (Bidla et al. 2005). Phenoloxidase changes the physical properties of the clot through cross-linking and melanization. Re-epithelization occurs under the scab by the proliferation and migration of the surrounding epithelial cells. In *Drosophila*, epidermal injury initiates epithelial remodeling and integument repair at wound sites via the transcription factor Grainyhead, the tyrosine kinase Stitcher, and extracellular-signal-regulated kinase ERK (Wang et al. 2009). ERK is activated in the early stages of regeneration in *Drosophila* as well as crickets (Nakamura et al. 2008a), suggesting that wound healing in the cricket and *Drosophila* is mediated by similar molecular mechanisms.

3.4 Blastema Formation

3.4.1 Blastema Formation in the Cricket

Blastema formation is a key step for tissue regeneration, and the ability to form a blastema is directly related to the insects' ability to regenerate lost legs. The blastema is a population of dedifferentiated proliferating cells or pluripotent stem cells that forms under the wound epidermis. Many regenerative organisms form blastema to repair lost part of tissues; however, the molecular mechanism in which blastema cells form is unclear. In the cricket, blastema cells are thought to be derived from differentiated tissue by a dedifferentiation process and not from stem cells. Two to 5 days after amputation of the leg, the blastema is formed at the distal end of the leg stump under the wound epidermis. Experiments have shown that BrdU (5-bromo-2-deoxyuridine) and EdU (5-ethynyl-2′-deoxyuridine) are incorporated into blastema cells, suggesting that blastema cells have highly proliferating activity (Bando et al. 2009; Mito et al. 2002).

3.4.2 Signal Transduction Pathway for Blastema Formation

To determine the molecular mechanism for blastema formation, a comparative transcriptome analysis between early blastema and the corresponding non-regenerative tissue was performed. The expression levels of various genes that are activated in the blastema are compared with non-regenerative tissue including apoptosis-related proteins, matrix metalloproteinases, and microRNA-binding proteins. These genes were also isolated from the blastema of other organisms undergoing regeneration suggesting that a comparative transcriptome approach is valid. Expressions of many components of signal transduction pathways including Wg/Wnt, Dpp/TGF-beta, Hh/Shh, EGF, IGF, Notch, Toll, Salvador/Warts/Hippo (Sav/Wts/Hpo), and JAK/STAT signaling pathways were also upregulated (Bando et al. 2013).

The JAK/STAT pathway transduces cytokine signals into the cell. When cytokines are received by the cytokine receptor Domeless (Dome), Janus kinase Hopscotch (Hop) is phosphorylated and in turn phosphorylates the transcription factors signal-transducer and activator of transcription (STAT) protein. The phosphorylated STAT is dimerized allowing it to transduce signals into the nuclei.

When dsRNA for a given gene is injected into a cricket nymph, in some cases, a phenotype appears after molting and lasts to the adult stage (nymphal RNAi). To examine the effect of RNAi silencing during leg regeneration, the metathoracic leg of third instar nymphs was amputated at the distal tibia immediately after injection of the dsRNA. In the cricket blastema, expressions of *Gb'dome*, *Gb'hop*, and *Gb'Stat* were upregulated temporally, and RNAi against *Gb'dome*, *Gb'hop*, or *Gb'Stat* inhibits leg regeneration. *Gb'Stat*^{RNAi} cricket did not regenerate the lost part of the leg; however, body morphology, size, and weight were similar to the control cricket (Fig. 3.4a, b). Proliferation of the blastema cells was suppressed when compared with the control, as shown by the relative quantification of *cyclin E* expression (Fig. 3.4c). RNAi against negative regulators of JAK/STAT signaling, such as *Socs2* or *Socs5*, leads to hyper activation of JAK/STAT signaling and caused lengthening in the regenerated leg when compared to the control. In the

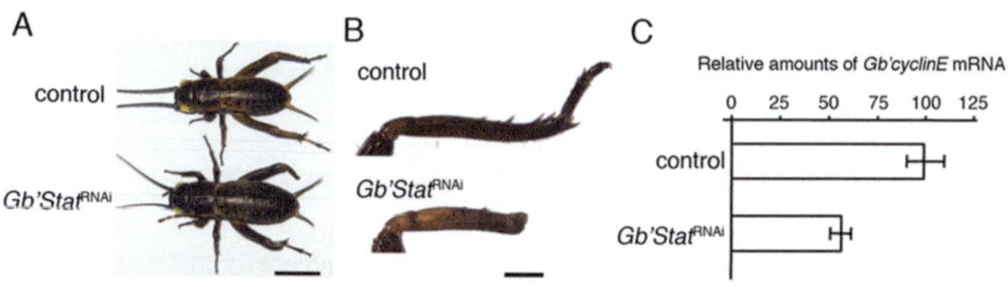

Fig. 3.4 Functional analyses of the role of *Gb'Stat* in blastema cell proliferation. (**a**) Morphologies of control and *Gb'Stat*^{RNAi} nymphs at sixth instar. (**b**) Morphologies of regenerated legs of control and *Gb'Stat*^{RNAi} crickets at sixth instar. (**c**) Relative *Gb'cyclinE* expression as revealed by q-PCR in the control (set at 100 %) and *Gb'Stat*^{RNAi} nymphs. *Error bars* indicate s.d. Scale bars: 5 mm in (**a**) and 1 mm in (**b**) (Bando et al. 2013)

longer regenerated legs, cell proliferation is activated by *cyclin E* expression. Similarly, both *Gb'Stat*^{RNAi} and *Gb'Socs2*^{RNAi} showed affects in leg regeneration and did not affect nymphal growth, suggesting JAK/STAT signaling is only involved in the proliferation of blastema cells during the leg regeneration process (Bando et al. 2013).

3.4.3 Signal Transduction Pathway for Suppression of Blastema Cell Proliferation

Recently the Sav/Wts/Hpo signaling pathway has been shown to be involved in tumor suppression (Harvey et al. 2013). In *Drosophila*, two large protocadherins, Dachsous (Ds) and Fat (Ft), transduce cell-cell contact into the nuclei via Sav/Wts/Hpo signaling to regulate cell proliferation. Hpo and Wts encode serine/threonine kinases and phosphorylate a transcriptional coactivator Yorkie (Yki). Unphosphorylated Yki locates in nuclei binding with the transcription factor Scalloped (Sd) and activates transcription of several genes including *Drosophila inhibitor of apoptosis protein* (*Diap1*), *cyclin E*, *expanded* (*ex*), and the microRNA *bantam*, to promote cell proliferation and inhibit apoptosis. Phosphorylated Yki is degraded by ubiquitin-proteasome machinery; hence, cell proliferation is suppressed (Harvey and Hariharan 2012; Staley and Irvine 2012).

The blastemas of *Gb'ds*^{RNAi}, *Gb'ft*^{RNAi} or *Gb'wts*^{RNAi} crickets were enlarged and the relative ratio of EdU incorporation increased compared to the control (Fig. 3.5a–c). Conversely, *Gb'yki*^{RNAi} suppressed blastema cell proliferation and caused a regeneration defective phenotype. RNAi against *Gb'wts* and *Gb'yki* was lethal and was not scored. *Gb'ds*^{RNAi} and *Gb'ft*^{RNAi} showed abnormal regeneration phenotypes but did not show abnormalities in other tissues, suggesting that Ds and Ft suppress blastema cell proliferation via Wts by phosphorylating Yki in the regenerating leg. Wts and Yki may regulate cell proliferation throughout the body based on the mortality from the RNAi experiments (Bando et al. 2009, 2011a).

3.4.4 Epigenetic Regulation During Blastema Formation

In addition to the signaling molecules, expressions of epigenetic factors including histone methyltransferases, histone demethylases, histone acetyltransferases, histone deacetylases, and components of chromatin remodeling SWI/SNF complex were all upregulated in the comparative transcriptome analysis (Bando et al. 2013). This is consistent with the current view that epigenetic regulation of cell fate-specific genes may result in cell differentiation. Epigenetic modifications could also

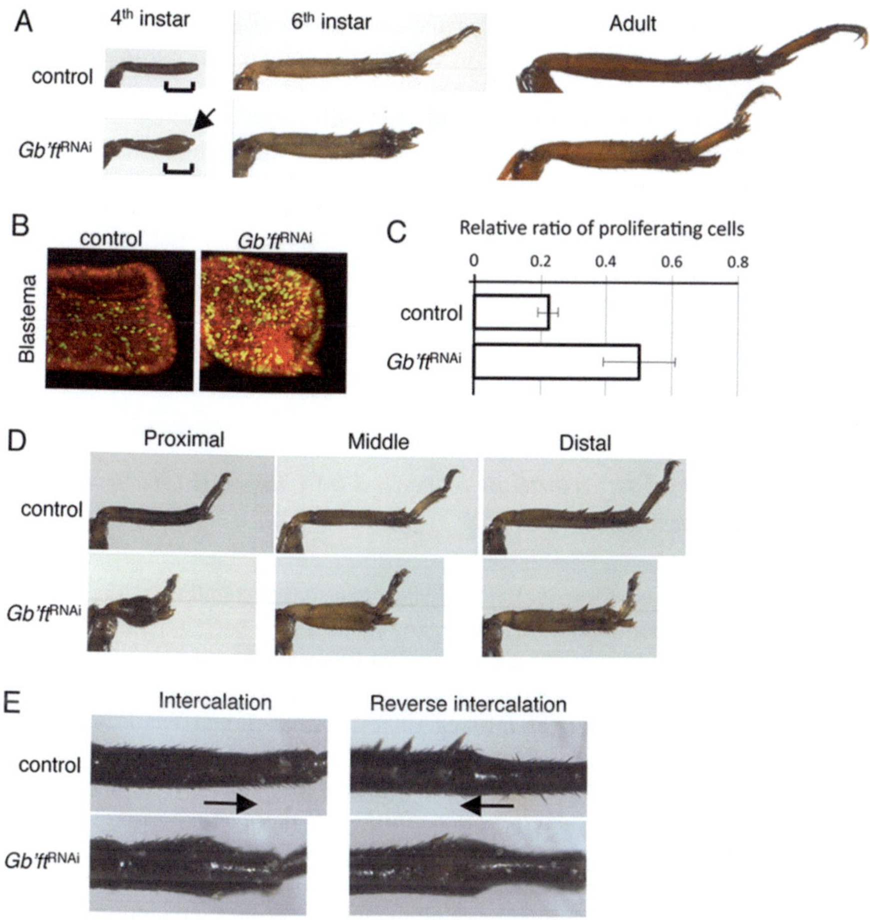

Fig. 3.5 Functional analyses of the role of *Gb′ft* in blastema cell proliferation and positional information. (**a**) Typical morphological changes of regenerating legs of control and *Gb′ft*^{RNAi} crickets. The regenerating legs after amputation at the third instar were observed for the fourth and sixth instar nymphs and adult. A normal regenerating leg is shown at the *top* as a control. The blastema and distal enlargement phenotype of *Gb′ft*^{RNAi} nymphs at the fourth instar is indicated in *bracket* and by *arrow*. (**b**) Effect of *Gb′ft*^{RNAi} on the proliferation of regenerating cells. PI staining (*red*) and EdU incorporation (*green*) of regenerating cells in control and *Gb′ft*^{RNAi} nymphs at 2 dpa. Merged signals appear *yellow*. (**c**) Relative ratio of proliferating cells/total cells in blastema of control and *Gb′ft*^{RNAi} nymphs. *Error bars* indicate s.d. (**d**) Effects of *Gb′ft*^{RNAi} on the size of regenerating legs at sixth instar. These regenerating legs were amputated at distal, middle, or proximal positions at the third instar. (**e**) Effect of *Gb′ft*^{RNAi} on intercalary regeneration. Normal (*left*) and reverse (*right*) intercalation in control and *Gb′ft*^{RNAi} nymphs at the fifth instar. Orientation of surface bristles is indicated by *leftward arrows* (reverse orientation) and *rightward arrows* (normal orientation) (Bando et al. 2009, 2011a)

promote dedifferentiation by changing the expressions of cell fate specification genes (Bando et al. 2013), as seen in the reprogramming process of iPS cells (Meissner 2010).

In conclusion, the cell fate of differentiated cells could be reprogrammed via epigenetic factors during the dedifferentiation process. Multipotent blastema cells are accumulated under the wound epidermis, and proliferation of blastema cells is activated via the JAK/STAT signaling pathway mediated by activation of *cyclin E* expression. Proliferation of blastema cells is suppressed by Ds and Ft protocadherins via Sav/Wts/Hpo signaling pathway.

3.5 Positional Information

3.5.1 Positional Information Revealed by Classical Grafting Experiments

Numerous grafting experiments have advanced our understanding of the mechanisms of pattern formation and tissue/organ regeneration. Such experiments have shown that each leg segment has a similar set of P/D and circumferential positional values; furthermore, when positional values are missing in the amputated or grafted legs, the legs can intercalate the missing values (Campbell and Tomlinson 1995; French 1981; Meinhardt 1983). *These studies lead to the formulation of the shortest intercalation rule, which states that the confrontation of cells with different positional values would result in the intercalation of intermediate values via the shortest route.*

Another interesting phenomenon in leg regeneration is the formation of supernumerary legs, which occurs after grafting the distal part of a leg onto the contralateral leg stump, a procedure that brings two structures with inverted AP axes in contact at the amputation plane (Campbell and Tomlinson 1995; French 1981; Meinhardt 1983). The formation of supernumerary legs in this context means that additional P/D axes were induced upon grafting. In 1976, the polar coordinate model was proposed to interpret the regeneration phenomena (French et al. 1976). This model has two main aspects. First, the rule of shortest intercalation states that the confrontation of cells with different positional values would result in the intercalation of intermediate values via the shortest route. Second, the full circle rule states that leg outgrowth and ultimately a P/D axis would be generated at the site where there was a full complement of circumferential positional values (Bryant et al. 1981). This model can account for the formation of supernumerary legs following grafting experiments, because a full circumference of positional values will be intercalated at the two positions of maximal circumferential misalignment (French et al. 1976).

Until 1986, scientists interested in the mechanisms of appendage regeneration, intercalary regeneration, and supernumerary leg formation used the nymphal leg system extensively, but very few groups are currently using this approach (mainly due to a lack of tools for functional analyses). However, the situation has changed in the past decade, as techniques for analyzing gene expression and function have become available. In this section, we will provide molecular mechanisms of positional information along P/D axis essential for appendage regeneration and intercalary regeneration.

3.5.2 Molecular Basis of Positional Information

When regenerative animals lose part of their tissue, they can recognize where the amputated position is and which part of the tissue is remained depending on positional information along P/D axis. This function allows them to epimorphically regenerate only the lost part and to repair the injured tissue to the same morphology as the normal one. If positional information would be lost during the regeneration process, the morphology of regenerated leg must be changed along P/D and/or circumferential axis. Identification of the molecular basis of positional information is a long-standing question. As stated above, positional information defines the longitudinal and circumferential position in each leg segment, suggesting that the molecule(s) required for positional information must be repeatedly expressed in each leg segment in a gradient manner along the longitudinal and circumferential axes.

Based on data from *Drosophila*, we speculated that Ds and Ft are candidates for the molecular basis of positional information. Ds and Ft encode large protocadherins located at the adherence junction of the cell membrane and interact with each other to sense cell-cell contact. In *Drosophila*, *ds* and *ft* are expressed in the eye, wing, and leg in a gradient manner along the P/D axis. Also, *ds* and *ft* mutants exhibit abnormal tissue morphologies. Recent studies for the function of Ds and Ft show that Ds and Ft form a heterodimer to transduce cell-cell contact information via the Sav/Wts/Hpo signaling pathway to regulate cell proliferation, apoptosis, and cell morphology. Ds and Ft regulate the pattern formation of tissues along the P/D axis via the unconventional myosin Dachs (D) and a novel dsRNA binding protein Lowfat (Lft). They also regulate planner cell polarity by the golgi kinase Four-jointed (Fj) and now multifunctional signaling based on Ds and Ft roles in Ds/Ft signaling (Harvey and Hariharan 2012; Staley and Irvine 2012).

We speculated that Ds/Ft signaling could be required for the regulation of positional information during leg regeneration in crickets. In the developing and regenerating legs of crickets, *Gb′ds* and *Gb′ft* were expressed in a distal and proximal region in each leg segment in a gradient manner. The expression pattern of *Gb′d* is similar to *Gb′ds*, and the expression patterns of *Gb′lft* and *Gb′fj* are similar to *Gb′ft*, suggesting that these molecules function coordinately (Bando et al. 2009, 2011a, b). To clarify the function of these genes, we performed RNAi against *Gb′ds*, *Gb′ft*, *Gb′fj*, *Gb′d*, and *Gb′lft* and compared the morphologies of the regenerated legs and patterns of intercalary regeneration. RNAi against *Gb'ds* or *Gb'ft* showed shortened and thick regenerated leg, while *Gb'd*^{RNAi} and *Gb'lft*^{RNAi} showed shortened regenerated leg phenotypes (Fig. 3.5a). The regenerated leg of *Gb'fj*^{RNAi} crickets is slightly shorter than those of *Gb'd*^{RNAi}. When we change the amputated position along the P/D axis on the leg, the length of regenerated leg is almost the same when compared to the contralateral normal leg in the control cricket. In RNAi crickets against *Gb'ds*, *Gb'ft*, *Gb'd*, and *Gb'lft*, the lengths of the regenerated legs were all shortened depending on the amputation position. When the leg was amputated at the proximal region of the tibia, the regenerated legs of RNAi crickets against these genes were very short; however, the shortened leg had both a regenerated tarsus and claw (Fig. 3.5d). In the distal region of the shortened regenerated tibia, the same numbers of spikes and tibial spurs were regenerated. Moreover, organization of the muscle structure in the short regenerated tibia is quite similar to the control, suggesting that the shortened regenerated legs in RNAi-treated crickets against *Gb'ds*, *Gb'ft*, *Gb'd*, and *Gb'lft* had positional information for not only proximal position but distal position as well, even though the tissue size is much smaller than normal. Next, RNAi against these genes was used to determine their roles in intercalary regeneration. In the control cricket, when a graft is transplanted into the host stump with discontinuous position, discontinuity can be recovered by intercalary regeneration of the missing region between graft and host stump. In RNAi cricket against *Gb'ds*, *Gb'ft*, and *Gb'd*, the wound healing between the graft and host stump occurred, but intercalary regeneration did not (Fig. 3.5e). This is different from *Gb'lft*^{RNAi} crickets where intercalary regeneration also occurred. These data indicate that positional information along the P/D axis is maintained by the Ds/Ft signaling pathway. This pathway is involved in the maintenance of positional information along P/D axis during longitudinal regeneration and intercalary regeneration but is not required for positional information along circumferential axis since supernumerary leg formation was occurred in RNAi crickets against *Gb'ds*, *Gb'ft*, *Gb'd*, and *Gb'lft* (Bando et al. 2009, 2011a, b).

3.6 Repatterning Process

3.6.1 Similarity Between Repatterning Process and Developmental Process

Lost part of tissue is recognized based on positional information and is reconstructed from blastema cells. Undifferentiated blastema cells can redifferentiate into several types of differentiated cells, which is a similar process to normal development. Therefore, repatterning is akin to a "redevelopment," and pattern formation genes (e.g., signaling pathways and tissue-specific transcription factors) could also be involved in this process.

3.6.2 Signaling Pathways and Patterning Genes During Repatterning Process

Lost parts of the leg are regenerated from blastema cells and depend on positional information to repair the leg. The morphology of the lost part of a leg is repatterned by the function of leg patterning genes, similar to pattern formation in limb development during embryogenesis. To verify that the molecular boundary model would apply during regeneration, expression patterns of *Gb'hh*, *Gb'wg*, and *Gb'dpp* were examined in regenerating legs (Mito et al. 2002) and shown to be expressed in the blastema of legs amputated at the distal tibia. These genes are expressed in a pattern comparable to that of the leg bud of the *Gryllus* embryo and the *Drosophila* leg imaginal disc (Mito et al. 2002; Niwa et al. 2000). This result suggests that the initiation mechanism for establishment of the P/D axis in regenerating legs is similar to that of normal leg development. In addition, the formation of a regenerating leg P/D axis is triggered at the site where ventral *wg*-expressing cells and dorsal *dpp*-expressing cells meet in the AP boundary, similar to *Drosophila* leg development (Campbell and Tomlinson 1995; Mito et al. 2002).

In nymphs silenced for *Gb'hh*, regeneration was abnormal, with the formation of a supernumerary leg, whereas no such phenotype was observed in the absence of amputation (Nakamura et al. 2008b). In nymphs silenced for *Gb'dpp* and *Gb'wg*, no phenotype was found either during development or during regeneration (Nakamura et al. 2008b). However, when expression of *Gb'arm* (the β-catenin ortholog) was knocked down through RNAi, no regeneration was observed, indicating that the canonical Wnt pathway may be involved in initiation of regeneration (Nakamura et al. 2007). During leg regeneration, *Gb'Egfr* is expressed in the blastema, as observed in the limb bud (Nakamura et al. 2008a). In the nymphs exposed to *Gb'Egfr* dsRNAs, the leg was formed normally; however, tibial amputation followed by *Gb'Egfr*^{RNAi} resulted in malformation of the tarsus and claw (Nakamura et al. 2008a). Furthermore, during formation of the supernumerary legs upon grafting, *Gb'hh*, *Gb'dpp*, and *Gb'wg* were expressed as predicted from the molecular boundary model (Mito et al. 2002). However, the *Gb'Egfr* RNAi nymphs were unable to form the distal structures of the supernumerary legs, indicating that EGFR is indeed involved in the regeneration of distal structures (Nakamura et al. 2008a, b). This is consistent with the role of EGF signaling during *Drosophila* leg development (Campbell 2002). Thus, these data indicate that the molecular boundary model is the most plausible hypothesis for explaining the various phenomena observed during the formation of the P/D axis of both the developing and regenerating cricket leg. In addition, *Gb'aristaless*, *Gb'Dll*, and *Gb'dac* were expressed in the regenerating leg along the P/D axis, similar to the expression patterns in the developing limb. *Gb'Dll*^{RNAi} nymphs were unable to regenerate distal structures including the distal region of the tibia, all tarsal segments, and claws (Ishimaru et al. 2015; Nakamura et al. 2008a), and *Gb'dac*^{RNAi} nymphs were unable to regenerate distal structures of the tibia and proximal structures of the tarsus (Ishimaru et al. 2015), as expected from the phenotype of *Drosophila Dll* and *dac* mutants (Cohen et al. 1989), suggesting that the repatterning process during regeneration could be a repetitive event of limb development during embryogenesis.

3.6.3 Epigenetic Regulation of Leg Patterning Genes

During regeneration, the already differentiated cells dedifferentiate into blastema cells, which in turn differentiate into several types of cells needed to reconstruct the amputated tissue. During this whole process, the global gene expression has to change as well, likely via the epigenetic mechanisms (Katsuyama and Paro 2011). The comparative transcriptome analysis in *Gryllus* revealed that the expressions of epigenetic factors were changed during leg regeneration (Bando et al. 2013). The observed changes can be grouped into two types: the methylation of cytosine in DNA and the chemical modifications (including methylation, acetylation, phosphorylation, and ubiquitylation) of specific amino acid residues of N-terminal tail of histones. Methylation of histone H3 lysine 27 (H3K27) residue leads to changes that transform euchromatin into heterochromatin, which in turn represses gene expression. *Enhancer of zeste* [*E*(*z*)] encodes methyltransferase of histone H3K27, and the expression of *Gb′E*(*z*) is upregulated in the blastema. The depletion of *Gb′E*(*z*) via RNAi also caused the reduction of methylated histone H3K27 (Fig. 3.6b). This, in turn, likely leads to the formation of euchromatin (from heterochromatin) and upregulation of expression of many target genes. In contrast, *Ubiquitously transcribed tetratricopeptide repeat gene on the X chromosome* (*Utx*) encodes demethylase on histone H3K27, and the expression of *Gb′Utx* is also upregulated in the blastema. RNAi against *Gb′Utx* caused the maintenance of trimethylation of histone H3K27 (Fig. 3.6b) and subsequent downregulation expression of target genes.

Regenerated legs of $Gb'E(z)^{RNAi}$ cricket showed extra tibial segment formation between the original tibia and tarsus (Fig. 3.6a). Interestingly, in the $Gb'E(z)^{RNAi}$ leg, methylation of histone H3K27 was reduced (Fig. 3.6b) and expression domain of *Gb′dac* was expanded to distal region in the regenerating leg compared with the control (Fig. 3.6c). Since expression of *Gb′dac* is required for tibia formation (Ishimaru et al. 2015), this observation suggests that the *Gb′E*(*z*) also regulates tibia development during repatterning process via the regulation of *Gb′dac* expression (Hamada et al. 2015). Regenerated legs of $Gb'Utx^{RNAi}$ cricket showed defect on tarsal joint formation (Fig. 3.6a). This observation may be explained by the fact that in $Gb'Utx^{RNAi}$ legs, the *Egfr* expression is also lost in the same location. Hence, the absence of *Gb′Egfr* may lead to defects in joint formation in the tarsus (Fig. 3.6c). Recent studies have shown that by positively activating *Egfr* expression, the *Gb′Utx* also regulates joint formation during repatterning process in regenerated legs (Nakamura et al. 2008a; Hamada et al. 2015).

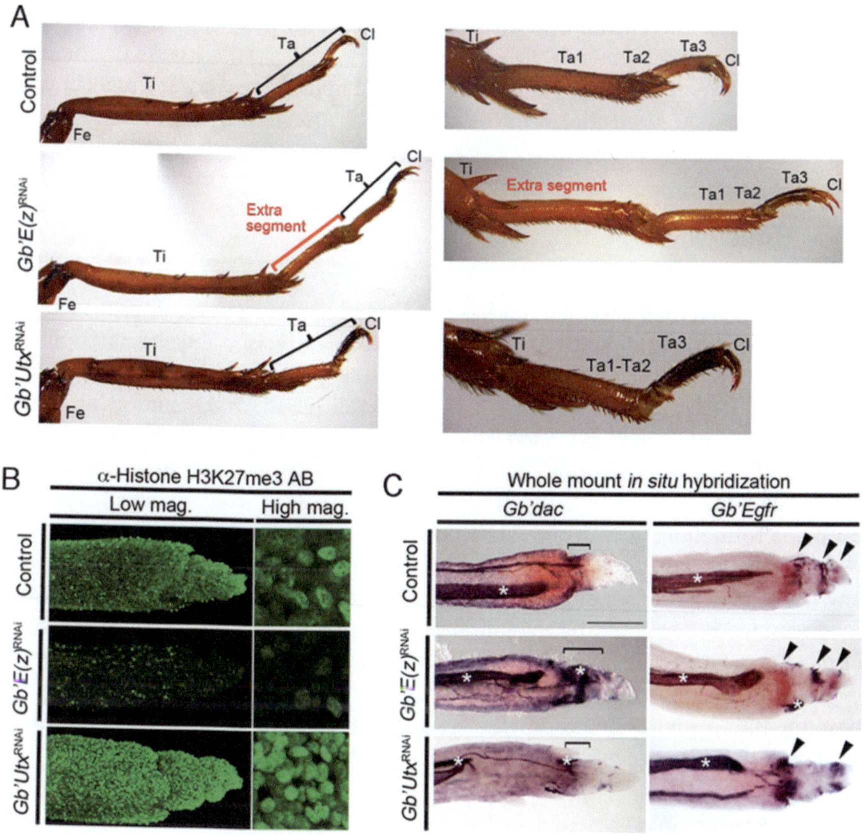

Fig. 3.6 The roles of epigenetic factors *Gb′E(z)* and *Gb′Utx* during repatterning process. (**a**) Typical morphological changes of regenerated adult legs in wild-type *Gb′E(z)*^{RNAi} and *Gb′Utx* ^{RNAi} crickets. Lateral views of low- and high-magnification images are shown on the *left* and *right*, respectively. The tarsus and ectopic tibial segments are indicated in *black* and *red brackets*, respectively. *Fe* femur, *Ti* tibia, *Ta* tarsus, *Cl* claw, *Ta* tarsomere. (**b**) The effects of *Gb′E(z)*^{RNAi} and *Gb′Utx*^{RNAi} on localization of the histone H3K27me3 (*green*) in regenerating legs. The distal portion of the regenerating leg is directed toward the *right*. (**c**) *Gb′dac* and *Gb′Egfr* expression patterns in wild-type *Gb′E(z)*^{RNAi} and *Gb′Utx*^{RNAi} adult regenerated legs. The expressions of *Gb′dac* and *Gb′Egfr* in the regenerated tarsi are indicated by *brackets* and *arrowheads*, respectively. *Asterisks* indicate non-specific staining (Hamada et al. 2015)

3.6.4 Common Features Between Longitudinal and Intercalary Regeneration

To identify the molecules that may be involved in intercalary regeneration, *Gb′hh* and *Gb′wg* expression patterns were examined during intercalary regeneration. While both genes are induced in the host when a tibia is amputated proximally, no such induction occurs in the reciprocal experiment when amputation was performed distally (Nakamura et al. 2007). This directional induction occurs even in the reversed intercalation. Furthermore, because no regeneration occurs when *Gb'arm* was knocked down by RNAi (Nakamura et al. 2007), it is likely that the canonical Wg/Wnt signaling pathway is involved in the process of leg regeneration and determination of positional information in the affected leg segment.

3.7 Conclusions

Studies of cricket leg regeneration have indicated that blastema cell proliferation, positional information, and repatterning are cooperatively regulated through the Wg/Wnt, JAK/STAT, Ds/Ft, and EGF signaling pathways (Bando et al. 2009, 2011b, 2013; Hamada et al. 2015; Ishimaru et al. 2015; Nakamura et al. 2007, 2008a). They have also provided clues for understanding the molecular mechanisms underlying regeneration that was first described some 300 years ago (Agata and Inoue 2012; Réaumur 1712). Many questions such as which molecules induce blastema formation, how differentiated cells are dedifferentiated to form blastema cells, and what is the molecular basis of positional information remain unanswered. However, recent studies suggest that tissue regeneration in both vertebrates and invertebrates is regulated by evolutionarily conserved signaling pathways (Dong et al. 2007; Pan 2010; Takeo et al. 2013; Zhao et al. 2011). By identifying the evolutionarily conserved molecular mechanisms involved in animal leg regeneration, the future studies could provide critical insights into the field of regenerative medicine and have potential human applications.

Acknowledgments We thank Hideyo Ohuchi, Taro Mito, Taro Nakamura, Yuko Maeda, Fumiaki Ito, Takuro Kida, Yuji Matsuoka, Yoshiyasu Ishimaru, Kazuki Kurita, Hidetaka Mizushima, Misa Okumura, and Yuki Bando to the cricket studies on leg development and leg regeneration described in this article. T.B. and S.N. were supported by a grant from the Ministry of Education, Culture, Sports, Science, and Technology of Japan.

References

Agata K, Inoue T (2012) Survey of the differences between regenerative and non-regenerative animals. Dev Growth Differ 54:143–152

Bando T, Mito T, Maeda Y, Nakamura T, Ito F, Watanabe T, Ohuchi H, Noji S (2009) Regulation of leg size and shape by the Dachsous/Fat signalling pathway during regeneration. Development 136:2235–2245

Bando T, Mito T, Nakamura T, Ohuchi H, Noji S (2011a) Regulation of leg size and shape: involvement of the Dachsous-fat signaling pathway. Dev Dyn 240:1028–1041

Bando T, Hamada Y, Kurita K, Nakamura T, Mito T, Ohuchi H, Noji S (2011b) Lowfat, a mammalian Lix1 homologue, regulates leg size and growth under the Dachsous/Fat signaling pathway during tissue regeneration. Dev Dyn 240:1440–1453

Bando T, Ishimaru Y, Kida T, Hamada Y, Matsuoka Y, Nakamura T, Ohuchi H, Noji S, Mito T (2013) Analysis of RNA-Seq data reveals involvement of JAK/STAT signalling during leg regeneration in the cricket Gryllus bimaculatus. Development 140:959–964

Basler K, Struhl G (1994) Compartment boundaries and the control of Drosophila limb pattern by hedgehog protein. Nature 368:208–214

Bidla G, Lindgren M, Theopold U, Dushay MS (2005) Hemolymph coagulation and phenoloxidase in Drosophila larvae. Dev Comp Immunol 29:669–679

Bryant SV, French V, Bryant PJ (1981) Distal regeneration and symmetry. Science (80-.).212:993–1002

Campbell G (2002) Distalization of the Drosophila leg by graded EGF-receptor activity. Nature 418:781–785

Campbell G, Tomlinson A (1995) Initiation of the proximodistal axis in insect legs. Development 121:619–628

Campbell G, Weaver T, Tomlinson A (1993) Axis specification in the developing Drosophila appendage: the role of wingless, decapentaplegic, and the homeobox gene aristaless. Cell 74:1113–1123

Cohen SM, Bronner G, Kuttner F, Jurgens G, Jaeckle H (1989) Distal-less encodes a homoeodomain protein required for limb development in Drosophila. Nature 338:432–434

Diaz-Benjumea FJ, Cohen B, Cohen SM (1994) Cell interaction between compartments establishes the proximal-distal axis of Drosophila legs. Nature 372:175–179

Dong J, Feldmann G, Huang J, Wu S, Zhang N, Comerford SA, Gayyed MF, Anders RA, Maitra A, Pan D (2007) Elucidation of a universal size-control mechanism in Drosophila and mammals. Cell 130:1120–1133

French V (1981) Pattern regulation and regeneration. Philos Trans R Soc Lond B Biol Sci 295:601–617

French V, Bryant PJ, Bryant SV (1976) Pattern regulation in epimorphic fields. Science (80-.). 193:969–981

Galindo MI, Bishop SA, Greig S, Couso JP (2002) Leg patterning driven by proximal-distal interactions and EGFR signaling. Science 297:256–259 (80–)

Galindo MI, Bishop SA, Couso JP (2005) Dynamic EGFR-Ras signaling in Drosophila leg development. Dev Dyn 233:1496–1508

Goto S, Hayashi S (1999) Proximal to distal cell communication in the Drosophila leg provides a basis for an intercalary mechanism of limb patterning. Development 126:3407–3413

Hamada Y, Bando T, Nakamura T, Ishimaru Y, Mito T, Noji S, Tomioka K, Ohuchi H (2015) Leg regeneration is epigenetically regulated by histone H3K27 methylation in the cricket Gryllus bimaculatus. Development 142:2916–2927

Harvey KF, Hariharan IK (2012) The hippo pathway. Cold Spring Harb Perspect Biol 4:a011288

Harvey KF, Zhang X, Thomas DM (2013) The Hippo pathway and human cancer. Nat Rev Cancer 13:246–257

Inoue Y, Mito T, Miyawaki K, Matsushima K, Shinmyo Y, Heanue TA, Mardon G, Ohuchi H, Noji S (2002) Correlation of expression patterns of homothorax, dachshund, and Distal-less with the proximodistal segmentation of the cricket leg bud. Mech Dev 113:141–148

Ishimaru Y, Nakamura T, Bando T, Matsuoka Y, Ohuchi H, Noji S, Mito T (2015) Involvement of dachshund and Distal-less in distal pattern formation of the cricket leg during regeneration. Sci Rep 5:8387

Katsuyama T, Paro R (2011) Epigenetic reprogramming during tissue regeneration. FEBS Lett 585:1617–1624

Kojima T (2004) The mechanism of Drosophila leg development along the proximodistal axis. Dev Growth Differ 46:115–129

Manjon C, Sanchez-Herrero E, Suzanne M (2007) Sharp boundaries of Dpp signalling trigger local cell death required for Drosophila leg morphogenesis. Nat Cell Biol 9:57–63

Meinhardt H (1983) Cell determination boundaries as organizing regions for secondary embryonic fields. Dev Biol 96:375–385

Meissner A (2010) Epigenetic modifications in pluripotent and differentiated cells. Nat Biotechnol 28:1079–1088

Mito T, Inoue Y, Kimura S, Miyawaki K, Niwa N, Shinmyo Y, Ohuchi H, Noji S (2002) Involvement of hedgehog, wingless, and dpp in the initiation of proximodistal axis formation

during the regeneration of insect legs, a verification of the modified boundary model. Mech Dev 114:27–35
Mito T, Shinmyo Y, Kurita K, Nakamura T, Ohuchi H, Noji S (2011) Ancestral functions of Delta/Notch signaling in the formation of body and leg segments in the cricket Gryllus bimaculatus. Development 138:3823–3833
Miyawaki K, Inoue Y, Mito T, Fujimoto T, Matsushima K, Shinmyo Y, Ohuchi H, Noji S (2002) Expression patterns of aristaless in developing appendages of Gryllus bimaculatus (cricket). Mech Dev 113:181–184
Nakamura T, Mito T, Tanaka Y, Bando T, Ohuchi H, Noji S (2007) Involvement of canonical Wnt/Wingless signaling in the determination of the positional values within the leg segment of the cricket Gryllus bimaculatus. Dev Growth Differ 49:79–88
Nakamura T, Mito T, Miyawaki K, Ohuchi H, Noji S (2008a) EGFR signaling is required for re-establishing the proximodistal axis during distal leg regeneration in the cricket Gryllus bimaculatus nymph. Dev Biol 319:46–55
Nakamura T, Mito T, Bando T, Ohuchi H, Noji S (2008b) Dissecting insect leg regeneration through RNA interference. Cell Mol Life Sci 65:64–72
Niwa N, Saitoh M, Ohuchi H, Yoshioka H, Noji S (1997) Correlation between Distal-less expression patterns and structures of appendages in development of the two-spotted cricket, Gryllus bimaculatus. Zool Sci 14:115–125
Niwa N, Inoue Y, Nozawa A, Saito M, Misumi Y, Ohuchi H, Yoshioka H, Noji S (2000) Correlation of diversity of leg morphology in Gryllus bimaculatus (cricket) with divergence in dpp expression pattern during leg development. Development 127:4373–4381
Pan D (2010) The Hippo signaling pathway in development and cancer. Dev Cell 19:491–505
Réaumur RAF (1712) Sur les diverses reproductions qui se font dans les Ecrevisse, les Omars, les Crabes, etc. et entr'autres sur celles de leurs Jambes et de leurs Ecailles. Mem Acad Roy Sci 223–245
Staley BK, Irvine KD (2012) Hippo signaling in Drosophila: recent advances and insights. Dev Dyn 241:3–15
Tajiri R, Misaki K, Yonemura S, Hayash S (2011) Joint morphology in the insect leg: evolutionary history inferred from Notch loss-of-function phenotypes in Drosophila. Development 138:4621–4626
Takeo M, Chou WC, Sun Q, Lee W, Rabbani P, Loomis C, Taketo MM, Ito M (2013) Wnt activation in nail epithelium couples nail growth to digit regeneration. Nature 499:228–232
Wang S, Tsarouhas V, Xylourgidis N, Sabri N, Tiklová K, Nautiyal N, Gallio M, Samakovlis C (2009) The tyrosine kinase stitcher activates grainy head and epidermal wound healing in Drosophila. Nat Cell Biol 11:890–895
Zhao B, Tumaneng K, Guan K-L (2011) The Hippo pathway in organ size control, tissue regeneration and stem cell self-renewal. Nat Cell Biol 13:877–883

Chapter 4
Eye Development and Photoreception of a Hemimetabolous Insect, *Gryllus bimaculatus*

Hideyo Ohuchi, Tetsuya Bando, Taro Mito, and Sumihare Noji

Abstract The hemimetabolous insect, *Gryllus bimaculatus*, has two compound eyes that begin to form in the embryo and increase in size five- to sixfolds during nymphal development. Retinal stemlike cells reside in the anteroventral proliferation zone (AVPZ) of the nymphal compound eye and proliferate to increase retinal progenitors, which then differentiate to form new ommatidia in the anterior region of the eye. Here, we introduce the morphology and development of the cricket eye first, and then we focus on the roles of retinal determination genes (RDGs) such as *eyes absent* (*eya*) and *sine oculis* (*so*) in *Gryllus* eye formation and growth. Since the principal function of the eye is photoreception, we finally summarize opsin photopigments in this species, broadening the roles of photoreception.

Keywords Compound eye • Eyes absent • Eye development • *Gryllus bimaculatus* • Hemimetabolous insect • Ocellus • Opsin • RNA interference • Sine oculis

4.1 Introduction: Morphology and Structure of the Cricket Eye

Crickets possess two kinds of visual organs: three dorsal ocelli innervated by the posterior part of the central brain and a pair of lateral compound eyes innervated by the optic lobes (Homberg 2004; Henze et al. 2012). Among the three dorsal ocelli, two lateral ones are positioned just dorsal to the antennal bases, whereas the median ocellus is located on the forehead (Fig. 4.1a). At present, the functions of these

H. Ohuchi (✉) • T. Bando
Department of Cytology and Histology, Okayama University Graduate School of Medicine, Dentistry and Pharmaceutical Sciences, Okayama City 700-8558, Japan
e-mail: hohuchi@okayama-u.ac.jp

T. Mito
Graduate School of Bioscience and Bioindustry, Tokushima University, Tokushima 770-8513, Japan

S. Noji
Tokushima University, Tokushima City 770-8501, Japan

H.W. Horch et al. (eds.), *The Cricket as a Model Organism*,
DOI 10.1007/978-4-431-56478-2_4

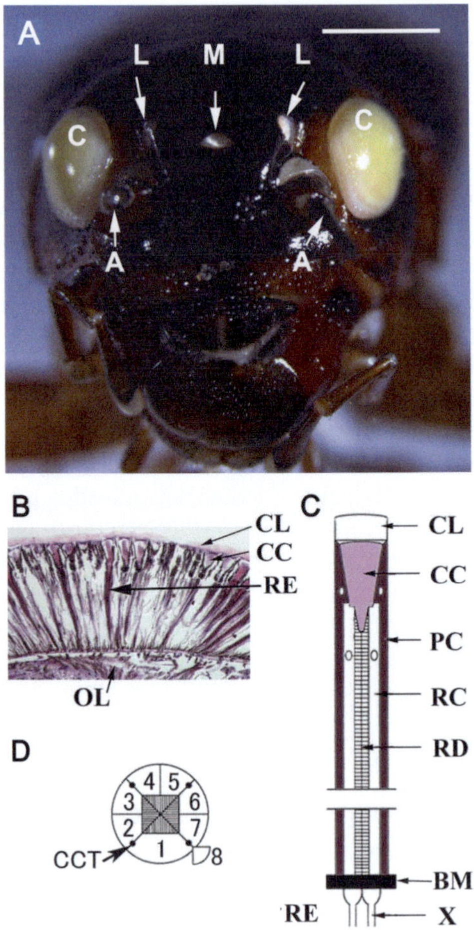

Fig. 4.1 Morphology of the cricket eye (Figure adapted from Takagi et al. 2012, except **a**). (**a**) Frontal view of a white-eye mutant cricket. *A* antenna (trimmed for better visualization), *C* compound eyes, *L* lateral ocelli, *M* median ocellus. *Scale bar*: 2 mm. (**b**) A longitudinal section of a cricket adult eye. (**c**) A diagrammatic illustration of an ommatidium of a cricket eye in a longitudinal section. (**d**) A diagrammatic illustration of an ommatidium of a cricket eye in a transverse section at the middle of the ommatidium. Retinular cells are numbered 1–8. *Rhabdoms* are indicated by striped patterns. Basement membrane (*BM*), axon (*X*), and optic lobe (*OL*) are evident in the eye. (**c**) and (**d**) are modified illustrations of Fig. 1A and D in Sakura et al. (2003) with permission. The scale of each image is arbitrary. *Abbreviations*: *CC* crystalline cones, *CCT* crystalline cone tracts, *CL* corneal lens, *PC* pigment cell, *RC* retinular cell, *RD* rhabdom, *RE* retina, *BM* basement membrane, *X* axon

ocelli are not well studied in most insect species. However, in the two-spotted cricket, it is the lateral compound eyes that are the main visual organs and on which we mainly focus here.

Each compound eye is made up of approximately 7000 repeated crystalline units called ommatidia (Burghause 1979), which is substantially larger than the 750 ommatidia present in *Drosophila* (Wolff and Ready 1993). Each *Gryllus* ommatidium is composed of precisely arranged photoreceptors and nonneural support cells (Fig. 4.1b, c; Takagi et al. 2012; Sakura et al. 2003). Inside the ommatidium in the midline region, there are rhabdoms composed of tightly packed tubular membranes (microvilli) that contain the phototransduction proteins (Sakura et al. 2003; Fig. 4.1c), and these are surrounded by eight retinular cells (Takagi et al. 2012; Sakura et al. 2003; Fig. 4.1d). Overall, the general aspects of the cricket ommatidium appear to be structurally similar to those observed in the *Drosophila* compound eye.

4.2 Developmental Processes of the Cricket Eye

Holometabolous insects like *Drosophila* proceed through two phases of visual system development (Friedrich 2006; Kumar 2011): the embryonic phase, which generates the simple eyes of the larva, and the postembryonic phase that produces the adult specific compound eyes during late larval development and pupation. In contrast, in the basal hemimetabolous insects such as crickets and grasshoppers, the development of the compound eyes is continuous throughout embryogenesis and postembryogenesis (Friedrich 2006).

The morphological changes of the developing retina in *Gryllus bimaculatus* are shown in Fig. 4.2. All eye components originate from the eye lobes, a pair of distinct embryonic anlagen derived from the lateral neuroectoderm of the embryonic procephalon. The main gene expressed in this region is *dachshund* (*Gb'dac*), a retinal determination transcription factor that will be discussed later (Inoue et al. 2004). During embryonic development, the eye primordia can be observed in the head at stage 11 (Mito and Noji 2008; Takagi et al. 2012; Fig. 4.2a, b). Note that distinct retinal structures are not yet visible at this stage (Fig. 4.2d). Later on, between stages 11 and 14, the morphogenetic furrow is formed (Takagi et al. 2012; Fig. 4.2c), similar to the previous reports in the grasshopper (Friedrich and Benzer 2000). Finally, at stage 14, corneal lenses, crystalline cones, and the retina consisting of pigment cells, retinular cells, and rhabdoms (Fig. 4.1b) can all be visualized in the differentiating primordium (Fig. 4.2e).

During postembryonic development, the *Gryllus* first instar nymphs hatch out with small compound eyes, which will continue to grow throughout the subsequent nymphal stages. Additional rows of ommatidia are formed during each molt. By the second nymphal stage, the main eye structures have become fairly well established (Fig. 4.2f). The compound eyes continue to develop at the

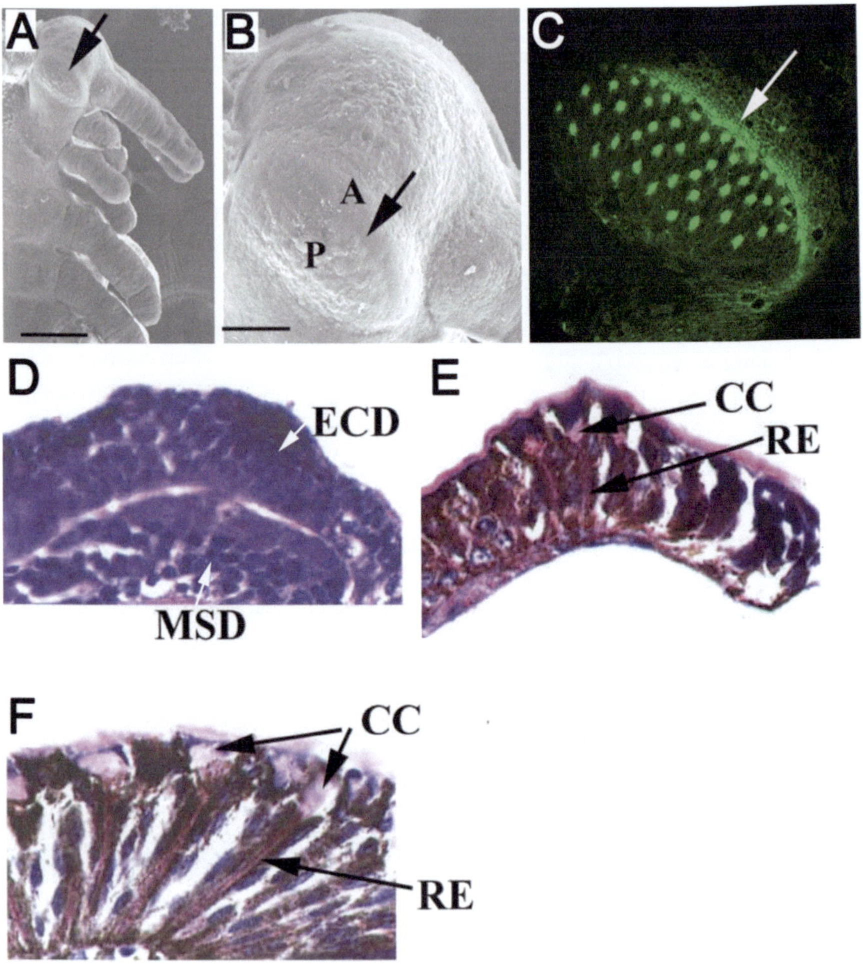

Fig. 4.2 Morphological changes during development of the cricket eye (Figure adapted from Takagi et al. 2012). (**a**) A scanning electron micrograph of a cricket embryo at stage 11. The *arrow* indicates a developing eye. Anterior is upward. (**b**) A high-magnification image of an embryonic head. The *arrow* indicates the morphogenetic furrow formed between *A* and *P*. *A* anterior, *P* posterior. (**c**) A developing eye stained with phalloidin to visualize F-actin in morphogenetically active cells. The *arrow* indicates the morphogenetic furrow. (**d**) A sagittal section of an embryonic eye at stage 11 (before formation of the morphogenetic furrow). *ECD* ectoderm, *MSD* mesoderm. (**e**) A section of an embryonic eye at stage 14, where ommatidia are formed. *CC* crystalline cones, *RE* retina. (**f**) A section of a nymph at the second instar. *Scale bars*: 1mm in (**a**), 250 μm in (**b**)

anterior margin during the third, fourth, and fifth instar nymphal stages (Takagi et al. 2012; Fig. 4.3a–c), similar to previous findings in other hemimetabolous insects (Anderson 1978). The anterior-most margin of nymphal eyes becomes filled with highly compacted cells, some of which may be retinal stem cells (Anderson 1978; Dong and Friedrich 2010). The thymidine analogue

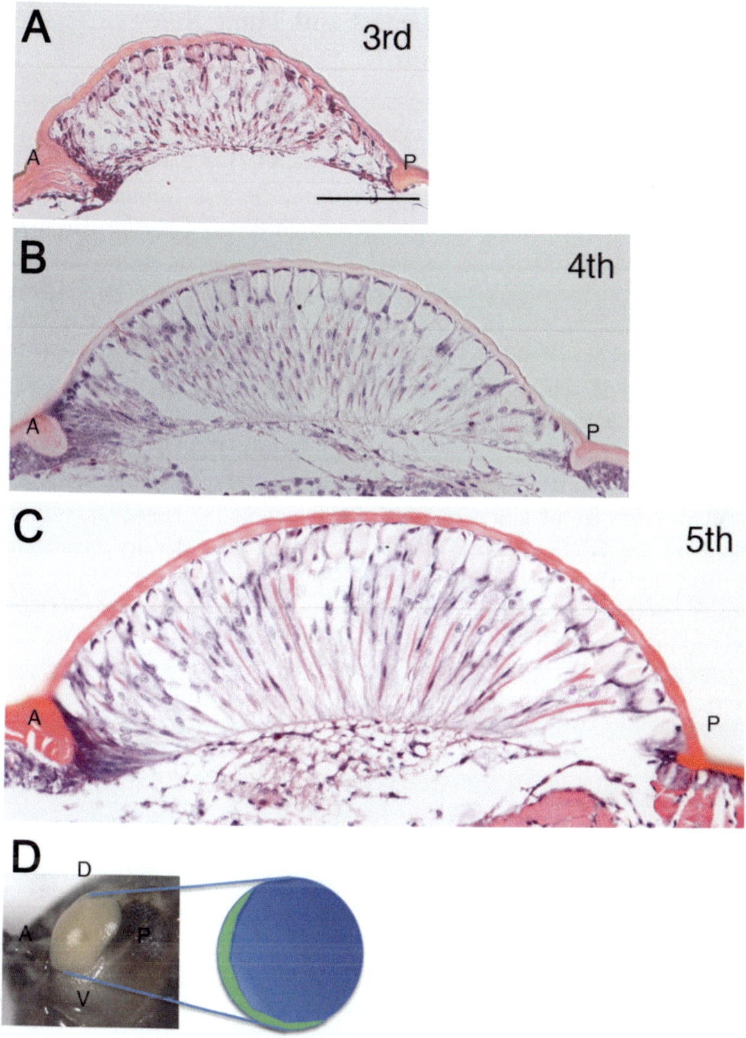

Fig. 4.3 Morphological changes of cricket nymphal eyes from the third to fifth instar (Figure adapted from Takagi et al. 2012). (**a**) A sagittal section of a nymphal eye at the third instar. (**b**) A sagittal section of a nymphal eye at the fourth instar. (**c**) A sagittal section of a nymphal eye at the fifth instar. (**d**) Illustration of the anteroventral proliferation zone (AVPZ, *green*) in the nymphal eye (*blue*) of a *white-eyed* cricket. *D* dorsal, *V* ventral, *A* anterior, *P* posterior. *Scale bar*: 100 μm in (**a-c**)

incorporation assays performed during third nymphal stage show that these cells are actually proliferating (Takagi et al. 2012). At this point, the anterior ventral proliferation zone (AVPZ), which is responsible for continuous growth of the compound eye, is formed at the anterior ventral margin of the nymphal eye primordium (Fig. 4.3d).

4.3 Retinal Determination Genes and Their Roles in Cricket Eye Formation

The extensive studies of *Drosophila* eye development have established the framework of the retinal determination gene (RDG) network (Fig. 4.4, Kumar 2011; Jemc and Rebay 2007). A crucial transcription factor for eye formation is Pax6, a member of the paired-box homeodomain proteins in vertebrates as well as in insects (Wawersik and Maas 2000). The *Drosophila* Pax6 homologs are *eyeless* (*ey*) and *twin of eyeless* (*toy*) that synergistically pattern the developing fly eye. Their targets include *sine oculis* (*so*), *eyes absent* (*eya*), and *dachshund* (*dac*, Fig. 4.4). Loss-of-function mutations of *ey*, *toy*, *so*, *eya*, and *dac* can result in a small eye or the lack of the eye in the adult, while ectopic expression of these genes, either alone or in combination, can induce ectopic eye formation, suggesting that these eye-forming genes act at the very early stages of eye development (Halder et al. 1995; Bonini et al. 1997; Chen et al. 1997; Pignoni and Zipursky 1997; Shen and Mardon 1997; Czerny et al. 1999; Kronhamn et al. 2002; Jang et al. 2003). However, very little is known about the RDG networks in basal insects. In order to understand the mechanisms underlying the continuous eye development in the hemimetabolous mode of development, it is critical to characterize the functions of the RDGs more precisely in these species.

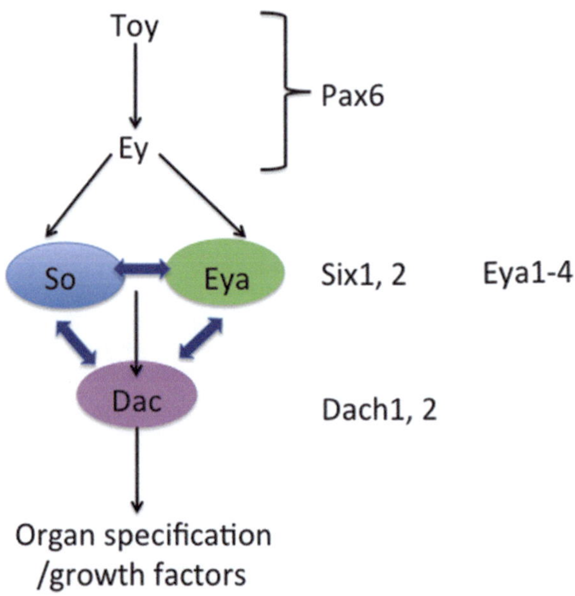

Fig. 4.4 Retinal determination gene (RDG) network as revealed by *Drosophila* studies (Figure modified from Kumar 2011 and Jemc and Rebay 2007). *Black arrows* represent direct transcriptional regulation or undetermined genetic regulatory pathways. *Blue arrows* depict protein-protein interactions. Mammalian homologs are shown on the *right*

4.3.1 Toy, Ey, *and* Dac

The *Gryllus* homologs of *toy* and *ey* have been cloned, and their expression patterns were localized in the eye primordium (S. N., H. O., unpublished data). However, neither targeting the individual genes with RNAi nor dual RNAi against both genes has any effects on eye formation and generates only wild-type morphology.

During embryonic eye development, *Gb'dac* expression is first observed in the lateral head region that corresponds to the eye primordium and a part of the deutocerebrum, the second part of the three-partite arthropod brain. Before the formation of compound eyes, the expression becomes restricted to the posterior region of the eye anlage. The expression then moves anteriorly as the morphogenetic furrow progresses (Inoue et al. 2004). This *Gryllus* expression pattern of *dac* corresponds very well to that observed in *Drosophila* eye imaginal disc, although these two insect species are markedly different in phylogeny (Takagi et al. 2012). Similar to the situation with *toy* and *ey*, no noticeable RNAi phenotype against *Gb'dac* has been observed (S. N., H. O., unpublished data). Recent work in *Tribolium* showed that only triple knockdowns (including *toy*, *ey*, and *dac*) result in the loss of compound eyes (Yang et al. 2009). Taken together, relationship among Toy, Ey, and Dac in embryonic eye development awaits further elucidation, but is likely modified in crickets as compared to that in *Drosophila* (Fig. 4.4).

4.3.2 Eya *and* So

Gryllus homologs of *sine oculis* (*Gb'so*) and *eyes absent* (*Gb'eya*) are expressed during embryonic eye development (Takagi et al. 2012). Their expression domains appear to overlap with the procephalic expression regions of the *Gryllus* homologs of *hedgehog* (*Gb'hh*) and *wingless* (*Gb'wg*) at stage 4 (Miyawaki et al. 2004). Parental RNAi (paRNAi) against *Gb'eya* results in smaller eye primordia or no embryonic eyes (see Fig. 4.5 from Takagi et al. 2012). Almost half of *Gb'eya* RNAi embryos have neither compound eyes nor ocelli. In contrast, the embryonic phenotypes of *Gb'so* RNAi are very subtle, possibly due to the insufficient depletion of its transcripts (Takagi et al. 2012).

During later embryogenesis, around stage 14, *Gb'eya and Gb'so* are expressed in the posterior region of the morphogenetic furrow (MF). These expression patterns are very similar to those observed in the *Drosophila* eye imaginal disc (Callaerts et al. 2006). Since it has been reported that Ey- and Toy-dependent activation of *eya*, *so*, and *dac* locks the cells in a retina-committed state (Chen et al. 1999; Curtiss and Mlodzik 2000; Bessa et al. 2002), it is tempting to postulate that *Gb'eya* and *Gb'so* in the posterior region of the cricket embryonic eye may be involved in keeping cells in a similar retinal-committed state.

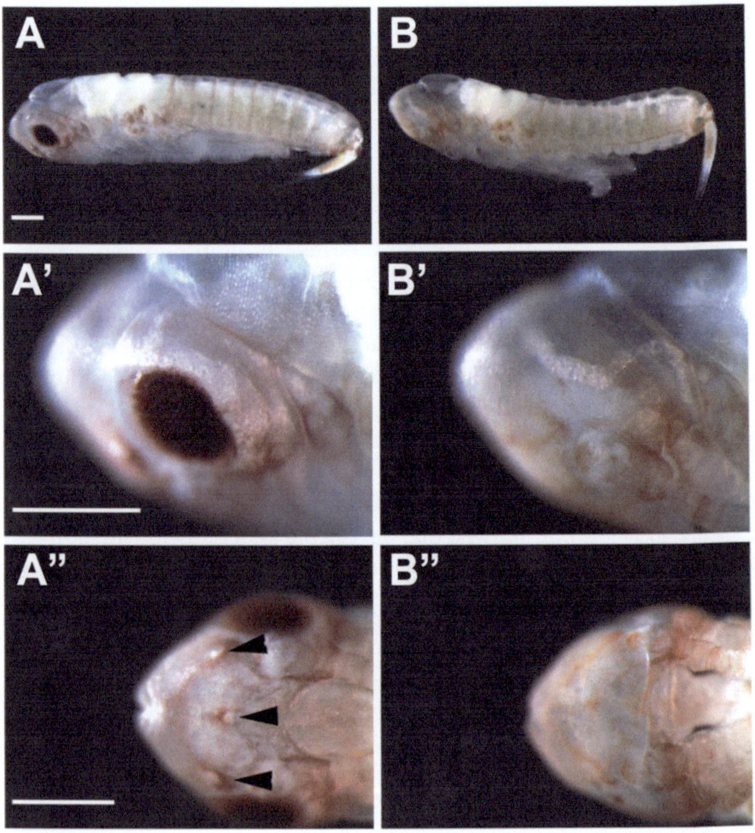

Fig. 4.5 *Gb'eya* is required for formation of embryonic eyes and ocelli of the hemimetabolous cricket as revealed by parental RNAi (Figure adapted from Takagi et al. 2012). (**a**, **a'** and **a"**) Control embryos. (**a**) Lateral view; (**a'**) High magnification of the head region of (**a**); (**a"**) Dorsal view. *Arrowheads* indicate three ocelli. (**b**, **b'** and **b"**) A *Gb'eya* RNAi embryo. Neither eye nor ocellus was observed. *Scale bars*: 300 μm

The expression patterns of *Gb'eya* and *Gb'so* were also examined in *Gryllus* nymphs. In situ hybridization of the fifth instar eye shows that *Gb'eya* is expressed intensely in the AVPZ of the nymphal eye primordium (Figs. 4.6a; compare with Fig. 4.6b) as well as in the pigment cells. In contrast, the *Gb'so* could not be detected (Takagi et al. 2012). These expression studies were followed up with the nymphal RNAi (nyRNAi) at the third or fourth instar stage using a white-eye mutant line (Fig. 4.6c). This line is highly inbred to reduce genetic background differences and to observe color changes in the eyes when making transgenic lines (Nakamura et al. 2010) or RNAi experiments (Takagi et al. 2012). In *Gb'eya* nyRNAi experiments with low amounts of dsRNA administered (2 μM conc.), the white eyes turn black in the immediate nymphal stage. The eyes of later nymphs and adults, though, revert back to white. When the higher dosage of *Gb'eya* dsRNA (20 μM) is injected into third instar nymphs, not only do their eyes turn black, but a *small-eye*

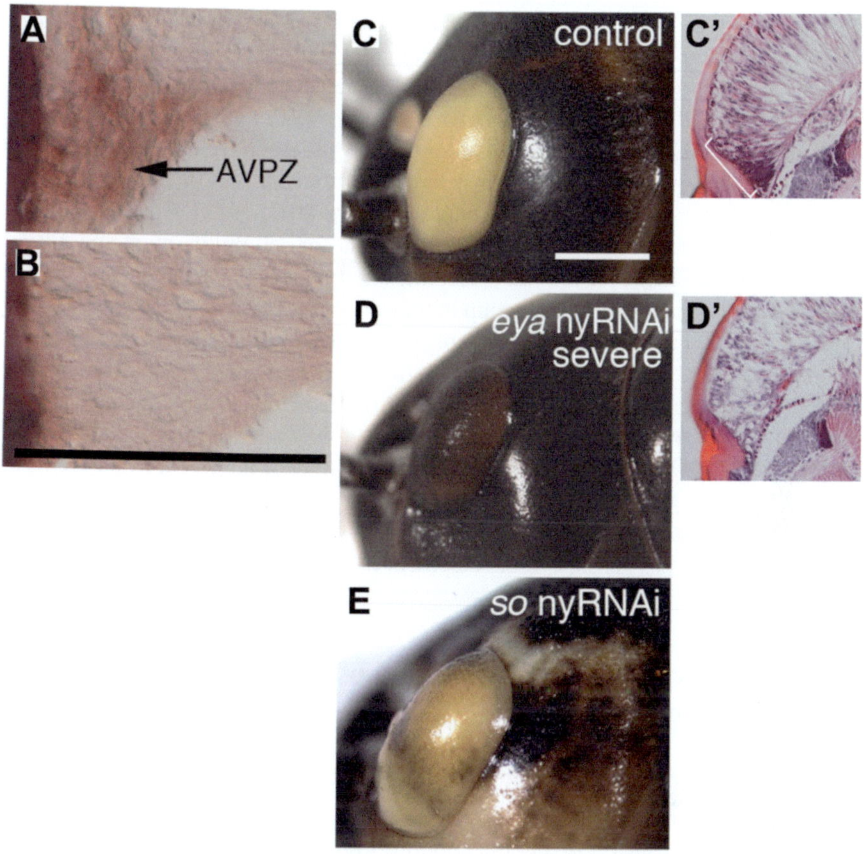

Fig. 4.6 Effects of *Gb'eya* or *Gb'so* RNAi on development of the nymphal eye (Figure adapted from Takagi et al. 2012). (**a**) Expression of *Gb'eya* in the AVPZ (*brown*, *arrow*) of the nymphal eye at the fifth instar as revealed by in situ hybridization. (**b**) A sense probe of *Gb'eya* does not give rise to *brown* staining in the AVPZ. (**c**) A control white adult eye. (**c'**) A sagittal section of a control nymphal eye at the fifth instar stained with hematoxylin and eosin (HE). AVPZ is shown by *bracket*. (**d**) A severe phenotype of *Gb'eya* nyRNAi nymph injected with dsRNA at the seventh instar. (**d'**) A sagittal section of the *Gb'eya* nyRNAi eye with a severe phenotype, stained with HE. A typical AVPZ is not observed and retinal structure is disrupted. (**e**) A severe phenotype of *Gb'so* nyRNAi nymph at the seventh instar. The same defects are observed in the mild *Gb'eya* nyRNAi eye. *Scale bars*: 100 μm in (**a, b**), 2 mm in (**c-e**)

phenotype can also be observed (Fig. 4.6d). Consistent with these results, when a 100-fold lower concentration of dsRNA is used (0.2 μM), the frequency of such severe phenotypes is greatly reduced. In *Gb'so* nyRNAi injections with the highest concentration of dsRNA (20 μM), there is a significant increase of black pigmentation in the eyes (Fig. 4.6e), resulting in dull-white eyes. However, the *small-eye* phenotype is never observed. Histological analysis shows that the darkening of the eyes is likely due to the formation of a brown cuticular layer at the surface of the eye, which is usually produced by the epidermal cells, instead of the cornea (Takagi et al. 2012). Since *eya* is normally expressed in the pigment cells abutting the corneal lens (CL; Fig. 4.1c), it is likely that reduction of *eya* expression in CL leads to transformation of the cornea to epidermal cells, which produce cuticle.

Severe cases of *Gb′eya* RNAi exhibit significant reduction in the size of compound eyes in addition (Fig. 4.6d; compare with Fig. 4.6c). Cell proliferation assays show that the cells in S phase are either completely absent in *Gb′eya* or have greatly reduced numbers in *Gb′so*, resulting in a dramatically smaller AVPZ in the fifth instar stage (Fig. 4.6d′; compare with Fig. 4.6c′) (Takagi et al. 2012). Combined, these results indicate that *Gb′eya* and *Gb′so* are essential for cell proliferation and AVPZ growth during postembryonic eye development. During eye development, *eya* and *so* are known to be components of a group of transcription factors that play an important role in the specification of the eye field not only in *Drosophila* but also in mammals as well (Oliver et al. 1995, 1996; Xu et al. 1999). Moreover, they function as a complex in both eye specification and differentiation (Pignoni et al. 1997; Kumar 2011; Fig. 4.4). Therefore, it is reasonable to assume that they might also act as a complex during eye development in *Gryllus bimaculatus*. In light of the RNAi experiments in *Gryllus* and *Tribolium* (Yang et al. 2009), the Eya/So complex is evolutionarily conserved and required for retinal differentiation, maintenance of the retinal region, and cornea formation during eye development in insects in general.

4.4 Photoreception and Expression of Opsins in the Cricket Eye

All eye types, irrespective of their level of morphological complexity or modification in regard to individual lifestyle and environmental conditions, have a primary function in animal vision or photoreception. Previous studies in *Gryllus* demonstrated the existence of three spectral classes of photoreceptors in its compound eyes with maximal absorption levels observed at 332 nm (ultraviolet; uv), 445 nm (blue), and 515 nm (green; Zufall et al. 1989). A different situation was observed in ocelli, with only two classes of photoreceptors being reported: green- and uv-sensitive at the peak absorbance of ~511 nm and ~351 nm, respectively (Henze et al. 2012). There are four opsin genes that are responsible for photoreception in crickets: one in the short-wavelength (SW), one in the middle-wavelength (MW), and two in the long-wavelength (LW) clade of the phylogenetic tree (Henze et al. 2012). The *Gb′*SW, MW, and two LW opsin genes are termed as UV, blue, green A, and green B, respectively. Of the four genes, only the green B is expressed in all eye regions except for the dorsal rim area (DRA) (Henze et al. 2012; Tamaki et al. 2013), while the others exhibit differential expressions. The DRA is a specialized region at the dorsal-frontal margin of the compound eye and is a non-imaging eye region (Labhart and Meyer 1999). The UV opsin is expressed in the DRA and in the remainder of the eye excluding a portion in the ventral retina. The blue opsin is expressed in the DRA and the ventral portion of the retina. Thus, UV, blue, and green B opsins are likely responsible for the three wavelengths that are absorbed by the cricket ommatidia. With the regard to ocelli, only green A is

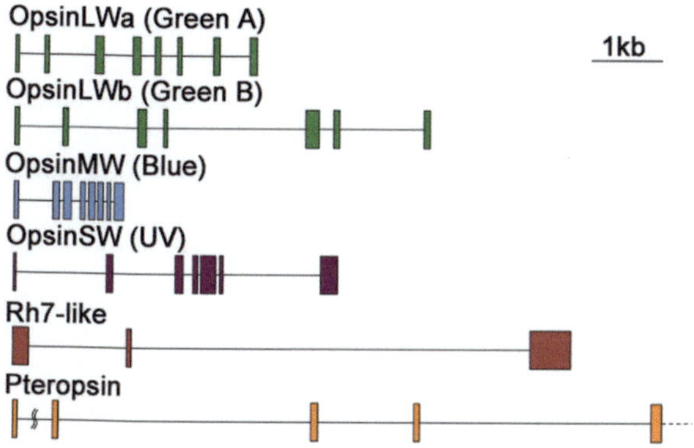

Fig. 4.7 Structure of *Gb'opsin* genes deduced from the cricket genome database. There are at least six opsin genes in the genome of *Gryllus bimaculatus*. OpsinLWa (*green A*) and OpsinLWb (*green B*) genes are located tail to tail in the reverse orientation, suggesting that they were produced by a recent gene duplication. The cricket Rh7-like gene has three exons, which is conserved among Rh7-like genes of other species. The genomic structure for a cricket pteropsin is preliminary, because the putative exon 1 and exon 2 are in different scaffolds and exon 5 does not contain a termination codon at present (2013.12 version)

expressed in all three ocelli types, while UV opsin is not. Therefore, it is presumed that an additional ocellus-specific opsin should be present in crickets. The preliminary analysis of the *Gryllus* genome has identified two additional opsin family members (TM and SN, unpublished data): an Rh7-like gene and pteropsin (Fig. 4.7), which is an invertebrate opsin 3 (Opn3) homolog. The photosensitivity of Rh7 has not been characterized in any invertebrate at present, whereas a single study of mosquito Opn3 homolog revealed it to be a green-sensitive photopigment (Koyanagi et al. 2013).

While the data is lacking in *Gryllus*, the functions of opsin genes have been examined via RNAi in another cricket, *Modicogryllus siamensis* (Tamaki et al. 2013). Interestingly, the loss of function of the UV opsin causes photoperiod-dependent changes in the durations of nymphal development in this species. Hence, some of the opsin genes may also play a role in photoperiodic responses in insects, although such photoperiodicity is not observed in *Gryllus bimaculatus*. It is also tempting to examine the role of opsins expressed in the ocelli by RNAi or genome-editing techniques developed recently (Chap. 21).

4.5 Conclusion

In this review, we summarized the morphology, development, and opsin photopigments of the *Gryllus* eye. Crickets have two types of eyes, ocelli and compound eyes, the latter of which are the main visual organs in this species. In the basal hemimetabolous insect, the development of the compound eyes continues from embryogenesis (patterning) to postembryogenesis (differentiation and growth). The anterior ventral proliferation zone (AVPZ), in which retinal stemlike cells reside, is responsible for continuous growth of the compound eye. The AVPZ in the cricket nymphal eye is reminiscent of the stem cell regions of some vertebrate retinas, the ciliary marginal zone (CMZ; for a review, see Kubo and Nakagawa 2008). Although retinal determination genes (RDGs) such as *ey*, *toy*, *so*, *eya*, and *dac* are expressed during cricket eye development, there seems to be divergence in their roles between crickets and holometabolous insects. *eya* and *so* are likely fundamental to cricket eye development rather than Pax6 homologs, which is different from that in *Drosophila*, where ey, toy, and dac functions at the top of the early eye patterning. In spite of its diversity in morphological complexity, the eye has a primary function in photoreception. Genomic analysis of *Gryllus bimaculatus* has revealed all the members of opsin photopigments. Their unidentified roles can now be elucidated by RNAi, transgenesis, or RNA-guided genome-editing systems in this species and will expand our understanding of vision in the hemimetabolous insects.

References

Anderson H (1978) Postembryonic development of the visual system of the locust. Schistocerca gregaria. II. An experimental investigation of the formation of the retina-lamina projection. J Embryol Exp Morphol 46:47–170

Bessa J, Gebelein B, Pichaud F, Casares F, Mann RS (2002) Combinatorial control of Drosophila eye development by eyeless, homothorax, and teashirt. Genes Dev 16:2415–2427

Bonini NM, Bui QT, Gray-Board GL, Warrick JM (1997) The Drosophila eyes absent gene directs ectopic eye formation in a pathway conserved between flies and vertebrates. Development 124:4819–4826

Burghause F (1979) Die structurelle spezialisierung des dosalen augenteils der Grillen (Orthoptera, Gryllidae). Zool Jb Physioll Bd 83:S502–S525

Callaerts P, Clements J, Francis C, Hens K (2006) Pax6 and eye development in Arthropoda. Arthropodan Struct Dev 35:379–391

Chen R, Amoui M, Zhang Z, Mardon G (1997) Dachshund and eyes absent proteins form a complex and function synergistically to induce ectopic eye development in Drosophila. Cell 91:893–903

Chen R, Halder G, Zhang Z, Mardon G (1999) Signaling by the TGF-beta homolog decapentaplegic functions reiteratively within the network of genes controlling retinal cell fate determination in Drosophila. Development 126:935–943

Curtiss J, Mlodzik M (2000) Morphogenetic furrow initiation and progression during eye development in Drosophila: the roles of decapentaplegic, hedgehog and eyes absent. Development 127:1325–1336

Czerny T, Halder G, Kloter U, Souabni A, Gehring WJ, Busslinger M (1999) Twin of eyeless, a second Pax-6 gene of Drosophila, acts upstream of eyeless in the control of eye development. Mol Cell 3:297–307

Dong Y, Friedrich M (2010) Enforcing biphasic eye development in a directly developing insect by transient knockdown of single eye selector genes. J Exp Zool B Mol Dev Evol 314:104–114

Friedrich M (2006) Continuity versus split and reconstitution: exploring the molecular developmental corollaries of insect eye primordium evolution. Dev Biol 299:310–329

Friedrich M, Benzer S (2000) Divergent decapentaplegic expression patterns in compound eye development and the evolution of insect metamorphosis. J Exp Zool 288:39–55

Halder G, Callaerts P, Gehring WJ (1995) New perspectives on eye evolution. Curr Opin Genet Dev 5:602–609

Henze MJ, Dannenhauer K, Kohler M, Labhart T, Gesemann M (2012) Opsin evolution and expression in arthropod compound eyes and ocelli: insights from the cricket Gryllus bimaculatus. BMC Evol Biol 12:163

Homberg U (2004) Multisensory processing in the insect brain. In: Christensen TA (ed) Methods in insect sensory neuroscience. CRC Press, Boca Raton, pp 3–25

Inoue Y, Miyawaki K, Terasawa T, Matsushima K, Shinmyo Y, Niwa N, Mito T, Ohuchi H, Noji S (2004) Expression patterns of dachshund during head development of Gryllus bimaculatus (cricket). Gene Expr Patterns 4:725–731

Jang CC, Chao JL, Jones N, Yao LC, Bessarab DA, Kuo YM, Jun S, Desplan C, Beckendorf SK, Sun YH (2003) Two Pax genes, eye gone and eyeless, act cooperatively in promoting Drosophila eye development. Development 130:2939–2951

Jemc J, Rebay I (2007) The eyes absent family of phosphotyrosine phosphatases: properties and roles in developmental regulation of transcription. Annu Rev Biochem 76:513–538

Koyanagi M, Takada E, Nagata T, Tsukamoto H, Terakita A (2013) Homologs of vertebrate Opn3 potentially serve a light sensor in nonphotoreceptive tissue. Proc Natl Acad Sci U S A 110:4998–5003

Kronhamn J, Frei E, Daube M, Jiao R, Shi Y, Noll M, Rasmuson-Lestander A (2002) Headless flies produced by mutations in the paralogous Pax6 genes eyeless and twin of eyeless. Development 129:1015–1026

Kubo F, Nakagawa S (2008) Wnt signaling in retinal stem cells and regeneration. Dev Growth Differ 50:245–251

Kumar JP (2011) My what big eyes you have: how the Drosophila retina grows. Dev Neurobiol 71 (12):1133–1152. doi:10.1002/dneu.20921

Labhart T, Meyer EP (1999) Detectors for polarized skylight in insects: a survey of ommatidial specializations in the dorsal rim area of the compound eye. Microsc Res Tech 47:368–379

Mito T, Noji S (2008) The two-spotted cricket Gryllus bimaculatus: an emerging model for developmental and regeneration studies. CSH protoc. 2008, pdb emo110

Miyawaki K, Mito T, Sarashina I, Zhang H, Shinmyo Y, Ohuchi H, Noji S (2004) Involvement of wingless/armadillo signaling in the posterior sequential segmentation in the cricket, Gryllus bimaculatus (Orthoptera), as revealed by RNAi analysis. Mech Dev 121:119–130

Nakamura T, Yoshizaki M, Ogawa S, Okamoto H, Shinmyo Y, Bando T, Ohuchi H, Noji S, Mito T (2010) Imaging of transgenic cricket embryos reveals cell movements consistent with a syncytial patterning mechanism. Curr Biol 20:1641–1647

Oliver G, Mailhos A, Wehr R, Copeland NG, Jenkins NA, Gruss P (1995) Six3, a murine homologue of the sine oculis gene, demarcates the most anterior border of the developing neural plate and is expressed during eye development. Development 121:4045–4055

Oliver G, Loosli F, Koster R, Wittbrodt J, Gruss P (1996) Ectopic lens induction in fish in response to the murine homeobox gene Six3. Mech Dev 60:233–239

Pignoni F, Zipursky SL (1997) Induction of Drosophila eye development by decapentaplegic. Development 124:271–278

Pignoni F, Hu B, Zavitz KH, Xiao J, Garrity PA, Zipursky SL (1997) The eye-specification proteins So and Eya form a complex and regulate multiple steps in Drosophila eye development. Cell 91:881–891

Sakura M, Takasuga K, Watanabe M, Eguchi E (2003) Diurnal and circadian rhythm in compound eye of cricket (Gryllus bimaculatus): changes in structure and photon capture efficiency. Zool Sci 20:833–840

Shen W, Mardon G (1997) Ectopic eye development in Drosophila induced by directed dachshund expression. Development 124:45–52

Takagi A, Kurita K, Terasawa T, Nakamura T, Bando T, Moriyama Y, Mito T, Noji S, Ohuchi H (2012) Functional analysis of the role of eyes absent and sine oculis in the developing eye of the cricket Gryllus bimaculatus. Dev Growth Differ 54:227–240

Tamaki S, Takemoto S, Uryu O, Kamae Y, Tomioka K (2013) Opsins are involved in nymphal photoperiodic responses in the cricket Modicogryllus siamensis. Physiol Entomol 38:163–172

Wawersik S, Maas RL (2000) Vertebrate eye development as modeled in Drosophila. Hum Mol Genet 9:917–925

Wolff T, Ready DF (1993) Pattern formation in the Drosophila retina. In: Bate M, Martinez-Arias A (eds) The development of Drosophila melanogaster. Cold Spring Harbor Laboratory Press, Cold Spring Harbor, New York, pp 1277–1325

Xu PX, Adams J, Peters H, Brown MC, Heaney S, Maas R (1999) Eya1-deficient mice lack ears and kidneys and show abnormal apoptosis of organ primordia. Nat Genet 23:113–117

Yang X, Zarinkamar N, Bao R, Friedrich M (2009) Probing the Drosophila retinal determination gene network in Tribolium (I): the early retinal genes dachshund, eyes absent and sine oculis. Dev Biol 333:202–214

Zufall F, Schmitt M, Menzel R (1989) Spectral and polarized-light sensitivity of photoreceptors in the compound eye of the cricket (Gryllus bimaculatus). J Comp Physiol A 164:597–608

Chapter 5
An Early Embryonic Diapause Stage and Developmental Plasticity in the Band-Legged Ground Cricket *Dianemobius nigrofasciatus*

Sakiko Shiga and Hideharu Numata

Abstract Egg diapause is found in a variety of insect orders and can occur in various stages, ranging from an early small germ band stage to a later stage in which an embryo exhibits appendages. Embryonic diapause is especially prominent in crickets inhabiting temperate regions, such as the band-legged ground cricket, *Dianemobius nigrofasciatus*. This species is common in the Japanese islands and exhibits a univoltine life cycle with obligatory egg diapause in northern regions, whereas a bivoltine life cycle with facultative egg diapause controlled by parental photoperiod and temperature is present in the southern regions. *D. nigrofasciatus* enters diapause during an early embryonic stage. Here we review the morphological aspects of early embryogenesis and how they relate to the various stages of diapause. By focusing on details of development, such as patterns of nuclear arrangement and the appearance of cell membranes, we can identify useful hallmarks for the study of diapause. In particular, the formation of a continuous thin layer of cells covering the egg's surface normally reveals the formation of the cellular blastoderm. In diapause eggs, however, that continuous layer of cells covers the interior egg surface without forming a germ band, indicating that diapause occurs at the cellular blastoderm stage. Furthermore, in a bivoltine strain, diapause can be induced by low temperatures and averted by high temperatures applied to the eggs just before formation of the cellular blastoderm. In conclusion, diapause in *D. nigrofasciatus* entails developmental arrest at the cellular blastoderm stage, which can be influenced by temperature before blastoderm formation.

S. Shiga (✉)
Department of Biological Science, Graduate School of Science, Osaka University, Toyonaka, Osaka 560-0043, Japan
e-mail: shigask@bio.sci.osaka-u.ac.jp

H. Numata
Department of Biological Science, Graduate School of Science, Kyoto University, Kyoto 606-8501, Japan

H.W. Horch et al. (eds.), *The Cricket as a Model Organism*,
DOI 10.1007/978-4-431-56478-2_5

Keywords Egg diapause • Cellular blastoderm • *Dianemobius nigrofasciatus* • Cellularization • Photoperiodic response • Maternal factors • Facultative diapause

5.1 Cricket Life Cycles and Diapause

Crickets have evolved various life cycles (Masaki and Walker 1987). Some species inhabiting tropical regions pass through all developmental stages at all times of the year. These crickets are virtually free of developmental arrest (i.e., diapause); this life cycle is called homodynamic. In the higher latitudes, however, the life cycle is usually synchronized with seasons so that the crickets overwinter at a fixed stage. Like many other insects, different developmental stages of crickets show different degrees of tolerance to environmental stressors, due to stage-specific aspects of size, structure, behavior, and/or physiology (Masaki and Walker 1987). Diapause is a physiological state of dormancy in which morphological development is suppressed (Danks 1987). Most insects enter diapause during a stage tolerant to adverse conditions, though some do not, e.g., the rice grasshopper *Oxya yezoensis* (Ando 2004). Crickets remain in diapause during adverse seasons but develop and reproduce in other seasons. This type of life cycle is called heterodynamic. The specific stage in which diapause occurs varies from species to species because each has evolved specific physiological adaptations to its environment. Masaki and Walker (1987) classified ten types of heterodynamic life cycles, according to the overwintering stage and voltinism. The overwintering stage can occur at any life-cycle stage, egg, nymph, or adult, although overwintering during the egg stage is the most common in crickets.

In Japan, both univoltine and bivoltine species are found. Diapause in the univoltine life cycle in the northern areas is obligatorily, while diapause in the bi- or multivoltine species inhabiting the southern regions is facultatively controlled by environmental factors. Even within the same species, both obligatory and facultative diapause can be found. The band-legged ground cricket, *Dianemobius nigrofasciatus* (formerly *Pteronemobius nigrofasciatus*), is commonly found in the Japanese islands between 30 and 44°N, where it enters diapause during the egg stage (Masaki 1972). This species shows a univoltine life cycle with obligatory diapause in the north, whereas a bivoltine life cycle with facultative diapause controlled by photoperiod and temperature is present in the south. Geographic variation is also found in the photoperiodic responses controlling nymphal developmental period. Nymphal development is retarded by a long photoperiod in the univoltine northern populations and by an intermediate photoperiod in the bivoltine southern population. These geographic variations in photoperiodism can be used to explain a gap in geographic cline regarding the adult size. The size of adults collected in autumn increases from north to south, but this trend is disrupted and reversed in the zone where the life cycle changes from univoltine to bivoltine. On the basis of the characteristic geographical trend in size and the photoperiodic control of development, it has been inferred that the bivoltine cycle may be the original type from which the univoltine cycle has evolved (Masaki 1972). Although the life history of this species has been well characterized, the exact stage at which diapause occurs and its control mechanisms are not well described. Here we describe a very early diapause stage, and photoperiodic and temperature control of egg diapause in a facultative strain of *D. nigrofasciatus*. Identification of both the stage at which development is arrested and the stimuli controlling the induction of diapause will help to elucidate the developmental plasticity in insect eggs.

5.2 Various Embryonic Diapause Stages in Orthopteran Insects

Umeya (1950) defined five types of embryonic diapause stages in insects, and orthopteran species (e.g., crickets, grasshoppers, and locusts) can enter diapause during different stages. According to his classification, the earliest diapause stage (i.e., type I) is the pyriform stage. During this stage, a germ band consists of only a tiny cap of cells without formation of the head or expansion of the tail, as is found in the cricket *Velarifictorus mikado* and the bell cricket, *Meloimorpha japonica* (Umeya 1950). In the type II stage, diapause occurs during a later period in egg development when a dumbbell-shaped germ band has formed prior to differentiation of the body segments. This type of embryonic diapause has been observed in the cricket *Teleogryllus emma* (Masaki 1960). A type III stage is an elongated embryo with the head and tail slightly expanded laterally and the entire abdomen elongated anteriorly and posteriorly without appendages. Some lepidopterans enter diapause at this stage but no examples have been reported in orthopterans. The type IV stage occurs when appendages on the head and thoracic regions are seen. This type of diapause has been observed in the migratory locust, *Locusta migratoria*, and the differential grasshopper, *Melanoplus differentialis* (Umeya 1950; Slifer 1932).

Some European species of Tettigoniidae, Orthoptera, such as *Decticus verrucivorus*, enter an initial diapause in round-shaped primordium, a simple structure compared with the type I stage, although these species enter an embryonic diapause again at the type IV stage (Ingrisch 1984). In several cricket species, *Polionemobius mikado*, *D. nigrofasciatus*, *Pteronemobius ohmachii*, and *Loxoblemmus aomoriensis*, no embryonic body has been found in diapause eggs (Masaki 1960; Fukumoto et al. 2006). Whole-egg observations suggest that these eggs enter diapause before germ band formation. It would be interesting to understand how development is arrested at such an early stage in the absence of organs, and this is a focus of the present work.

5.3 Early Development of Nondiapause Eggs and the Diapause Stage in *D. nigrofasciatus*

Experiments investigating early embryonic development in *D. nigrofasciatus* were performed at 25 °C under long-day conditions of LD 16:8 h. Observations of egg whole mounts under a light microscope identified about ten nuclei that were equally

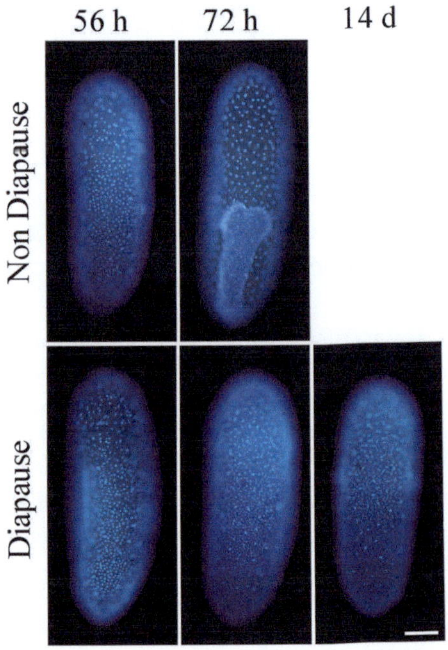

Fig. 5.1 Early development of nondiapause and diapause eggs of *Dianemobius nigrofasciatus*. Whole eggs were stained with DAPI (4,6-diamidino-2-phenylindole) 56 h, 72 h, and 14 days after oviposition at 25 °C under LD 16:8 h. Although a germ band formed 72 h after oviposition in the nondiapause eggs, it was not seen in the diapause eggs even at 14 days after oviposition. The *lower tip* of the eggs is the posterior pole. Scale = 200 μm (Modified from Tanigawa et al. 2009)

distributed within the yolk mass of the nondiapause eggs 12 h after oviposition. As both dividing and nondividing nuclei were observed, mitosis appeared to occur asynchronously in the yolk mass. Twenty-four hours after oviposition, the nuclei had moved from the yolk mass area to the superficial region of the egg where subsequent mitosis occurs. Nuclear density in the superficial region uniformly increased 40 h after oviposition. Fifty-six h after oviposition, the nuclear density was higher in the posterior region than the anterior region. Seventy-two hours after oviposition, an elongated heart-shaped germ band was observed in the posterior region (Fig. 5.1).

In semi-thin sections of the regions near the posterior pole of the egg, at 12 h after oviposition no periplasmic space in the surface region was observed, and a small number of nuclei surrounded by the cytoplasm was observed among yolk granules and lipid droplets. Twenty-four hours after oviposition, nuclei surrounded by the cytoplasm, known as energids, appeared at the egg surface. Forty hours after oviposition, the number of nuclei at the egg surface increased, but the surface was not entirely covered by these nuclei (Fig. 5.2A). Fifty-six hours after oviposition, a layer of nuclei was observed beneath the vitelline membrane surrounding the egg (Fig. 5.2B). A continuous thin layer of nuclei completely covered the egg surface at this time, indicating that the eggs of *D. nigrofasciatus* reached the blastoderm stage between 40 and 56 h after oviposition. The embryonic rudiment was observed as a thick layer in the posterior ventral region of the blastoderm 56 h after oviposition. Using electron

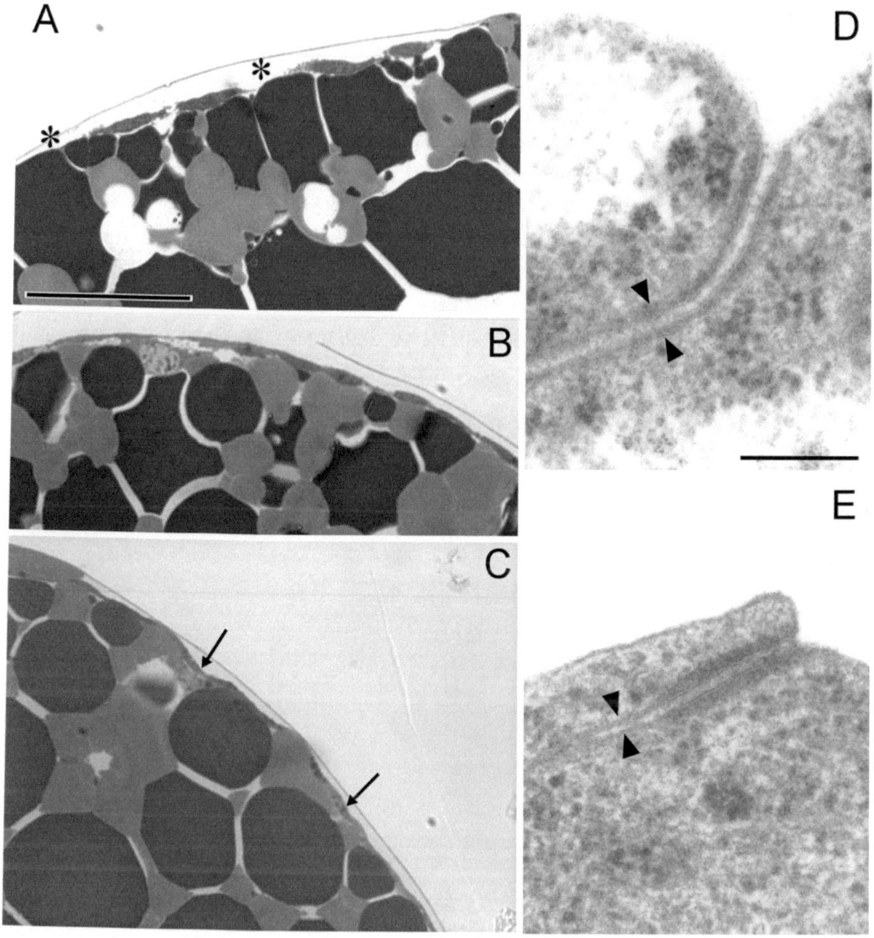

Fig. 5.2 Nuclear distribution near the posterior pole (–**C**) and cell membrane (**D**, **E**) of nondiapause and diapause eggs in *D. nigrofasciatus* at 25 °C. Panels illustrate toluidine blue staining of semi-thin sections and electron microphotographs of ultrathin sections of the surface region of the posterior pole. (**A**) At 40 h after oviposition of a nondiapause egg, nuclei were not arranged in a continuous layer, and some gaps between energids were observed (*asterisks*). (**B**) At 56 h after oviposition of a nondiapause egg, nuclei were found in a continuous layer. (**C**) A continuous layer (*arrows*) of the nuclei and cytoplasm was seen at the surface region of the diapause eggs at 14 days after oviposition. (**D**) A nondiapause egg at 40 h after oviposition. (**E**) A diapause egg 14 days after oviposition. *Triangles* indicate the cell membrane. Scale = 50μm in A-C and 200 nm in D and E (Modified from Tanigawa et al. 2009)

microscopy, the plasma membrane was observed between the nuclei 40 h after oviposition even before the blastoderm stage was reached (Fig. 5.2D). Fifty-six hours after oviposition, the presence of the plasma membrane in the blastoderm was also confirmed. At this stage, large nuclei surrounded by cytoplasm were found just beneath the cell layer of the blastoderm.

The timing of cellularization was also confirmed by DAPI staining of the nuclei by injection of rhodamine-dextran dye into the anterior region of the egg. If rhodamine stained the cytoplasm surrounding the nuclei, that would indicate that the plasma membrane had not yet formed and that the nuclei were present within the cytoplasm as energids. When rhodamine was injected into eggs 12 h after oviposition, rhodamine surrounded all nuclei. Injecting rhodamine into the eggs 24 h after oviposition showed that only a few energids were stained (Tanigawa et al. 2009). The results indicate the presence of some cell membranes prior to or at 24 h after oviposition, when the nuclei appear at the egg surface. These results indicate that cellularization in *D. nigrofasciatus* occurs before blastoderm formation, which occurs at 40–56 h after oviposition (see above and Fig. 5.3).

The timing of cellularization with respect to blastoderm formation is different from *Gryllus bimaculatus*, in which cellularization occurs during the blastoderm

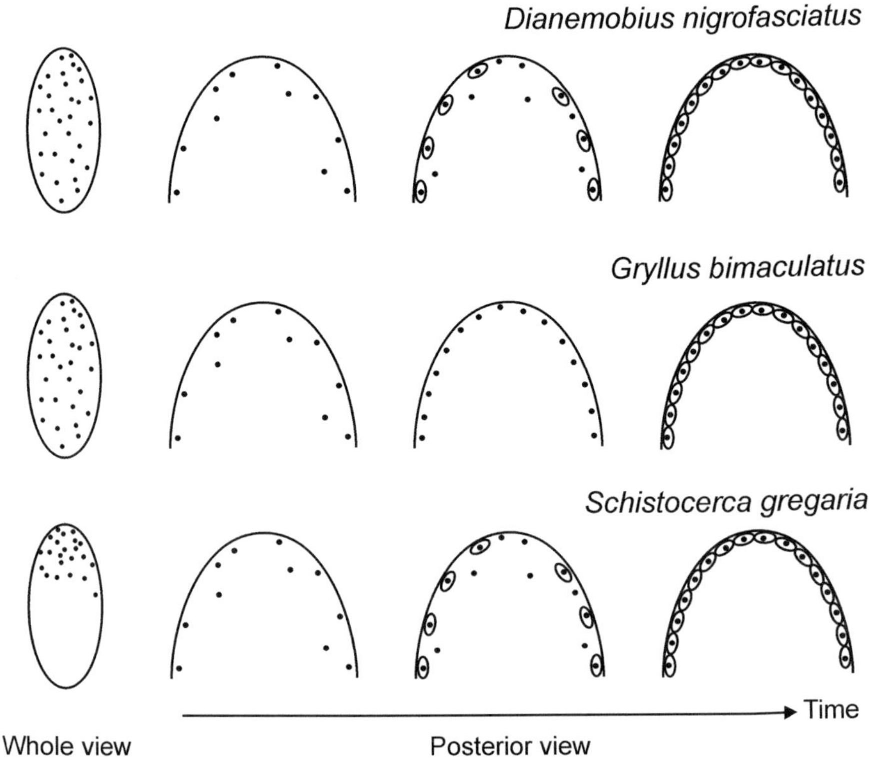

Fig. 5.3 Schematic illustrations of developmental patterns in the early embryos of *D. nigrofasciatus*, *Gryllus bimaculatus*, and *Schistocerca gregaria*. *Left* schemas show the whole view. Posterior regions are shown in the enlarged view. *D. nigrofasciatus* and *G. bimaculatus* have a uniform blastoderm formation, whereas *S. gregaria* has a differentiating blastoderm formation. Each *dot* indicates a nucleus. *Circles* surrounding the nucleus are plasma membrane. See the text for further explanation

stage, indicating that the cellular blastoderm follows a syncytial blastoderm (Fig. 5.3; Sarashina et al. 2005). Both *G. bimaculatus* and *D. nigrofasciatus* showed uniform blastoderm formation in which the cleavage energids emerged in a uniform array over the entire surface of the yolk mass, but the timing of cellularization differed. The mode of blastoderm formation has also been well studied in the desert locust *Schistocerca gregaria*. This species undergoes differentiating blastoderm formation. After several mitoses, a small number of energids appear in the surface region of the egg near the posterior pole, and these energids gradually spread anteriorly as asynchronous mitosis proceeds. Cellularization occurs shortly after the nuclei move to the egg's surface, before they completely enclose the whole egg, indicating that it occurs before blastoderm formation (Fig. 5.3; Ho et al. 1997). Thus, blastoderm formation is variable even within the Orthoptera, and the timing of cellularization even differs between the two crickets, *G. bimaculatus* and *D. nigrofasciatus*.

No distinct differences were apparent between the diapause and nondiapause eggs at 12, 24, and 40 h after oviposition. However, the development of diapause and nondiapause eggs differed after this time. In nondiapause eggs, from 40 to 56 h after oviposition, the number of nuclei appeared to increase, while the nuclear distribution was unchanged among diapause eggs at 56 h, 72 h, and 14 days after oviposition (Fig. 5.1). The periplasmic layer completely covered the egg's surface, but the embryonic rudiment was not observed even 14 days after oviposition in diapause eggs, and the cell membrane in these eggs was observed (Fig. 5.2c, e). These observations indicate that *D. nigrofasciatus* enters diapause at the cellular blastoderm stage. Among various possible diapause stages during embryogenesis, diapausing during the cellular blastoderm stage should have some advantages for the small ground crickets. Masaki (1986) suggested that for eggs laid only several millimeters in the ground, a simple organization of a uniform cell layer at the cellular blastoderm stage might be more tolerable to desiccation and temperature changes than a more differentiated and complex embryo.

5.4 Voltinism and Photoperiodic Response in the Osaka Strain of *D. nigrofasciatus*

The seasonal occurrence of embryonic diapause in Osaka, Japan (34°50′N 135°28′E) is shown in Fig. 5.4. Some females laid only nondiapause eggs (i.e., nondiapause egg producers), some laid both nondiapause and diapause eggs (i.e., mixed egg producers), and others laid only diapause eggs (diapause egg producers). In the summer months, from June to the beginning of August, nondiapause eggs were laid by nondiapause egg producers and mixed egg producers. During summer, diapause, eggs were also produced to some extent (Fig. 5.4). In September and October, females mostly consisted of diapause egg producers. Therefore, diapause eggs were almost exclusively produced during the autumn. Population size decreased from the middle of August to the beginning of September. This suggests

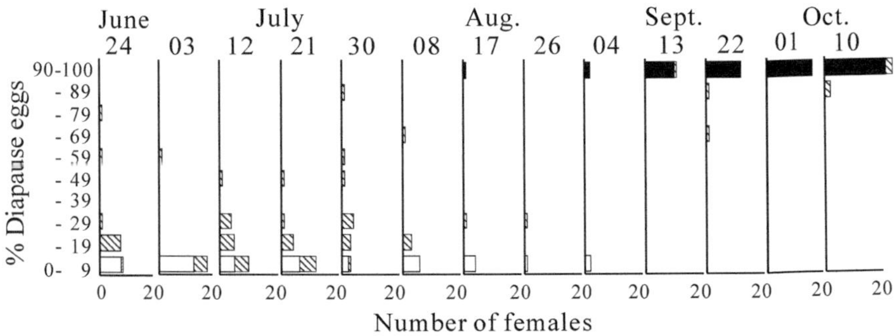

Fig. 5.4 Seasonal changes in the incidence of diapause eggs in *Dianemobius nigrofasciatus* in Osaka, Japan. Approximately 20 pairs of adults were collected in the field every 9 days. Adults were kept outdoors in Osaka (34°41′N, 135°31′E). Their eggs were collected every day from 6 to 9 or 10 days after the adults were set outdoors. The eggs were maintained in the laboratory (25 ± 1 °C, LD 16:8 h). Females were classified into 10 grades in terms of the percentage of diapause eggs (ordinate) that they oviposited. *Closed*, *hatched*, and *open columns* designate diapause egg, mixed egg, and nondiapause egg producers, respectively. Results for all of the individuals that laid ≥10 eggs are shown. Note that a few females produced diapause eggs from June to August (Modified from Shiga and Numata 1997)

that the adults present from June to August represent the overwintering generation, and the adults present in September and October represent the first generation from the current year. That is, *D. nigrofasciatus* probably has a bivoltine life cycle in Osaka.

In the Osaka strain, crickets from the overwintering versus the first generation produced different percentages of diapause eggs (Fig. 5.4). This shows that *D. nigrofasciatus* undergoes facultative diapause, which is controlled by environmental factors. The effect of maternal photoperiod on egg production and the incidence of diapause were examined. At 25 °C, females laid fewer eggs under short days than under long days. This is consistent with fecundity changes under different photoperiods reported in other nondiapause insects (Philogène and McNeil 1984). Figure 5.5 shows the effect of maternal photoperiod on the incidence of diapause. The longer the photophase, the smaller the incidence of diapause egg producers and the larger the incidence of nondiapause egg and mixed egg producers. The critical day length was between 13 and 14 h. The natural day length from late August to the middle of September in Osaka is about 13–14 h, which corresponds to the critical day length at 25 °C measured in the laboratory. The combined laboratory and field studies are consistent with the overwintering generation experiencing long-day photoperiods and thus developing into mixed egg producers or nondiapause egg producers and the first generation experiencing short-day photoperiods and developing into diapause egg producers.

Seasonal changes in egg production by individual females in bivoltine species were examined in *L. migratoria* (Tanaka 1994) and the Australian plague locust *Chortoicetes terminifera* (Wardhaugh 1986); in both of these locusts, all of the

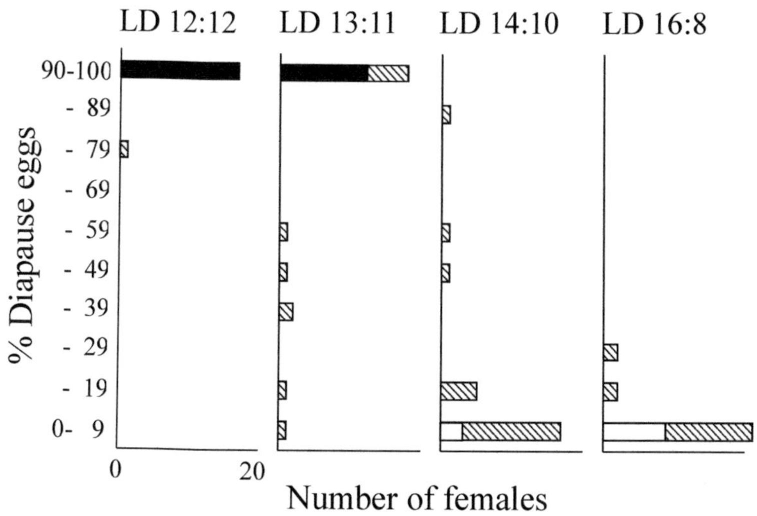

Fig. 5.5 Effects of maternal photoperiod on the induction of embryonic diapause in *D. nigrofasciatus* at 25 °C. Female adults were classified into 10 grades in terms of the incidence of diapause eggs (ordinate). The number of females in each grade is plotted (abscissa, maximum = 20). *Closed*, *hatched*, and *open columns* designate diapause egg, mixed egg, and nondiapause egg producers, respectively $n = 18$–25 (Modified from Shiga and Numata 1997)

females in the summer were nondiapause egg producers. In *D. nigrofasciatus*, however, mixed egg producers occurred in the summer, which suggests that the *D. nigrofasciatus* population in Osaka may partly consist of a univoltine, as well as a bivoltine, population (Shiga and Numata 1997). In addition, the occurrence of diapause eggs in the field during the summer may depend on natural temperature changes within a given year, because the incidence of diapause in *D. nigrofasciatus* was also affected by the temperature to which the mothers and their eggs were exposed (Fukumoto et al. 2006). In different years, the same Osaka strain reared in outdoor conditions showed very few diapause eggs during the summer (Fukumoto et al. 2006).

5.5 Effects of Egg Temperature on Diapause Induction in *D. nigrofasciatus*

Diapause induction in *D. nigrofasciatus* is also influenced by the temperature at which eggs are maintained. In an experiment, nymphs and adults were reared at a fixed temperature at 25 °C under long- or short-day conditions, and their eggs were incubated at different temperatures, which ranged from 5 to 30 °C. At 5–20 °C, all short- and long-day eggs entered diapause. At temperatures of ≥20 °C, the higher

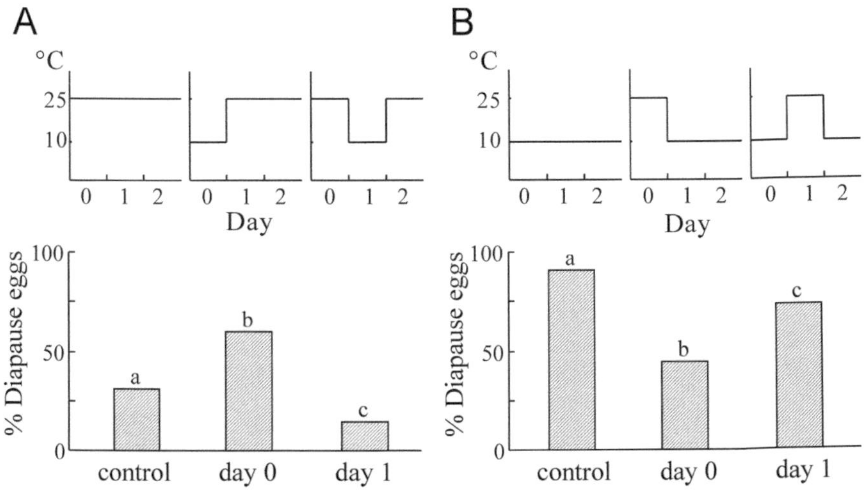

Fig. 5.6 Effects of temperature pulses applied to the eggs upon the induction of embryonic diapause in *D. nigrofasciatus*. **A** ($n = 36$–134) and **B** ($n = 26$–73) show the results with eggs laid by adults reared under LD 16:8 h at 25 °C or LD 12:12 h at 25 °C, respectively. The temperature pulses were 10 °C in (**A**) and 35 °C in (**B**). Eggs were maintained at 25 °C, except for the temperature pulses. The abscissas of the *lower graphs* show the time point at which the temperature pulses were applied. Different letters at the *tops of the columns* denote significant differences between temperature treatments (Tukey-type multiple comparison test for proportions) (Modified from Fukumoto et al. 2006)

the temperature at which the eggs were maintained, the lower was the incidence of diapause; this occurred in eggs laid by females under long- and short-day conditions (Fukumoto et al. 2006). At 22.5–30.0 °C, eggs laid by long-day adults exhibited a lower incidence of diapause than those laid by short-day adults. In eggs laid by adults under short- and long-day conditions, the higher the temperature at which the eggs were maintained, the lower the incidence of diapause.

It is interesting that the temperature experienced by the egg before the diapause stage (cellular blastoderm) could change the developmental program (i.e., nondiapause or diapause). This was shown by experiments in which eggs were exposed to high- or low-temperature pulses before the blastoderm stage. Considering that the cellular blastoderm forms within 40–56 h after oviposition (Fig. 5.2A, B), a temperature pulse was given on days 0 or 1. A low-temperature pulse applied to long-day eggs on day 0 was found to increase the incidence of diapause. However, when the pulse was applied on day 1, the incidence of diapause was reduced (Fig. 5.6A). A high-temperature pulse applied to short-day eggs reduced the incidence of diapause when applied to the eggs either on day 0 or on day 1 (Fig. 5.6B). This shows that temperature sensitivity in *D. nigrofasciatus* eggs occurs at 25 °C on days 0 or 1 when organs or distinct tissues have not yet formed.

5.6 Future Perspective

It would be interesting to define the molecular mechanisms underlying developmental arrest at the cellular blastoderm stage. Zygotic gene expression occurs at or after the formation of the blastoderm in most insects (Sander et al. 1985). Suppression of initial zygotic transcription might be involved in developmental arrest at the cellular blastoderm stage in diapause eggs in *D. nigrofasciatus*. Future studies should determine the point at which embryonic transcription is first initiated. Considering that diapause induction is controlled by the maternal photoperiod, maternal temperature, and egg temperature and that diapause occurs at the cellular blastoderm stage, key molecular factors influencing development of the nondiapause or diapause program may include maternal factors. Such factors may control the suppression of zygotic transcription and may also be under the control of the maternal photoperiod and/or temperature.

Several maternal RNAs are known to be important for early embryonic development. As an example, in *Drosophila melanogaster*, *cau*dal is transcribed both maternally and zygotically and plays a crucial role as a transcription factor in posterior embryonic patterning after the cellular blastoderm stage (Mlodzik et al. 1985). Maternal *caudal* mRNA is transported from the nurse cells to the oocytes from mid-oogenesis onward. To date, *caudal* has been identified in a variety of animals (Benton et al. 2013). Posttranscriptional modification of maternal mRNA, as in *caudal*, or other encoding transcription factors for early embryogenesis, might be another way to arrest further development after the cellular blastoderm stage. It would be interesting to investigate the molecular machinery underlying diapause control.

References

Ando Y (2004) Egg diapause and seasonal adaptations in rice grasshoppers. In: Tanaka S, Higaki M, Kotaki T (eds) Insect diapause: mechanisms and evolution. Tokai University Press, Hatano, pp 4–15 (in Japanese)

Benton MA, Akam M, Pavlopoulos A (2013) Cell and tissue dynamics during *Tribolium* embryogenesis revealed by versatile fluorescence labeling approaches. Development 140:3210–3220

Danks HV (1987) Insect dormancy: an ecological perspective. Biological Survey of Canada, Ottawa

Fukumoto E, Numata H, Shiga S (2006) Effects of temperature of adults and eggs on the induction of embryonic diapause in the band-legged ground cricket, *Dianemobius nigrofasciatus*. Physiol Entomol 31:211–217

Ho K, Dunin-Borkowski OM, Akam M (1997) Cellularization in locust embryos occurs before blastoderm formation. Development 124:2761–2768

Ingrisch S (1984) Embryonic development of *Decticus verrucivorus* (Orthoptera: Tettigoniidae). Entomol Gen 10:1–9

Masaki S (1960) Thermal relations of diapause in the eggs of certain crickets (Orthoptera; Gryllidae). Bull Fac Agric Hirosaki Univ 6:5–20

Masaki S (1972) Climatic adaptation and photoperiodic response in the band-legged ground cricket. Evolution 26:587–600

Masaki S (1986) Significance of ovipositor length in life cycle adaptations of crickets. In: Taylor F, Karban R (eds) The evolution of insect life cycles. Springer, New York, pp 20–34

Masaki S, Walker TJ (1987) Cricket life cycles. In: Hecht MK, Wallace B, Prance GT (eds) Evolutionary biology, vol 21. Plenum, New York, pp 349–423

Mlodzik M, Fjose A, Gehring WJ (1985) Isolation of *caudal*, a *Drosophila* homeo box-containing gene with maternal expression, whose transcripts form a concentration gradient at the pre-blastoderm stage. EMBO J 4:2961–2969

Philogène BJR, McNeil JN (1984) The influence of light on the non-diapause related aspects of development and reproduction in insects. Photochem Photobiol 40:753–761

Sander K, Gutzeit HO, Jäckle H (1985) Insect embryogenesis: morphology, physiology, genetical and molecular aspects. In: Kerkut GA, Gilbert LI (eds) Comprehensive insect physiology biochemistry and pharmacology, vol 1. Pergamon, Oxford, pp 319–385

Sarashina I, Mito T, Saito M, Uneme H, Miyawaki K, Shinmyo Y, Ohuchi H, Noji S (2005) Location of micropyles and early embryonic development of the two-spotted cricket *Gryllus bimaculatus* (Insecta, Orthoptera). Dev Growth Differ 47:99–108

Shiga S, Numata H (1997) Seasonal changes in the incidence of embryonic diapause in the band-legged ground cricket, *Dianemobius nigrofasciatus*. Zool Sci 14:1015–1018

Slifer EH (1932) Insect development. IV. External morphology of grasshopper embryos of known age and with a known temperature history. J Morph 53:1–21

Tanaka H (1994) Embryonic diapause and life cycle in the migratory locust, *Locusta migratoria* L. (Orthoptera: Acrididae). Appl Entomol Zool 29:179–191

Tanigawa N, Matsumoto K, Yasuyama K, Numata H, Shiga S (2009) Early embryonic development and diapause stage in the band-legged ground cricket *Dianemobius nigrofasciatus*. Dev Genes Evol 219:589–596

Umeya Y (1950) Studies on embryonic hibernation and diapause in insects. Proc Jpn Acad 26:1–9

Wardhaugh KG (1986) Diapause strategies in the Australian plague locust (*Chortoicetes terminifera* Walker). In: Taylor F, Karban R (eds) The evolution of insect life cycles. Springer, New York, pp 89–104

Part II
Physiology, Nervous System, and Behavior

Chapter 6
Molecular Approach to the Circadian Clock Mechanism in the Cricket

Kenji Tomioka, Outa Uryu, Yuichi Kamae, Yoshiyuki Moriyama, ASM Saifullah, and Taishi Yoshii

Abstract Circadian rhythms are physiological and behavioral changes that follow a roughly 24-h period, responding primarily to daily cycles in an organism's environment. Crickets have provided a good model to study the neural mechanisms controlling the circadian rhythm, because they have a large central nervous system. Neurobiological studies revealed that the circadian clock is located in the optic lobe and the photoreceptors necessary for light entrainment are in the compound eye. Recent progress in molecular technology enabled us to use crickets for dissection of the circadian system at a molecular level. The oscillatory mechanism of the circadian clock has been studied in *Drosophila* and a few higher order insect species, but the results from those insects are often inconsistent. We employed a reverse genetic approach to the cricket clock. We first obtained clock genes, *period* (*per*), *timeless* (*tim*), and *Clock* (*Clk*) with molecular cloning and then analyzed their functions with RNAi technology. The obtained results could be only partially explained by the *Drosophila* model. The central oscillatory mechanism of the cricket clock will be discussed together with the peripheral oscillators and the involvement of pigment-dispersing factor as a neurotransmitter in regulating the locomotor rhythm.

Keywords *Gryllus bimaculatus* • Circadian rhythms • *period* • *timeless* • Clock gene • Molecular oscillation

K. Tomioka (✉) • O. Uryu • Y. Kamae • Y. Moriyama • T. Yoshii
Graduate School of Natural Science and Technology, Okayama University, Okayama 700-8530, Japan
e-mail: tomioka@cc.okayama-u.ac.jp

A. Saifullah
Bangladesh Atomic Energy Commission, Dhaka 1212, Bangladesh

H.W. Horch et al. (eds.), *The Cricket as a Model Organism*,
DOI 10.1007/978-4-431-56478-2_6

6.1 Circadian Rhythms in the Cricket

Most animals show daily rhythms in their lives. The rhythm is thought to be acquired as an adaptive mechanism to daily environmental cycles associated with the Earth's rotation (Dunlap et al. 2004). The rhythm is generated by an endogenous mechanism called the circadian clock. Crickets have been a good model for studying the mechanisms underlying the circadian rhythm (Loher 1972, 1974; Tomioka et al. 2001; Tomioka and Abdelsalam 2004). They show clear circadian rhythms in a variety of physiological functions such as locomotor and stridulatory activity, spermatophore formation, retinal photosensitivity, and responsiveness of visual interneurons (Loher 1972, 1974; Tomioka and Chiba 1982a, b, 1992; Tomioka et al. 1994; Abe et al. 1997). It would be advantageous to search for the mechanisms underlying these rhythms in the cricket.

6.2 The Optic Lobe Is the Circadian Clock Locus

The relatively large size of crickets prompted the search for the tissue that generates the circadian behavioral rhythms. Experimental lesions revealed that the optic lobe is the candidate tissue that generates the activity rhythm: removal of the optic lobe or transection of the optic tract abolishes the locomotor or stridulatory rhythm in *Teleogryllus commodus*, *Gryllus bimaculatus*, *Gryllodes sigillatus*, and *Dianemobius nigrofasciatus* (Loher 1972; Sokolove and Loher 1975; Tomioka and Chiba 1984; Shiga et al. 1999; Abe et al. 1997). The optic lobe was later unequivocally shown to generate the circadian rhythm by measuring its efferent electrical activity in isolated and culture conditions (Tomioka and Chiba 1992). The isolated optic lobe maintains an electrical activity rhythm with a peak during the subjective night.

The photoreceptive area necessary for synchronization to environmental light dark cycle is the compound eye. Disruption of the photic input through the compound eye via bilateral optic nerve transection abolishes the photic entrainment (Tomioka and Chiba 1984). The photoreception in the compound eye is also necessary for mutual synchronization between the bilaterally paired circadian clocks; the dorso-caudal region of the eye plays the most important role since surgical lesion of this area disrupts the synchronization (Tomioka and Yukizane 1997). The photic information is most likely mediated by a group of neurons called medulla bilateral neurons. They show a circadian rhythm in their responsiveness to light stimuli presented to the compound eye, and sectioning of their axonal tract prevents the mutual entrainment of the circadian pacemaker in the optic lobe (Yukizane and Tomioka 1995; Yukizane et al. 2002).

6.3 Molecular Oscillatory Mechanism of the Circadian Clock

The circadian clock machinery has been studied in several insects such as *Drosophila*, monarch butterflies, honeybees, and mosquitoes. The basic clock mechanism is believed to be composed of transcriptional/translational feedback loops. In *Drosophila*, the major players in the loop are *period* (*per*), *timeless* (*tim*), *Clock* (*Clk*), and *cycle* (*cyc*; Fig. 6.1). In brief, transcriptional activators CLOCK (CLK) and CYCLE (CYC), which are protein products of the *Clk* and *cyc* genes, heterodimerize and bind to the promoter regions of *per* and *tim* to activate their transcription (Allada et al. 1998; Darlington et al. 1998; Rutila et al. 1998). The resultant PER and TIM proteins accumulate during the night, form PER/TIM heterodimers, and translocate to the nucleus at late night to repress their own transcription through inactivation of CLK/CYC (Curtin et al. 1995; Saez and Young 1996; Darlington et al. 1998; Lee et al. 1998; Rutila et al. 1998). PER and TIM are subsequently subjected to a degradation process. These processes eventually reduce the PER and TIM levels and release the CLK/CYC from the PER/TIM-dependent inactivation. Thus, the loop goes to the next cycle.

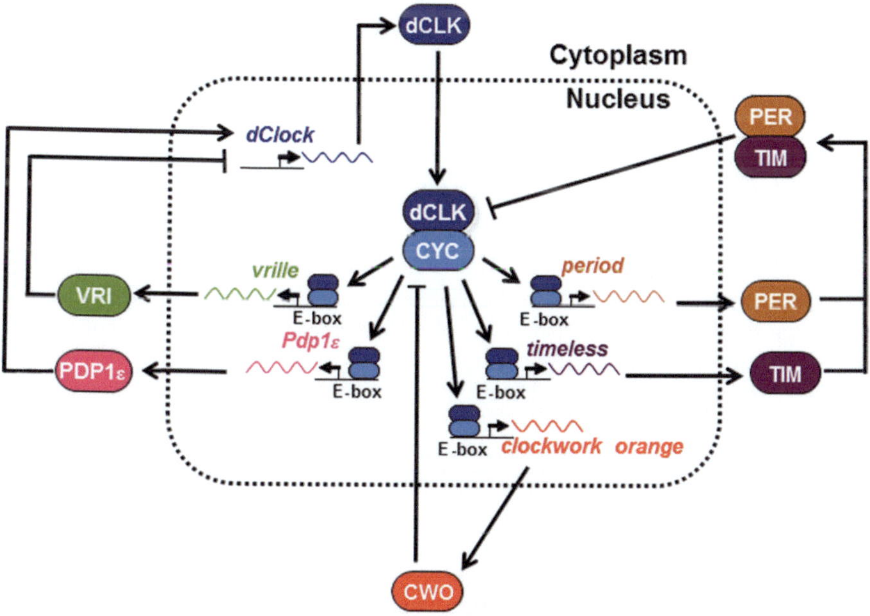

Fig. 6.1 The molecular oscillatory mechanism of the *Drosophila* circadian clock. It includes three transcriptional/translational loops consisting of *period* (*per*), *timeless* (*tim*), *Clock* (*Clk*), *cycle* (*cyc*), *Par domain protein 1ε* (*Pdp1ε*), *vrille* (*vri*), and *clockwork orange* (*cwo*). For details, see text

Transcription of the *Clk* gene is also under circadian regulation. There are two other factors, *vrille* (*vri*) and *Par domain protein 1ε* (*Pdp1ε*), involved in this mechanism (Cyran et al. 2003; Glossop et al. 2003). Their transcription is activated by CLK/CYC during the early night, but VRI accumulates earlier than PDP1ε and represses the *Clk* transcription through binding to a V/P-box, which lies on the regulatory region of the *Clk* gene. Later, accumulating PDP1ε then activates *Clk* transcription by competitively binding to the V/P-box. These processes result in the oscillation of *Clk* mRNA levels, which increase during the late night to early day.

The third loop involves *clockwork orange* (*cwo*), a transcription factor belonging to bHLH *orange* family (Lim et al. 2007; Matsumoto et al. 2007). *cwo* mRNA levels oscillate through a negative feedback of its product protein CWO and regulates the oscillation amplitude of other clock genes such as *per* and *tim*.

However, the *Drosophila* clock model is not completely supported in most of the other insects that have been examined. For example, there is no evidence that PER enters the nucleus in moths, cockroaches, or bloodsucking bugs (Sauman and Reppert 1996; Vafopoulou et al. 2007; Wen and Lee 2008). In moths, it is hypothesized that antisense and sense RNAs are transcribed simultaneously from the *per* gene and that antisense RNA is involved in the rhythmic expression of the PER protein (Sauman and Reppert 1996). *tim* does not exist in the hymenopteran genome, which includes honeybees (Rubin et al. 2006; Zhan et al. 2011). Instead of *tim*, mammalian-type *cryptochrome* (*cry2*) is thought to act as a negative component together with *per*. However, most of the studies in these non-model insects did not make rigorous functional analyses of the clock genes because of the limitation of available genetic techniques.

6.4 Molecular Approach to the Circadian System

The greatest advantage of using crickets is that RNAi is quite effective in analyzing the gene function. We thus use the cricket *Gryllus bimaculatus* to dissect the molecular oscillatory mechanism of the circadian clock as well as its output system regulating the overt activity rhythm.

6.4.1 Dissection of the Molecular Oscillatory Mechanism of the Circadian Clock

We first completed molecular cloning of the clock genes by a degenerate PCR strategy. The primers were designed based on known insects' clock genes. We have succeeded in obtaining cDNA fragments of the clock genes, *per*, *tim*, and *Clk*. 5′ and 3′ RACEs were performed for each of these genes and their full length cDNAs

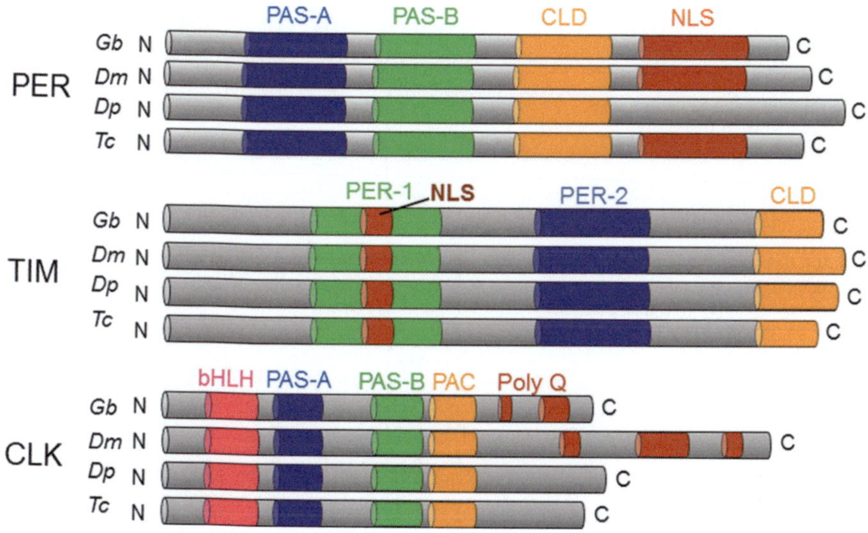

Fig. 6.2 Molecular structure of the protein products of *per*, *tim*, and *Clk* in the cricket *Gryllus bimaculatus* (*Gb*) aligned with those of *Drosophila melanogaster* (*Dm*), *Danaus plexippus* (*Dp*), and *Tribolium castaneum* (*Tc*). PER includes PAS-A, PAS-B, nuclear localization signal (NLS), and cytoplasm localization domains (CLD). TIM includes PER-1, PER-2, NLS, and CLD, and CLK is characterized by bHLH, PAS-A, PAS-B, PAC, and poly-Q. For details, see text

were obtained. The structural analyses revealed their resemblance to those of *Drosophila* and other insects (Fig. 6.2) (Moriyama et al. 2008, 2012; Danbara et al. 2010). *per* is characterized by four functional domains: PAS-A and PAS-B domains that are involved in the protein-protein interaction (Allada et al. 1998; Darlington et al. 1998), a cytoplasmic localization domain (CLD), and a nuclear localization signal (NLS). *tim* has two regions for dimerization with PERIOD (PER-1, PER-2) (Gekakis et al. 1995; Myers et al. 1995; Saez and Young 1996), NLS, and CLD. *Clk* includes a basic helix-loop-helix (bHLH) domain for binding to DNA (Allada et al. 1998; Darlington et al. 1998), PAS-A, PAS-B, and a polyglutamine repeat (poly-Q) in the C-terminal region that is implicated in transcriptional activity in *Drosophila* (Allada et al. 1998; Darlington et al. 1998). These structural similarities suggest that the clock genes *per*, *tim*, and *Clk* may have similar roles to those that they play in the *Drosophila* clock.

We measured the mRNA levels of those clock genes. Figure 6.3 summarizes the expression profiles of the clock gene mRNAs in the optic lobe. The mRNA levels of *per* and *tim* oscillate with a peak at early night. The profiles are similar to those reported for *Drosophila* and other insects. Interestingly, unlike in *Drosophila*, *Clk* is not rhythmically expressed. This is more similar to the mammalian clock than that of the *Drosophila* clock.

To investigate their importance in rhythm generation, we performed RNA interference. We synthesized dsRNA from the cDNA of the clock genes each of

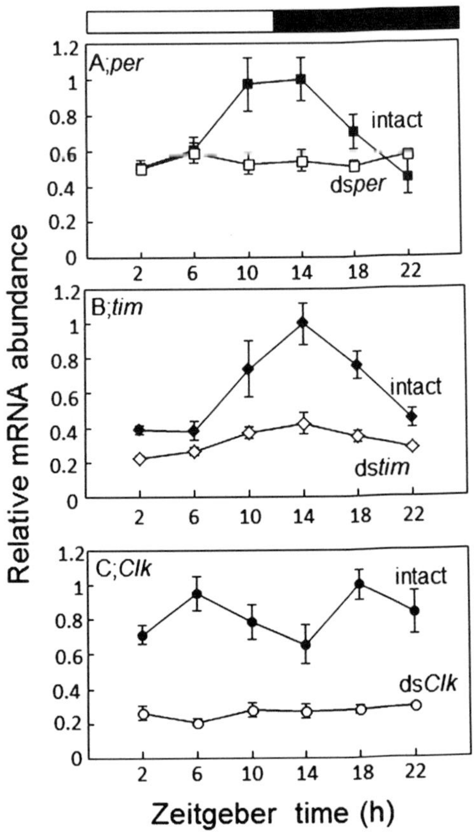

Fig. 6.3 mRNA expression patterns of *per* (**A**), *tim* (**B**), and *Clk* (**C**) and the effect of specific dsRNA on the mRNA levels of respective genes in the optic lobe of the cricket *Gryllus bimaculatus*. In control crickets (*filled symbols*), *per* and *tim* show a rhythmic expression with a peak during the late day to early night, while *Clk* shows rather constitutive expression. Systemic treatment with dsRNA (*open symbols*) effectively knocked down mRNA levels of the respective clock gene and abolished the rhythmic expression in *per* and *tim*. *White* and *black bars* indicate light and dark phases, respectively. **C**: after Moriyama et al. (2012)

which was injected into the abdomen of nymphal or adult crickets. The mRNA levels were gradually reduced over several days after dsRNA injection and a substantial, significant reduction was observed 7 days after the treatment (Uryu et al. 2013). The dsRNA injection successfully reduced the levels of respective clock gene mRNAs and abolished the rhythm in *per* and *tim* as shown in Fig. 6.3. Thus, RNAi against the clock genes has the desired effects in the cricket *G. bimaculatus*.

We then examined the role of clock genes in rhythm generation by measuring locomotor activity of the RNAi-treated crickets. The RNAi against *per* and *Clk* resulted in a loss of rhythmicity in constant darkness (Fig. 6.4). The arrhythmicity persisted for more than 30 days until the end of the experiment. Only a small fraction of the treated crickets maintained the rhythm but the power of the rhythm was significantly reduced. *tim* RNAi crickets maintained the rhythm but with shortened free-running periods (Fig. 6.4) (Danbara et al. 2010), when compared with those of control crickets injected with dsRNA against *DsRed2*, which is derived from a coral species.

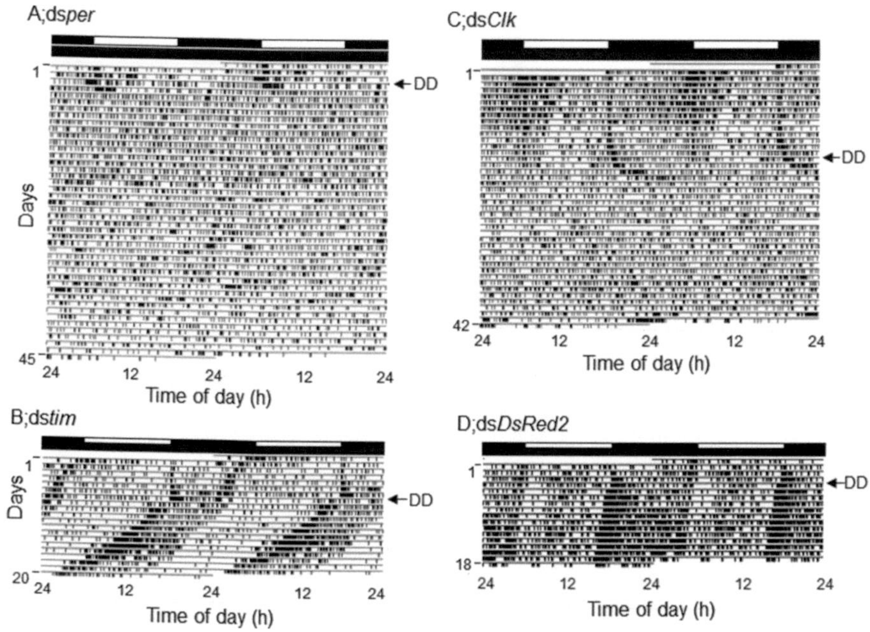

Fig. 6.4 Effects of *per* (**A**), *tim* (**B**), or *Clk* (**C**) RNAi on the locomotor rhythm of the cricket *Gryllus bimaculatus*. The locomotor rhythm disappeared in *per* and *Clk* RNAi crickets in DD, while it persisted with a shortened free-running period in the *tim* RNAi cricket. Control crickets treated with ds*DsRed2* showed a rhythm similar to that of intact crickets (**D**). *White* and *black bars* above actograms indicate light and dark phases, respectively. **A**, after Moriyama et al. (2008); **B** and **D**, after Danbara et al. (2010); **C**, after Moriyama et al. (2012)

6.4.2 *The Role of Clock Genes in Molecular Oscillatory Mechanism*

To know the role of clock genes in the molecular clock machinery, we examined the effects of knocking down one clock gene on the expression of other clock genes. In *Clk* RNAi crickets, both *per* and *tim* transcripts were significantly reduced, and the daily expression was lost, suggesting that it may act as a transcriptional activator as in *Drosophila*. In *per* RNAi, *tim* and *Clk* transcripts were both significantly reduced and rhythmic expression of *tim* was abolished. Similarly, in *tim* RNAi crickets, expression levels of both *per* and *Clk* were significantly reduced. These data suggest that the expression of clock genes is regulated by a complex mechanism, probably through some complex gene network. Based on the results, we assume that the cricket circadian clock functions as follows (Fig. 6.5). Since we have already confirmed the existence of the *cycle* (*cyc*) gene and its rhythmic expression under LD in the cricket optic lobe (Uryu et al. 2013), we assume that *Clk* and *cyc* work together as transcriptional activators as in *Drosophila*, even though their expression profiles are more similar to those of their counterparts in mammals.

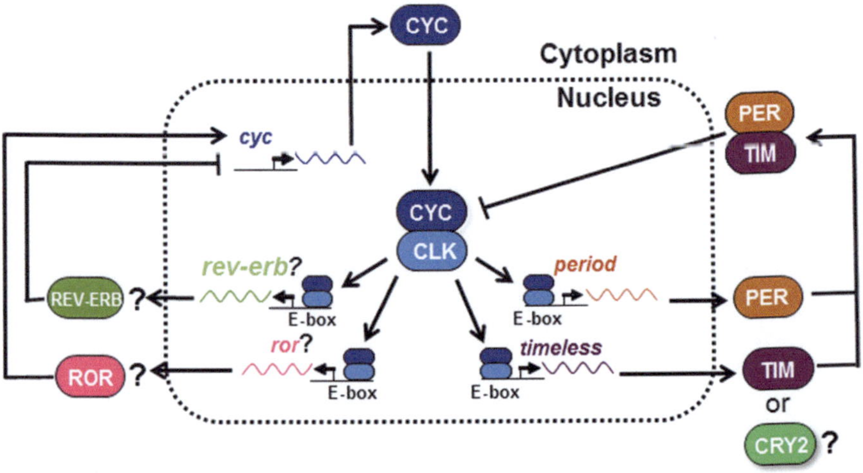

Fig. 6.5 A schematic model of the cricket circadian clock. It probably consists of at least two feedback loops for rhythmic expression of *per* and *tim* and for *cyc*. In analogy to the mammalian clock, *rev-erb* and *ror* may be involved in regulation of *cyc* expression. For further explanation, see text

Probably, *per* and *tim* act as negative factors to repress their own transcription through their negative effect on CLK/CYC, although there may be an additional complex pathway(s) that regulates the rhythmic expression of the *cyc* gene. *tim* RNAi knocked down *per* and *Clk* mRNA levels and stopped the oscillation in *per*; nevertheless, a locomotor rhythm persisted. We speculate that some other loop involving *cwo* or *cryptochrome* might retain its oscillation, independent of the oscillation of *per* and *tim*. The diversity and commonality of the circadian oscillatory mechanism in insects is a challenging issue to understand.

6.5 Peripheral Oscillators

In addition to the central clock tissues regulating activity rhythms, peripheral tissues also show circadian rhythms in crickets. The compound eyes have a circadian rhythm in their responsiveness to light (Tomioka and Chiba 1982a) and spermatophore formation occurs in a rhythmic, circadian manner (Loher 1974). Peripheral rhythms are also known in other insects. Cuticle secretion is known to occur in a rhythmic manner in cockroaches and flies (Wiedenmann et al. 1986; Ito et al. 2008), and antennal odor sensitivity is under a regulation of the circadian clock in cockroaches and flies (Krishnan et al. 2001; Page and Koelling 2003). The controlling mechanism for these peripheral rhythms has been investigated only in a few species. For example, in *Drosophila* the rhythms are controlled by circadian clocks located in those related tissues (Plautz et al. 1997), while in cockroaches at least the rhythms in the compound eyes and antennae are under the control of the central clock that is located in the optic lobe (Wills et al. 1985; Page and Koelling 2003). However, the oscillatory mechanism of the peripheral oscillator and the relationship between the central and peripheral oscillators remain to be elucidated.

Crickets again provide a good model to address these issues because the central clock has been localized and some peripheral tissues show circadian rhythms (Tomioka and Abdelsalam 2004). We first analyzed the peripheral oscillations in various tissues by measuring circadian expression of the clock genes *per* and *tim* in the cricket *Gryllus bimaculatus*. mRNA levels of both *per* and *tim* genes showed a circadian rhythmic expression in the brain, terminal abdominal ganglion (TAG), anterior stomach, midgut, and testis in DD, suggesting that they include a circadian oscillator (Uryu and Tomioka 2010). However, the amplitude and the levels of the mRNA rhythms varied among those tissues. Removal of the optic lobe, the central clock tissue, differentially affected the rhythms. The rhythm of both*per*and *tim* was lost in the anterior stomach, while in the midgut and TAG, *per* maintained rhythmic expression but *tim* expression became arrhythmic. In the brain, both *per* and *tim* mRNA retained rhythmic expression but with a shifted phase (Uryu and Tomioka 2010). These data suggest that rhythms outside the optic lobe are controlled by the optic lobe to varying degrees and that the oscillatory mechanism may be different from that of the central clock in the optic lobe.

6.6 Molecules Involved in the Output Pathway

Molecules involved in the output pathway regulating insect activity rhythms are still largely unknown except for a few neurotransmitters. The pigment-dispersing factor (PDF), an octadeca-neuropeptide, is one such molecule and is believed to be a principal neurotransmitter regulating locomotor rhythms in various insects including *Drosophila* and cockroaches (Renn et al. 1999; Helfrich-Forster et al. 2000; Park et al. 2000; Lee et al. 2009). In the cricket, however, PDF seems unessential for locomotor rhythms, because partial removal of the optic lobe, which spares the PDF cells in the medulla, abolishes the locomotor rhythm (Okamoto et al. 2001). But PDF is apparently involved in the regulation of the responsiveness of the circadian rhythms of the visual system (Saifullah and Tomioka 2003).

We thus used RNA interference to study the role of the *pigment-dispersing factor* (*pdf*) gene in the regulation of circadian locomotor rhythms in the cricket, *Gryllus bimaculatus* (Hassaneen et al. 2011a). Injections of *pdf* dsRNA effectively knocked down the *pdf* mRNA and PDF peptide levels. The treated crickets maintained the rhythm both under LD and DD, confirming our previous assumption that PDF is not an essential neurotransmitter in the cricket circadian system (Okamoto et al. 2001). However, the RNAi-treated crickets showed reduced nocturnal activity and synchronized to the shifted LD more quickly than control crickets. The free-running periods of the *pdf* RNAi crickets were shorter than those of control crickets in DD. Molecular oscillations of *per* and *tim* in the optic lobe showed a reduced amplitude and an advanced phase, corresponding to the weak and shorter free-running rhythms in the *pdf* RNAi crickets (Hassaneen et al. 2011b). These results suggest that PDF is not essential for rhythm generation but plays an important role in the control of nocturnality, photic entrainment, and fine-tuning of the free-running period of the circadian clock. To further understand the role of PDF at a molecular level, we need to investigate the role of PDF receptors, which have been studied in detail in *Drosophila* (Im and Taghert 2010), in our future studies.

6.7 Future Perspective

Crickets provide a unique model to study important aspects of the circadian rhythm. They have clocks in peripheral tissues (Uryu and Tomioka 2010) and are suitable for studies investigating the relationship between the central and peripheral clocks, in addition to the oscillatory mechanism of the peripheral clocks. They also show ontogenetic changes from nymphal diurnal to adult nocturnal rhythm (Tomioka and Chiba 1982b), providing a model to study phase regulation by the clock, which has been a question for years. With molecular techniques, we will be able to approach these questions. The key will be to clarify the molecular oscillatory mechanism of the central circadian clock. As described in this chapter, we have cloned some clock and clock-related genes in the cricket *Gryllus bimaculatus* and investigated their roles in the clock machinery (Moriyama et al. 2008, 2012; Danbara et al. 2010). However, for complete elucidation of the circadian clock machinery, we need to obtain many other clock genes known in other species, such as *Pdp1ε*, *vrille*, *rev-erb*, and *ror*, and to reveal their functional roles in the circadian clock. Elucidation of the cricket clock would lead to understanding how the insect circadian clock diversified evolutionarily, since the cricket is phylogenetically more primitive than flies, butterflies, and moths and has a clock that at least partially resembles the mammalian clock.

Acknowledgement This study was supported in part by grants from Japan Society for Promotion of Science. We are grateful of the members of Chronobiology Laboratory of Okayama University for their continuous discussion and technical assistance.

References

Abe Y, Ushirogawa H, Tomioka K (1997) Circadian locomotor rhythm of the cricket *Gryllodes sigillatus*. I. Localization of the pacemaker and the photoreceptor. Zool Sci 14:719–727

Allada R, White NE, So WV, Hall JC, Rosbash M (1998) A mutant *Drosophila* homolog of mammalian *Clock* disrupts circadian rhythms and transcription of *period* and *timeless*. Cell 93:791–804

Curtin KD, Huang ZJ, Rosbash M (1995) Temporally regulated nuclear entry of the *Drosophila* Period protein contributes to the circadian clock. Neuron 14:365–372

Cyran SA, Buchsbaum AM, Reddy KL, Lin M-C, Glossop NRJ, Hardin PE, Young MW, Storti RV, Blau J (2003) *vrille*, *Pdp1* and *dClock* form a second feedback loop in the *Drosophila* circadian clock. Cell 112:329–341

Danbara Y, Sakamoto T, Uryu O, Tomioka K (2010) RNA interference of *timeless* gene does not disrupt circadian locomotor rhythms in the cricket *Gryllus bimaculatus*. J Insect Physiol 56:1738–1745

Darlington TK, Wager-Smith K, Ceriani MF, Staknis D, Gekakis N, Steeves TDL, Weitz CJ, Takahashi JS, Kay SA (1998) Closing the circadian loop: CLOCK-induced transcription of its own inhibitors *per* and *tim*. Science 280:1599–1603

Dunlap JC, Loros J, DeCoursey PJ (2004) Chronobiology: biological timekeeping. Sinauer, Sunderland

Gekakis N, Saez L, Delahaye-Brown AM, Myers MP, Sehgal A, Young MW, Weitz CJ (1995) Isolation of *timeless* by PER protein interaction: defective interaction between TIMELESS protein and long-period mutant PER^{L}. Science 270:811–815

Glossop NR, Houl JH, Zheng H, Ng FS, Dudek SM, Hardin PE (2003) VRILLE feeds back to control circadian transcription of *Clock* in the *Drosophila* circadian oscillator. Neuron 37:249–261

Hassaneen E, Sallam A, Abo-Ghalia A, Moriyama Y, Karpova S, Abdelsalam S, Matsushima A, Shimohigashi Y, Tomioka K (2011a) Pigment-dispersing factor affects nocturnal activity rhythms, photic entrainment and the free-running period of the circadian clock in the cricket *Gryllus bimaculatus*. J Biol Rhythms 26:3–13

Hassaneen E, Sallam AE-D, Abo-Ghalia A, Moriyama Y, Tomioka K (2011b) Pigment-dispersing factor affects circadian molecular oscillations in the cricket *Gryllus bimaculatus*. Entomol Sci 14:278–282

Helfrich-Förster C, Tauber M, Park JH, Muhlig-Versen M, Schneuwly S, Hofbauer A (2000) Ectopic expression of the neuropeptide pigment-dispersing factor alters behavioral rhythms in drosophila melanogaster. J Neurosci 20:3339–3353

Im SH, Taghert PH (2010) PDF receptor expression reveals direct interactions between circadian oscillators in *Drosophila*. J Comp Neurol 518:1925–1945

Ito C, Goto SG, Shiga S, Tomioka K, Numata H (2008) Peripheral circadian clock for the cuticle deposition rhythm in *Drosophila melanogaster*. Proc Natl Acad Sci U S A 105:8446–8451

Krishnan B, Levine JD, Lynch MK, Dowse HB, Funes P, Hall JC, Hardin PE, Dryer SE (2001) A new role for cryptochrome in a *Drosophila* circadian oscillator. Nature 411:313–317

Lee C, Bae K, Edery I (1998) The *Drosophila* CLOCK protein undergoes daily rhythms in abundance, phosphorylation, and interactions with the PER-TIM complex. Neuron 21:857–867

Lee C-M, Su M-T, Lee H-J (2009) Pigment dispersing factor: an output regulator of the circadian clock in the german cockroach. J Biol Rhythms 24:35–43

Lim C, Chung BY, Pitman JL, McGill JJ, Pradhan S, Lee J, Keegan KP, Choe J, Allada R (2007) *clockwork orange* encodes a transcriptional repressor important for circadian-clock amplitude in *Drosophila*. Curr Biol 17:1082–1089

Loher W (1972) Circadian control of stridulation in the cricket *Teleogryllus commodus* Walker. J Comp Physiol 79:173–190

Loher W (1974) Circadian control of spermatophore formation in the cricket *Teleogryllus commodus* Walker. J Insect Physiol 20:1155–1172

Matsumoto A, Ukai-Tadenuma M, Yamada RG, Houl J, Uno KD, Kasukawa T, Dauwalder B, Itoh TQ, Takahashi K, Ueda R, Hardin PE, Tanimura T, Ueda HR (2007) A functional genomics strategy reveals *clockwork orange* as a transcriptional regulator in the *Drosophila* circadian clock. Gene Dev 21:1687–1700

Moriyama Y, Sakamoto T, Karpova SG, Matsumoto A, Noji S, Tomioka K (2008) RNA interference of the clock gene *period* disrupts circadian rhythms in the cricket *Gryllus bimaculatus*. J Biol Rhythms 23:308–318

Moriyama Y, Kamae Y, Uryu O, Tomioka K (2012) *Gb'Clock* is expressed in the optic lobe and required for the circadian clock in the cricket *Gryllus bimaculatus*. J Biol Rhythms 27:467–477
Myers MP, Wager-Smith K, Wesley CS, Young MW, Sehgal A (1995) Positional cloning and sequence analysis of the *Drosophila* clock gene, *timeless*. Science 270:805–858
Okamoto A, Mori H, Tomioka K (2001) The role of optic lobe in generation of circadian rhythms with special reference to the PDH immunoreactive neurons. J Insect Physiol 47:889–895
Page TL, Koelling E (2003) Circadian rhythm in olfactory response in the antennae controlled by the optic lobe in the cockroach. J Insect Physiol 49:697–707
Park JH, Helfrich-Förster C, Lee G, Liu L, Rosbash M, Hall JC (2000) Differential regulation of circadian pacemaker output by separate clock genes in *Drosophila*. Proc Natl Acad Sci U S A 97:3608–3613
Plautz JD, Kaneko M, Hall JC, Kay SA (1997) Independent photoreceptive circadian clocks throughout *Drosophila*. Science 278:1632–1635
Renn SCP, Park JH, Rosbash M, Hall JC, Taghert PH (1999) A *pdf* neuropeptide gene mutation and ablation of PDF neurons each cause severe abnormalities of behavioral circadian rhythms in *Drosophila*. Cell 99:791–802
Rubin EB, Shemesh Y, Cohen M, Elgavish S, Robertson HM, Bloch G (2006) Molecular and phylogenetic analyses reveal mammalian-like clockwork in the honey bee (*Apis mellifera*) and shed new light on the molecular evolution of the circadian clock. Genome Res 16:1352–1365
Rutila JE, Suri V, Le M, So WV, Rosbash M, Hall JC (1998) CYCLE is a second bHLH-PAS clock protein essential for circadian rhythmicity and transcription of *Drosophila period* and *timeless*. Cell 93:805–814
Saez L, Young MW (1996) Regulation of nuclear entry of the *Drosophila* clock proteins PERIOD and TIMELESS. Neuron 17:911–920
Saifullah ASM, Tomioka K (2003) Pigment-dispersing factor sets the night state of the medulla bilateral neurons in the optic lobe of the cricket, *Gryllus bimaculatus*. J Insect Physiol 49:231–239
Sauman I, Reppert SM (1996) Circadian clock neurons in the silkmoth *Antheraea pernyi*: novel mechanisms of Period protein regulation. Neuron 17:889–900
Shiga S, Numata H, Yoshioka E (1999) Localization of the photoreceptor and pacemaker for the circadian activity rhythm in the band-legged ground cricket, Dianemobius nigrofasciatus. Zool Sci 16:193–201
Sokolove PG, Loher W (1975) Role of the eyes, optic lobes and pars intercerebralis in locomotory and stridulatory circadian rhythms of *Teleogryllus commodus*. J Insect Physiol 21:785–799
Tomioka K, Abdelsalam SA (2004) Circadian organization in hemimetabolous insects. Zool Sci 21:1153–1162
Tomioka K, Chiba Y (1982a) Persistence of circadian ERG rhythms in the cricket with optic tract severed. Naturwissenschaften 69:355–356
Tomioka K, Chiba Y (1982b) Post-embryonic development of circadian rhythm in the cricket, *Gryllus bimaculatus*. J Comp Physiol A 147:299–304
Tomioka K, Chiba Y (1984) Effects of nymphal stage optic nerve severance or optic lobe removal on the circadian locomotor rhythm of the cricket, *Gryllus bimaculatus*. Zool Sci 1:385–394
Tomioka K, Chiba Y (1992) Characterization of optic lobe circadian pacemaker by in situ and in vitro recording of neuronal activity in the cricket *Gryllus bimaculatus*. J Comp Physiol A 171:1–7
Tomioka K, Yukizane M (1997) A specific area of the compound eye in the cricket Gryllus bimaculatus sends photic information to the circadian pacemaker in the contralateral optic lobe. J Comp Physiol A 180:63–70
Tomioka K, Nakamichi M, Yukizane M (1994) Optic lobe circadian pacemaker sends its information to the contralateral optic lobe in the cricket *Gryllus bimaculatus*. J Comp Physiol A 175:381–388

Tomioka K, Saifullah ASM, Koga M (2001) The circadian clock system of hemimetabolous insects. In: Denlinger DL, Giebultowicz JM, Saunders DS (eds) Insect timing: circadian rhythmicity to seasonality. Elsevier, Amsterdam, pp 43–54

Uryu O, Tomioka K (2010) Circadian oscillations outside the optic lobe in the cricket *Gryllus bimaculatus*. J Insect Physiol 56:1284–1290

Uryu O, Kamae Y, Tomioka K, Yoshii T (2013) Long-term effect of systemic RNA interference on circadian clock genes in hemimetabolous insects. J Insect Physiol 59:494–499

Vafopoulou X, Steel CGH, Terry KL (2007) Neuroanatomical relations of prothoracicotropic hormone neurons with the circadian timekeeping system in the brain of larval and adult *Rhodnius prolixus* (Hemiptera). J Comp Neurol 503:511–524

Wen C-J, Lee H-J (2008) Mapping the cellular network of the circadian clock in two cockroach species. Arch Insect Biochem Physiol 68:215–231

Wiedenmann G, Lukat R, Weber F (1986) Cyclic layer deposition in the cockroach endocuticle: a circadian rhythm? J Insect Physiol 32:1019–1027

Wills SA, Page TL, Colwell CS (1985) Circadian rhythms in the electroretinogram of the cockroach. J Biol Rhythms 1:25–37

Yukizane M, Tomioka K (1995) Neural pathways involved in mutual interactions between optic lobe circadian pacemakers in the cricket *Gryllus bimaculatus*. J Comp Physiol A 176:601–610

Yukizane M, Kaneko A, Tomioka K (2002) Electrophysiological and morphological characterization of the medulla bilateral neurons that connect bilateral optic lobes in the cricket, Gryllus bimaculatus. J Insect Physiol 43:631–641

Zhan S, Merlin C, Boore JL, Reppert SM (2011) The monarch butterfly genome yields insights into long-distance migration. Cell 147:1171–1185

Chapter 7
Hormonal Circadian Rhythm in the Wing-Polymorphic Cricket *Gryllus firmus*: Integrating Chronobiology, Endocrinology, and Evolution

Anthony J. Zera, Neetha Nanoth Vellichirammal, and Jennifer A. Brisson

Abstract Hormonal circadian rhythms, an important aspect of endocrinology, have been extensively studied in vertebrates, but much less so in insects. Moreover, adaptive evolutionary change of these rhythms has been barely investigated in any species. In the wing-polymorphic cricket, *Gryllus firmus*, the key insect hormone, juvenile hormone (JH), exhibits a large-amplitude (ca. tenfold change) circadian rhythm in the flight-capable (long-winged, LW(f)) morph that delays reproduction, but a much shallower (twofold change) rhythm in the flightless, reproductive morph (SW, short-winged). The morph-specific cycle is a genetic polymorphism and occurs in both the laboratory and field in *G. firmus*, as well as in several other cricket species. The cycle is primarily driven by a circadian rhythm in the rate of JH biosynthesis. The function of the morph-specific cycle is unknown at present, but possibly regulates some aspect of flight, which is also circadian in crickets (only occurs at night), and which first occurs when the morph-specific JH titer cycle is also first manifest. Endocrine circadian cycles are likely common and functionally important in insects but have largely been unstudied. Recent transcriptome profiling has identified numerous morph-specific diel changes in transcript abundance that are more common in the LW(f) than in the SW morph and thus are correlated with the greater circadian change in the JH titer in the LW(f) morph. Thus, morph-specific daily change in the JH titer and global transcript abundance appear to be key features of wing polymorphism in *Gryllus*. These studies in *Gryllus* are contributing significantly to the synthesis of the previously independent subdisciplines of chronobiology, endocrinology, and life history evolution.

A.J. Zera (✉) • N.N. Vellichirammal
School of Biological Sciences, University of Nebraska, Lincoln, NE 68588, USA
e-mail: azera1@unl.edu

J.A. Brisson
School of Biological Sciences, University of Nebraska, Lincoln, NE 68588, USA

Department of Biology, University of Rochester, Rochester, NY 14627, USA

H.W. Horch et al. (eds.), *The Cricket as a Model Organism*,
DOI 10.1007/978-4-431-56478-2_7

Keywords Hormone circadian rhythm • JH, juvenile hormone • Wing polymorphism • Transcriptome profiling • *Gryllus firmus*

7.1 Introduction

Many aspects of endocrine regulation, such as blood hormone titers, rates of hormone biosynthesis and release, receptor expression, etc., exhibit circadian rhythms (Nelson 1995; Norris 1997; Hadley and Levine 2007). These daily cycles in endocrine control coordinate the expression of various aspects of physiology (e.g., metabolism, immunology) and behavior with functions that occur during specific times of the day or night. For decades, hormonal circadian rhythms have been extensively studied in humans and other vertebrates (Nelson 1995; Norris 1997; Dunlap et al. 2004). By contrast, with a few notable exceptions, this phenomenon has been much less studied in insects (reviewed in Vafopoulou and Steel 2005; Bloch et al. 2013). In addition, evolutionary aspects of hormone circadian rhythms have been largely unexplored (Zera et al. 2007; Zera and Zhao 2009). Like any trait, a circadian rhythm is subject to evolutionary change and adaptive modification, resulting initially in genetically based variation of the rhythm within a species. Yet, until very recently almost nothing was known about the existence, characteristics, or functional significance of adaptive variation for endocrine circadian rhythms within or among species.

In the course of investigating the endocrine control of morph specialization for flight versus reproduction in the wing-polymorphic cricket, *Gryllus firmus*, an unexpected, high-amplitude, and genetically variable circadian rhythm was observed for the key insect hormone, juvenile hormone (JH) (Zera and Cisper 2001; Zhao and Zera 2004a). This and subsequent studies, discussed below, have now made *Gryllus* a powerful experimental model for investigating not only hormonal circadian rhythms but, more importantly, adaptive evolutionary modification of these rhythms. The present chapter will review recent and ongoing investigations of this topic in *Gryllus*, mainly focusing on *Gryllus firmus*.

7.2 Background on Wing Polymorphism

Because studies of endocrine circadian rhythms in *Gryllus* were undertaken in the context of wing polymorphism, some background on this trait will first be given (for more detailed information on this topic, see Harrison (1980), Zera and Denno (1997), and Zera (2009); also see Chap. 15 of this volume). Wing polymorphism is a common feature of many insect species and essentially involves adaptive specialization of distinct phenotypes (morphs) within a population for flight at the expense of reproduction and vice versa. One morph is capable of flight, having fully

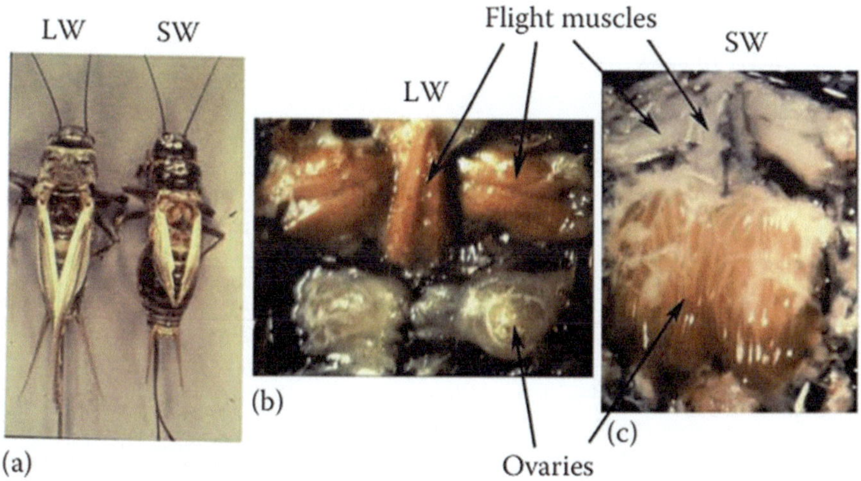

Fig. 7.1 Flight-capable, *LW* (referred to as *LW*(f) in text) and *SW* female morphs of *Gryllus firmus*. In panel (**a**), the forewings have been removed to show variation in the hind wings used for flight. Panels (**b**) and (**c**) illustrate dissections of the same-aged (day 5 of adulthood) morphs showing much larger flight muscles and much smaller ovaries in the LW compared with the *SW* morph (Figure from Zera 2009)

developed wings and flight muscles (termed LW(f), long wings with functional flight muscles), but delays egg production. An obligately flightless morph (SW, short-winged) has vestigial wings and flight muscles but exhibits enhanced reproductive output (e.g., greater egg production in females; Fig. 7.1). Many wing-polymorphic species also contain another flightless morph that has fully developed wings but degenerated flight muscles (termed LW(h), long wings with histolyzed flight muscles) (Zera et al. 1997). This morph is derived from the LW(f) morph during adulthood by muscle histolysis and also has enhanced reproductive output similar to the SW morph. A central feature of wing polymorphism is the diverse set of traits that differ between the morphs, such as the length of wings, size of flight muscles, size of ovaries, aspects of intermediary metabolism (lipid and protein biosynthesis), and numerous behaviors (Harrison 1980; Zera 2009; see Chap. 15 of this volume). Many of the traits that differ between morphs are regulated by juvenile hormone in non-polymorphic species. This has given rise to the classic hypothesis of morph regulation in which the level of juvenile hormone (JH), chronically above or below a threshold value, is thought to control the expression of many traits that differ between the flight-capable and flightless morphs. Although widely accepted, until recently, very few studies have directly tested the JH-morph-induction hypothesis by measuring the blood JH titer in alternate morphs (Zera and Denno 1997; Zera 2009). This lack of direct information was the original motivation for measuring JH titers in adult morphs of the wing-polymorphic cricket, *Gryllus firmus* (Zera and Cisper 2001).

7.3 Morph-Specific JH Titer Circadian Rhythm in *G. firmus*

7.3.1 Characterization of the Morph-Specific JH Titer Cycle

Initial investigations of the JH titer in LW(f) and SW morphs of *G. firmus* identified an unexpected, high-amplitude, morph-specific daily rhythm for the blood concentration of JH starting on about the fifth day after adult eclosion (Fig. 7.2, top left panel; Zera and Cisper 2001; Zhao and Zera 2004a). In the LW (f) morph, the JH titer rose about tenfold starting about 6 h before lights off (16 light:8 dark photoregime) and dropped down to baseline levels within about 2 h after lights off (Fig. 7.2, top left panel). By contrast, in the SW morph, the hemolymph JH titer exhibited only a very shallow daily rhythm (about twofold change). This shallow rhythm appeared to be primarily, if not exclusively, due to the daily contraction of whole-body hemolymph volume, which occurs to a similar degree in both the SW and LW(f) morphs (Zhao and Zera 2004a). Importantly, the high-amplitude JH titer rhythm only occurred in LW individuals with functional flight muscles (i.e., the LW(f) morph). Flightless LW (h) individuals, with histolyzed flight muscles, exhibited the shallow JH titer cycle characteristic of the SW morph (Zera and Cisper 2001). Thus, the JH titer differs substantially between flight-capable and flightless morphs, but in a manner completely different from that proposed by the classic model of JH regulation of wing polymorphism. Henceforth, the "high-amplitude JH titer cycle" and "shallow or low-amplitude JH titer cycle" will simply be referred to as "the cyclic" and "acyclic" JH titer, respectively.

The onset of the JH titer cycle in the LW(f) morph of *G. firmus* is correlated with the onset of flight ability which also begins around day 5 of adulthood and which is also circadian (flight only occurs at night). This and other observations mentioned above collectively suggest that the JH titer cycle may directly regulate some aspect of flight or some trait correlated with flight. Alternatively, other as yet unidentified morph-specific trait(s) may vary diurnally and may also be the main target of the JH titer cycle. For example, excitability and locomotor activity varies diurnally in the long-winged cricket *Gryllus bimaculatus* (Lorenz et al. 2004), with increased activity just before lights off (Fassbold et al. 2010). By contrast, the blood ecdysteroid titer exhibited no large-amplitude, morph-specific cycle, but is chronically elevated in SW vs. LW(f) individuals (Zhao and Zera 2004a; Zera et al. 2007). Finally, the high-amplitude JH titer cycle occurred in each of three LW(f) genetic stocks and is absent in each of three SW genetic stocks, thus demonstrating that the morph-specific pattern is a genetic polymorphism (Zera and Cisper 2001).

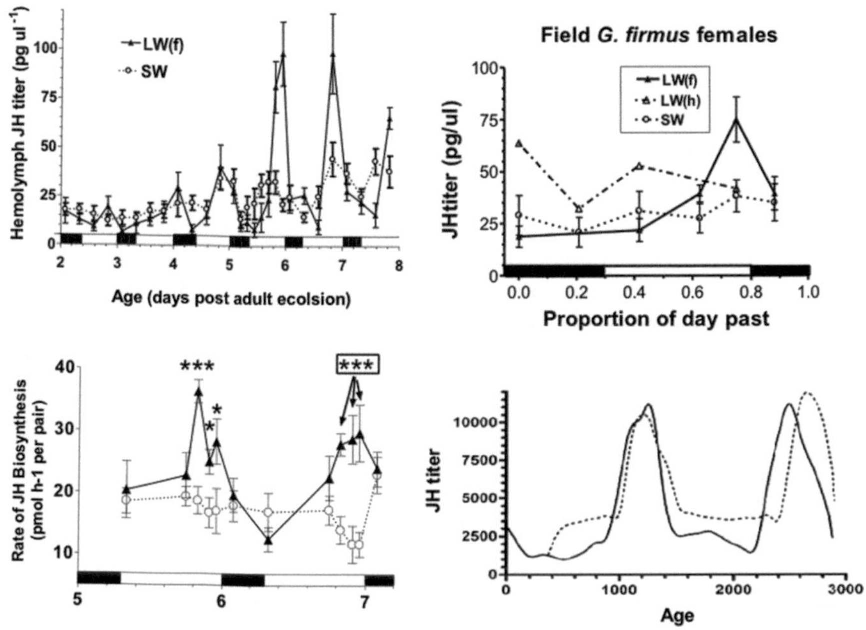

Fig. 7.2 *Top left panel*. Hemolymph JH titer in *LW*(f) and SW laboratory stocks of *G. firmus*. On the x-axis, numbers refer to days of adulthood, and *black* and *white bars* designate, respectively, scotophase (*dark*) and photophase (*light*) of 16 light:8 *dark* cycle (data from Zhao and Zera (2004a)). *Bottom left panel*. In vitro rate of JH biosynthesis by corpora allata from laboratory stocks of *LW*(f) and *SW G. firmus*. Same symbols as *top left panel*. *Asterisks* indicate significant difference between morphs (* $= P < 0.05$; *** $= P < 0.001$) (data from Zhao and Zera (2004b)). *Top right panel*. Hemolymph JH titer in field-collected *G. firmus* females. X axis = proportion of the day passed (0.0, 1. 0 = midnight); *dark and light bars* on x-axis refer, respectively, to night and day (sunrise to sunset) (data from Zera et al. (2007)). *Bottom right panel*. Fit between predicted JH titer circadian rhythm (*dashed line*) in the *LW*(f) morph based on measured rates of JH biosynthesis and JH degradation using the mathematical model described in the text and actual JH titer measurements (*solid line*). For the predicted JH titer (*dashed line*), rates of JH biosynthesis above the baseline (i.e., above 20 pmol h^{-1} per pair) had been increased by 1.8-fold, and in vivo rate of JH degradation had been increased by twofold. Without these adjustments, the predicted position of the JH titer peaks from the model was the same as in this panel, but the baseline occurs at 5,800 units. See text for additional details and references in the text for sample sizes (for more extensive discussion of the model and analysis, see Zera (2013)). "Y" axis = relative, whole-body JH concentration; "X" axis refers to age of adulthood; the *dotted line* starts near the beginning of the photophase of day 5 and ends in the early scotophase of day 7 of adulthood

7.3.2 *Proximate Mechanisms Regulating the Morph-Specific Cycle*

There are a number of important questions regarding the morph-specific JH titer cycle such as the following:

1. What are the proximate mechanisms causing the generation of the morph-specific cycle?
2. Is the daily cycle a true endogenous circadian rhythm, as opposed to being caused exclusively by external change in light intensity?

3. What are the specific function(s) of the cyclic/acyclic JH titer in morphs specialized for dispersal vs. reproduction?
4. What is the phylogenetic extent of the morph-specific cycle? That is, is the cycle found only in laboratory stocks of *G. firmus*? In field populations of this species? In other *Gryllus* species? In insect groups other than orthopterans?

To begin to address the first question, we focused on two important regulators of the JH titer: rate of JH biosynthesis and rate of JH degradation by juvenile hormone esterases (JHE) (Goodman and Cusson 2012; Zera 2009). The in vitro JH biosynthetic rate exhibited a large-amplitude, morph-specific cycle that strongly paralleled the morph-specific blood JH titer cycle during days 5–7 of adulthood (Fig. 7.2, bottom left panel; Zhao and Zera 2004b). By contrast, hemolymph JHE activity only exhibited a weak cycle (onefold increase) that was equivalent in the morphs except for a brief period when the JH titer is declining, just after lights off (Zhao and Zera 2004b). Previous studies have indicated that an approximately threefold to sixfold change in the in vitro activity of blood JHE is required to alter in vivo JH degradation (reviewed in Zera 2006, 2013). Thus, the morph-specific JH titer pattern appears to be regulated primarily if not exclusively by change in the rate of JH biosynthesis. Rate of JH biosynthesis itself is strongly regulated (inhibited) by allatostatins, neuropeptides that are produced in the brain and directly delivered via neuronal axons to the corpora allata, the paired glands in which JH biosynthesis occurs (Stay and Tobe 2007; Goodman and Cusson 2012). An anti-allatostatin monoclonal antibody tagged with a fluorescent label was used to estimate, via confocal microscopy, the concentration of allatostatin within the corpora allata of LW(f) and SW morphs (Stay and Zera 2010). During the early portion of the photophase, when the blood JH titer was low and equivalent in the morphs, the concentration of the inhibitory allatostatin was high and equivalent in the morphs. However, late in the photophase, when the blood JH titer rose to a higher level in the LW(f) compared with the SW morph, the concentration of allatostatin dropped to a lower concentration in the LW(f) morph compared to the SW morph. This change is consistent with a morph-specific daily cycle in allatostatin concentration, at least partially, regulating the dramatic morph-specific daily cycle in the blood JH titer.

7.3.3 Mathematical Modeling of the JH Titer Rhythm and Its Circadian Basis

The extent to which the experimentally measured temporal change in in vitro rate of JH biosynthesis, on a background of constant in vivo rate of JH degradation, could account for the measured temporal change in the blood JH titer in the LW(f) morph was assessed quantitatively, using a standard differential equation: $d[\mathrm{JH}]/dt = s(t) - k[\mathrm{JH}]$. In this equation, expected change in the JH titer is a function of "s," the measured JH biosynthetic rate, and "k," the in vivo rate of hormone degradation derived from its in vivo half-life (see Zera (2013) for additional details). Data on the JH titer and rates of JH biosynthesis on day 5-to-early day 7 of adulthood (Fig. 7.2, left panels) were used in the model. Because the in vivo rate constant of JH degradation had not been measured in adult *G. firmus*, the rate constant estimated from the congener *G. rubens*, early in the photophase of day 5 of adulthood (Zera et al. 1993), was used. *G. rubens* and *G. firmus* have similar blood JHE activities during the juvenile and adult stages (Zera and Huang 1999; Zera 2006, 2013). In this initial study, the in vivo rate of JH degradation was assumed to be temporally constant, based on the only minor temporal fluctuation in blood JHE activity, discussed above.

The model produced an expected daily JH titer cycle in the LW(f) morph that reasonably, but not completely, accounted for the experimentally measured JH titer cycle reported in Zhao and Zera (2004a), Zera (2013), and Zera et al. (unpublished data). The model accurately predicted the position of the JH titer peaks but underestimated the amplitude of the peak by about two- to 2.5-fold and overestimated the baseline JH titer by the same amount (see Fig. 3.11 in Zera 2013). A near perfect fit between experimentally measured and predicted JH titer cycles only required that the rates of JH biosynthesis during the titer peaks (see legend of Fig. 7.2) and rate of JH degradation each be increased by 1.8- to twofold (Fig. 7.2, bottom right panel). Given the magnitude of errors inherent in estimating physiological parameters such as in vivo rates of hormone biosynthesis and degradation, as well as the JH titer itself, this is a fairly modest adjustment of the rates of JH biosynthesis and degradation. Results of the modeling indicate that the cyclic rate of JH biosynthesis appears to be the main driver of the blood JH titer cycle in the LW(f) morph. The lack of fit between observed and expected profiles provides insights as to which aspects of JH titer regulation should be the focus of future study. For example, the model suggests that the in vitro assay of JH biosynthesis may underestimate the in vivo rate of JH biosynthesis, when biosynthesis is elevated, perhaps due to the absence of in vivo regulators, or severing the neural connections between the corpora allata and the brain, which is required to perform the assay (see extensive discussion in Zera 2013). In addition, some factors that will improve the fit between predicted and observed JH titers have yet to be incorporated into the preliminary analysis. For example, there is a 30 % daily reduction in whole-body hemolymph volume from early to late in the photophase in both *Gryllus* morphs (Zhao and Zera 2004a) that will increase the predicted amplitude of the JH titer by 1.4-fold when included in the model.

Although JH titer regulation has been a prominent area of research in insect endocrinology since the 1960s, there have been relatively few quantitative investigations of the extent to which measured parameters of JH titer regulators, specifically, rates of JH biosynthesis and degradation, can account for the observed JH titer (Tobe and Stay 1985; Zera 2013). Thus quantitative modeling of JH titer regulation in *Gryllus* represents an important contribution to the study of JH titer regulation in insects in general, over and above its importance in the study of the JH titer circadian rhythm. Finally, JH titer measurements over short time intervals (3–6 h) in *G. firmus* morphs currently represent the most detailed characterization of temporal change in the JH titer to date in insects. The unexpected finding of a morph-specific JH titer cycle in *G. firmus* indicates that temporally intensive measurement of the JH titer may identify other unanticipated temporal patterns of JH titer variation in insects (Goodman and Cusson 2012).

A daily rhythm is not necessarily an endogenous circadian rhythm: it may be *caused* by the daily cycling of an environmental signal, and the rhythm might not persist in the absence of the environmental signal (Giebultowicz 2000; Saunders 2002). Three key pieces of information are required to provide strong evidence that a daily cycle is an endogenous circadian rhythm: First, the rhythm must persist under constant environmental conditions (e.g., constant darkness in the absence of other daily varying environmental signals). Second, the rhythm must be temperature compensated, exhibiting the same period under different temperatures. Third, the rhythm must shift in response to a shift in the environmental signal that entrains the rhythm to a 24 h period (Saunders 2002; Bloch et al. 2013). Indeed, the morph-specific JH titer cycle and rate of JH biosynthesis over a 2-day period persisted when crickets were transferred from the 16 light:8 dark photoperiod to constant darkness (Zera and Zhao 2009). Moreover, the same morph-specific JH titer cycle was observed when crickets were raised at 20 °C instead of the standard 28 °C (temperature compensated), and there was a concordant 6 h shift in the onset and peak of the JH titer cycle when the onset of the photophase was shifted 6 h earlier. In addition, the JH titer cycle was completely lost under constant light (Zera and Zhao 2009), a common characteristic of many physiological circadian cycles (Vafopoulou and Steel 2001; Giebultowicz 2000; Saunders 2002), which is thought to be caused by the sensitivity of some clock gene products, such as *timeless*, to light. These data collectively provide strong evidence that the morph-specific cycle is indeed an endogenous morph-specific circadian rhythm. Finally, the persistence of the morph-specific rhythm in multiple LW(f) and SW lines demonstrated that it is a genetic polymorphism for a circadian rhythm (Zera and Zhao 2009).

7.3.4 Field Studies and Comparisons with Other Species

Several studies were conducted to determine the extent to which the morph-specific JH titer daily cycle, which had been characterized exclusively in the laboratory in *G. firmus*, also occurs under field conditions (Zera et al. 2007; Zera and Zhao 2009). Laboratory stocks of LW(f) and SW *G. firmus* raised in the field (Gainesville, Florida, September to October, 2002–2004) exhibited a similar morph-specific cycle as was observed in the laboratory, although the amplitude of the cycle was not as great in the field. A similar cycle was observed in field-collected *G. firmus*, bled in the field (Fig. 7.2, top right panel). The morph-specific cycle was observed in youngest (8.4 ± 0.2 days), intermediate (12.0 ± 0.1 days), and oldest (15.0 ± 0.2 days) adults sampled (numbers refer to days after adult eclosion measured by counting daily cuticular growth layers). Thus, the cycle occurs in natural populations, throughout the adult stage of *G. firmus*. Finally, the JH titer cycle was found in LW(f) individuals of three other cricket species, while no cycle was found in SW individuals of a different SW-monomorphic species (i.e., only contains SW individuals). Thus, the JH titer cycle is strongly associated with the LW(f) morph in comparisons between species as well as within species of crickets.

Only a few well-documented cases of diurnal cycles of circadian rhythms of hormones have been reported in insects other than *Gryllus*. Most notable are extensive studies on the circadian rhythm of ecdysteroid and prothoracicotropic hormone titers in the bedbug, *Rodnius prolixus*, and the pheromone-regulating hormone PBAN in lepidopterans (reviewed in Vafopoulou and Steel 2005; Bloch et al. 2013). In a few other cases, diel cycles in aspects of hormonal regulation, including JH regulation, have been reported in insects (Ramaswamy et al. 2000; Elekonich et al. 2001; reviewed in Zera and Zhao 2009). One case is of particular relevance to the morph-specific JH titer cycle in *Gryllus firmus*: the developmental change in adults from an acyclic to a cyclic hemolymph JH titer in worker honeybees (*Apis mellifera*) (Elekonich et al. 2001). This change occurs when adults change their behavior from hive maintenance to foraging via flight outside the hive. This pattern is the developmental analogue of the genetic polymorphism of the acyclic vs. cyclic JH titer in morphs of *G. firmus*, which provides additional evidence suggesting that a cyclic JH titer functions in some aspect of flight. To our knowledge, the circadian basis of these diel rhythms has only been investigated in one case (Kondrik et al. 2005). The limited data obtained to date suggest that circadian rhythms for aspects of hormonal regulation may be a functionally important and widespread phenomenon in insects, but understudied (Zera et al. 2007; Zera and Zhao 2009; Bloch et al. 2013). The existence of unidentified daily cycles in JH titer regulators may account for several previously puzzling cases in which a negative correlation or a lack of a correlation was observed between the rate of JH biosynthesis and JH titer (Bloch et al. 2000; and references therein).

7.3.5 Functional and Regulatory Aspects

The morph-specific JH titer cycle raises many fascinating issues regarding JH regulation of morph-specific traits. Because of space limitations, only one issue is briefly mentioned here (for detailed discussion of this topic, see Zera et al. (2007) and Zera and Zhao (2009)). An elevated JH titer per se has long been thought to regulate enhanced ovarian growth and histolysis of flight muscles in insects (Zera and Denno 1997). However, this cannot be the case in the LW(f) morph of *G. firmus*, which is capable of flight, but which also exhibits a repeated daily increase in the JH titer that is much higher than that in the SW or LW(h) morphs (Zera and Cisper 2001; Zera et al. 2007). Thus, JH regulation of morph-specific traits, at least in *Gryllus*, must be more complex than previously suspected. For example, rather than being regulated only by an elevated JH titer, ovarian growth in *Gryllus* might be regulated by the duration of time that the JH titer is elevated (JH titer is slightly higher over all time points in SW vs. LW(f) females; Zhao and Zera 2004a). In other words, the short duration of the elevated JH titer in the LW(f) morph may be insufficient to induce ovarian growth and flight muscle histolysis, while, at the same time, it might be elevated for a sufficient time to regulate the expression of other traits. Alternatively, in *Gryllus*, hormones other than JH may regulate aspects of ovarian growth, freeing an elevated JH titer to regulate morph functions related specifically to flight. The ecdysteroid titer is chronically elevated in the hemolymph of SW vs. LW(f) morphs throughout adulthood in both the laboratory and field (Zhao and Zera 2004a; Zera et al. 2007). In addition, a recent transcriptome study (see below) has identified an insulin-like peptide transcript that is substantially and chronically elevated in SW vs. LW(f) fat body of *G. firmus* (Vellichirammel et al. 2014). The homologue of this gene is also expressed to a much greater degree in the fat body of the gregarious phase of a phase-polymorphic locust, which exhibits earlier egg maturation (Chen et al. 2010). This situation would be similar to the reduced role of JH in reproduction, but greater role in regulating caste functions, in more advanced vs. primitive bees (Bloch et al. 2000). Alternatively, JH receptors or JH inhibitors may also cycle during the day in a tissue-specific manner, preventing the ovaries or flight muscles from responding to the elevated JH titer.

7.3.6 Chrono-transcriptomics

The recent development of microarrays and high throughput sequencing (e.g., RNA-Seq) has allowed assessment of the degree to which morphs of wing- or phase-polymorphic species differ in global transcript abundance and gene expression (Brisson et al. 2007; Chen et al. 2010). Circadian rhythms in global gene expression have been identified in a wide variety of organisms (Dunlap et al. 2004; Winjinen and Young 2006). However, no published study has yet reported daily changes in global gene expression in morphs of a polymorphic species. To determine the extent to which the morph-specific JH titer cycle is associated with and possibly regulates global changes in morph-specific gene expression, a *G. firmus* de novo reference transcriptome was sequenced and assembled using RNA collected from LW(f) and SW *G. firmus* fat body, the major organ of intermediary metabolism, as well as from the flight muscles of LW(f) and LW(h) morphs (Vellichirammal et al. 2014; unpublished data). Fat body samples were taken early (within a few h after lights on (AM), low JH titer in both morphs) or late in the photophase (within a few hours before lights off (PM), elevated JH titer in the LW(f) morph). Data analyses are currently in progress and only a brief description of a few major findings is reported here.

When the AM and PM fat body transcripts were compared for each morph, fat body of the LW(f) morph showed a much greater (50 %) number of transcripts that changed in abundance between the two sampling periods compared to SW fat body samples. Indeed, the percentage of transcripts that exhibited diel change was much greater than the percentage that differed between the morphs during either the AM or PM samples. Thus morph-specific diel change in transcript abundance is a prominent feature of wing-polymorphic *G. firmus*. The extent to which the morph-specific diel change in transcript abundance is regulated by the morph-associated change in the JH titer is not known at present. Interestingly, the greater change in transcript abundance in the LW(f) fat body from AM to PM samples was primarily due to a decrease in abundance of transcripts, many of which are involved in aerobic metabolism. These included subunits of various electron-transport enzymes (e.g., NADH dehydrogenase, cytochrome oxidase, succinate cytochrome b, NADH-cytochrome b5 reductase, NADH:ubiquinone dehydrogenase), Krebs cycle and associated enzymes (e.g., ATP-citrate synthase, acetyl CoA synthase), enzymes of aerobic glucose metabolism (pyruvate dehydrogenase complex), as well as a variety of proteins involved in the transport of metabolites into and out of the mitochondria. The morph-specific decrease in transcripts of proteins involved in aerobic metabolism is consistent with preliminary results of an ongoing fat body metabolomics analysis, which also indicates a decrease in aerobic metabolism in PM vs. AM in LW(f) fat body (D. Renault personal communication). Clearly, the morph-specific circadian change in the JH titer is only an indicator of a much more widespread morph-specific diel change in gene expression in *G. firmus*.

7.4 Summary, Conclusions, and Future Studies

The morph-associated JH titer circadian rhythm in *Gryllus* is currently one of the few thoroughly investigated hormonal circadian rhythms in insects. Moreover it is the most thoroughly studied polymorphic rhythm in any organism, the only genetic polymorphism for a hormonal circadian rhythm identified to date, and one of the only hormonal rhythms studied in the laboratory as well as in the field. Finally, it is the only hormonal rhythm strongly correlated with a life history polymorphism in which morphs that differ in flight ability and egg production also differ in cyclic vs. acyclic hormone levels. All of these factors make the morph-specific JH titer rhythm in *Gryllus* a premier experimental model for investigating adaptive variation and evolution of hormone circadian rhythms. There are several especially fruitful areas for future research on this topic. For example, very little is known about the morph-specific interaction between the circadian clock and JH biosynthesis in the corpora allata (e.g., do morphs differ in aspects of the circadian clock as has been demonstrated in honeybee castes (Toma et al. 2000)? In the output of the clock? In differential reception by LW(f) vs. SW corpora allata to equivalent signals from the clock?). The specific point of the JH biosynthetic pathway that is modulated in a morph-specific manner to dramatically increase JH production also represents a fascinating issue in flux control (i.e., control of metabolite flow) through pathways of metabolism, a central topic in biochemical systems biology (Fell 1997). Nor is much known about the function of the morph-specific JH titer cycle. Does this cycle regulate some direct aspect of flight, a correlated aspect of flight, or some trait independent of flight, but which varies in a morph-specific circadian manner? The recent finding of global, morph-dependent changes in transcript abundance allows a broadening of studies of morph-specific circadian adaptation to include genes regulated by JH, as well as those independent of this hormone. Clearly, the morph-specific JH titer circadian rhythm represents a powerful experimental model to investigate a variety of issues at the interface of chronobiology, endocrinology, and life history evolution and to synthesize these previous independent subdisciplines.

Acknowledgments A. J. Zera gratefully acknowledges support of his research on insect wing polymorphism by the National Science Foundation since 1990 (most recently awards IOS-1122075, IOS-0516973, and IBN-0212486). J. A. Brisson is grateful for start-up funds from the School of Biological Sciences, University of Nebraska.

References

Bloch GL, Borst DW, Huang Z-Y, Robinson GE, Cnaani J, Hefetz A (2000) Juvenile hormone titers, juvenile hormone biosynthesis, ovarian development and social environment in *Bombus terrestris*. J Insect Physiol 46:47–57

Bloch GL, Hazan E, Rafaeli A (2013) Circadian rhythms and endocrine functions in adult insects. J Insect Physiol 59:56–69

Brisson JA, Davis GK, Stern DL (2007) Common genome-wide patterns of transcript accumulation underlying the wing polyphenism and polymorphism in the pea aphid (*Acyrthosiphon pisum*). Evol Dev 9:338–346

Chen S, Yang P, Jiang F, Wei Y, Ma Z et al (2010) De novo analysis of transcriptome dynamics in the migratory locust during the development of phase traits. PLoS One 5:e15633

Dunlap JC, Loros JL, DeCoursey PJ (2004) Chronobiology. Biological timekeeping. Sinauer, Sunderland

Elekonich MM, Schultz DJ, Bloch GL, Robinson GE (2001) Juvenile hormone levels in honey bee (*Apis mellifera* L.) foragers: foraging experience and diurnal variation. J Insect Physiol 47:1119–1125

Fassbold K, El-Damanhouri H, Lorenz MW (2010) Age-dependent cyclic locomotor activity in the cricket, *Gryllus bimaculatus*, and the effect of adipokinetic hormone on locomotion and excitability. J Comp Physiol A 196:271–283

Fell D (1997) Understanding the control of metabolism. Portland Press, London

Giebultowicz J (2000) Molecular mechanism and cellular distribution of insect circadian clocks. Annu Rev Entomol 45:769–793

Goodman WG, Cusson M (2012) The juvenile hormones. In: Gilbert LI (ed) Insect endocrinology. Elsevier, Amsterdam, pp 310–365

Hadley M, Levine J (2007) Endocrinology. Pearson Prentice Hall, Upper Saddle River

Harrison RG (1980) Dispersal polymorphisms in insects. Annu Rev Ecol Syst 11:95–118

Kondrik D, Socha R, Syrova Z, Zemek R (2005) The effect of constant darkness on the content of adipokinetic hormone, adipokinetic response and walking activity in macropterous females of *Pyrrhocoris apterus* (L.). Physiol Entomol 30:248–255

Lorenz MW, Zemek R, Kodik D, Socha R (2004) Lipid mobilization and locomotor stimulation in *Gryllus bimaculatus* by topically applied adipokinetic hormone. Physiol Entomol 29:146–151

Nelson RJ (1995) An introduction to behavioral endocrinology. Sinauer, Sunderland

Norris DO (1997) Vertebrate endocrinology. Academic, San Diego

Ramaswamy SB, Shengqiang S, Mbata GN, Rachinsky A, Park YI, Crigler L, Donald S, Srinivasan A (2000) Role of juvenile hormone-esterase in mating-stimulated egg development in the moth *Heliothis virescens*. Insect Biochem Mol Biol 30:785–791

Saunders DS (2002) Insect clocks. Elsevier, Amsterdam

Stay B, Tobe SS (2007) The role of allatostatins in juvenile hormone synthesis in insects and crustaceans. Annu Rev Entomol 52:277–299

Stay B, Zera AJ (2010) Morph-specific diurnal variation in allatostatin immunostaining in the corpora allata of *Gryllus firmus*: implications for the regulation of a morph-specific circadian rhythm in JH biosynthetic rate. J Insect Physiol 56:266–270

Tobe SS, Stay B (1985) Structure and regulation of the corpora allata. Adv Insect Physiol 18:305–433

Toma DP, Bloch GL, Moore D, Robinson GE (2000) Changes in period mRNA levels in the brain and division of labor in honey bee colonies. Proc Natl Acad Sci U S A 97:6914–6919

Vafopoulou X, Steel CGH (2001) Induction of rhythmicity in prothoracicotropic hormone and ecdysteroids in *Rhodnius prolixus*: roles of photic and neuroendocrine zeitgebers. J Insect Physiol 47:935–941

Vafopoulou X, Steel CGH (2005) Circadian organization of the endocrine system. In: Gilbert LI, Kostas I, Gill SS (eds) Comprehensive molecular insect science. Elsevier, Amsterdam, pp 551–650

Vellichirammel NN, Zera AJ, Schilder RJ, Wehrkamp C, Riethoven J-JM, Brisson J (2014) De novo transcriptome assembly and morph-specific gene expression profiling of the wing-polymorphic cricket, *Gryllus firmus*. PLoS ONE 10.1371/journal.pone. 0082129.

Winjnen H, Young MJ (2006) Interplay of circadian clocks and metabolic rhythms. Annu Rev Genet 40:409–448

Zera AJ (2006) Evolutionary genetics of juvenile hormone and ecdysteroid regulation in *Gryllus*: a case study in the microevolution of endocrine regulation. Comp Biochem Physiol A 144:365–379

Zera AJ (2009) Wing polymorphism in *Gryllus* (Orthoptera: Gryllidae): Proximate endocrine, energetic and biochemical mechanisms underlying morph specialization for flight vs. reproduction. In: Whitman DW, Ananthakrishnan T (eds) Phenotypic plasticity of insects. Mechanisms and consequences. Science Publishers, Enfield, pp 609–653

Zera AJ (2013) Morph-specific JH titer regulation in wing-polymorprhic *Gryllus* crickets. Proximate mechanisms underlying adaptive genetic modification of JH regulation. In: Devillers J (ed) Juvenile hormone and juvenoids: modeling biological effects and environmental fate. CRC Press, Boca Raton, pp 31–64

Zera AJ, Cisper G (2001) Genetic and diurnal variation in the juvenile hormone titer in a wing-polymorphic cricket: implications for the evolution of life histories and dispersal. Physiol Biochem Zool 74:293–306

Zera AJ, Denno RF (1997) Physiology and ecology of dispersal polymorphism in insects. Annu Rev Entomol 42:207–231

Zera AJ, Huang Y (1999) Evolutionary endocrinology of juvenile hormone esterase: functional relationship with wing polymorphism in the cricket, *Gryllus firmus*. Evolution 53:837–847

Zera AJ, Zhao Z (2009) Morph-associated JH titer diel rhythm in *Gryllus firmus*: experimental verification of its circadian basis and cycle characterization in artificially-selected lines raised in the field. J Insect Physiol 55:450–458

Zera AJ, Borcher C, Gaines SB (1993) Juvenile hormone degradation in adult wing morphs of the cricket, *Gryllus rubens*. J Insect Physiol 39:845–856

Zera AJ, Sall J, Grudzinski K (1997) Flight-muscle polymorphism in the cricket *Gryllus firmus*: muscle characteristics and their influence on the evolution of flightlessness. Physiol Zool 70:519–529

Zera AJ, Zhao Z, Kaliseck K (2007) Hormones in the field: evolutionary endocrinology of juvenile hormone and ecdysteroids in field populations of the wing-dimorphic cricket *Gryllus firmus*. Physiol Biochem Zool 80:592–606

Zhao Z, Zera AJ (2004a) The hemolymph JH titer exhibits a large-amplitude, morph-dependent, diurnal cycle in the wing-polymorphic cricket, *Gryllus firmus*. J Insect Physiol 50:93–102

Zhao Z, Zera AJ (2004b) A morph-specific daily cycle in the rate of JH biosynthesis underlies a morph-specific daily cycle in the hemolymph JH titer in a wing-polymorphic cricket. J Insect Physiol 50:965–973

Chapter 8
Plasticity in the Cricket Central Nervous System

Hadley Wilson Horch, Alexandra Pfister, Olaf Ellers, and Amy S. Johnson

Abstract The auditory system of the cricket shows a remarkable level of anatomical plasticity in response to injury. Removal of the auditory organ deafferents several types of auditory neurons of the central nervous system. These neurons respond to the loss of input by sending dendrites across the midline, a boundary they typically respect, and forming synapses with the auditory afferents of the contralateral ear. This compensatory growth and synapse formation reinstates neuron-specific frequency tuning curves. Growth and branching after deafferentation are sexually dimorphic, with male growth rates being linear and female growth rates being nonlinear. Female dendrites stop growing and branching after only a few days, while male dendrites continue to grow steadily, becoming twice as long as females by 20 days after deafferentation. Exploration of the molecular basis of this compensatory plasticity has revealed a number of possible candidates, including semaphorins and a VAMP family member, neuronal synaptobrevin.

Keywords *Gryllus bimaculatus* • Auditory system • Compensatory plasticity • Deafferentation • *Teleogryllus oceanicus* • Semaphorins • Neuronal synaptobrevin • Sexual dimorphism

8.1 Introduction

The nervous system of the cricket has been studied for many decades. Early neuroethologists recognized the advantages to studying an animal that could generate interesting, robust behaviors with a relatively simple nervous system. Over the years, the plasticity of such simple nervous systems has been debated numerous times in the literature. While some have claimed that the short life spans and the high level of similarity among individual nervous systems mean that competition plays an inconsequential role in these systems (c.f. Easter et al. 1985), others have roundly criticized this idea, countering with multiple, convincing examples of

H.W. Horch (✉) • A. Pfister • O. Ellers • A.S. Johnson
Departments of Biology and Neuroscience, Bowdoin College, 6500 College Station, Brunswick, ME 04011, USA
e-mail: hhorch@bowdoin.edu

H.W. Horch et al. (eds.), *The Cricket as a Model Organism*,
DOI 10.1007/978-4-431-56478-2_8

invertebrate plasticity (Palka 1984; Murphey 1986; Meinertzhagen 2001; Groh and Meinertzhagen 2010).

In fact, the cricket demonstrates some impressive examples of nervous system plasticity. For example, the transplantation of cerci has revealed how axons find targets and form synapses under unusual conditions (McLean and Edwards 1976; Murphey et al. 1981, 1983a, b). Furthermore, the compensatory recovery of giant interneurons in response to deprivation has been explored after amputation of cerci (Palka and Edwards 1974; Murphey et al. 1975). More recently, several experiments have explored the characteristics of learning and memory in this animal (Chap. 9 and Matsumoto and Mizunami 2000, 2002; Matsumoto et al. 2003, 2006; Scotto-Lomassese et al. 2003; Matsumoto 2004). The auditory system of the cricket demonstrates robust plasticity in response to injury both during pre-eclosion maturation and in the adult. The deafferentation-induced compensatory plasticity of the auditory system has been explored since the mid-1980s and will be the main topic of this chapter.

8.2 Compensatory Plasticity in the Auditory System

The auditory system is the subject of additional chapters in this book (c.f. Chaps. 11 and 19), but a basic description of the anatomy will be included here. Briefly, the cricket's tympanal organs are on its first foreleg. Each prothoracic leg has both an anterior and a posterior tympanal membrane. The transduction of sound occurs nearby in the tympanal organ, and the auditory information is conveyed into the central nervous system via nerve 5. Approximately 70 auditory receptor neurons project into the first ganglion of the thorax (the prothoracic ganglion) and arborize in a claw-shaped structure that mainly respects the midline (Fig. 8.1a). There, auditory information is transferred to several mirror-image pairs of auditory neurons including ascending neuron-1 (AN-1; which responds best to cricket song), ascending neuron-2 (AN-2; which responds best to ultrasound), and the omega neurons (which act to sharpen directionality through reciprocal inhibition). These several pairs of auditory neurons in *Gryllus bimaculatus* have been anatomically and physiologically described in detail (Wohlers and Huber 1982). It is important to note that homologous cells in different cricket species have different names, for example, AN-2 in *G. bimaculatus* is homologous to Int-1 in *Teleogryllus oceanicus*. Regardless of species, however, the dendrites of these auditory neurons tend to respect the midline of the prothoracic ganglion and receive auditory input exclusively from the ipsilateral afferents (Selverston et al. 1985). This organization is important for the generation of both positive and negative phonotaxis, behaviors that are critical for a cricket's survival.

In the mid-1980s, several labs explored the effect of input removal on the anatomy and physiology of these central auditory neurons. The first description was published by Ron Hoy and colleagues (1985) and was completed in *Teleogryllus oceanicus*. In these experiments, removing one prothoracic leg a day

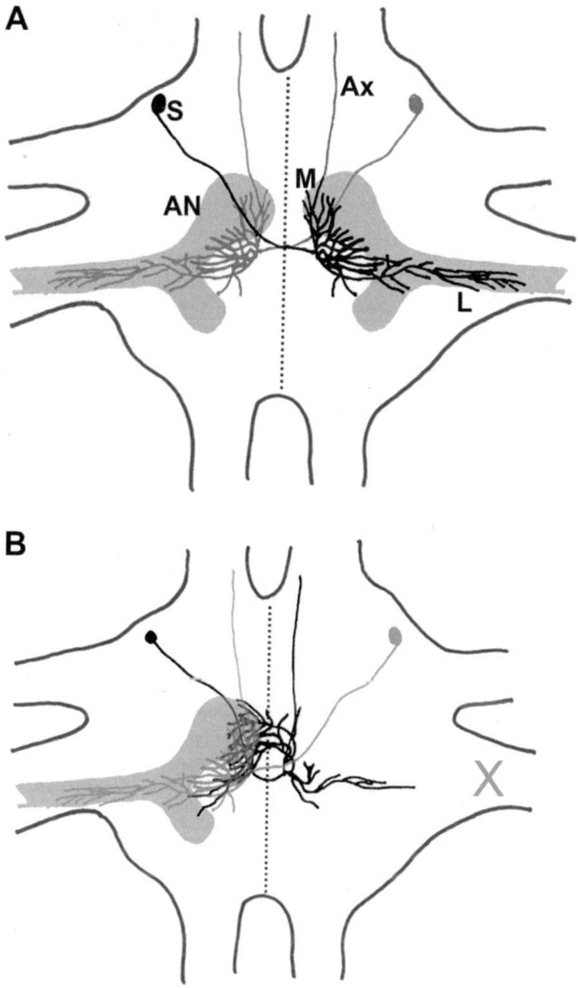

Fig. 8.1 Chronic deafferentation causes profound reorganization of the dendrites of central auditory neurons in the adult. (**a**) Mirror-image pairs of AN-2 (*black* and *gray*) are shown within the outline of the prothoracic ganglion. Auditory afferents project from the ears on both sides and end in mirror-image "claw"-shaped arbors (*light gray*). The midline of the prothoracic ganglion is represented as a dotted line. Relevant parts of the right AN-2 are labeled: *M* medial dendrites, *L* lateral dendrites, *Ax* axon. The soma (S) is contralateral to the main dendritic arbor. (**b**) Chronic deafferentation of the right AN-2, via unilateral ear removal throughout larval development (X), results in loss of the right auditory nerve and reorganization of the AN-2 dendritic arbor. The contralateral axons and AN-2 remain intact (Figure adapted from Horch et al. 2011)

or two after hatching deprives these crickets of the sensory input from the ear on that side. Once these amputated crickets molted into adults, the morphology of the deafferented ascending neuron (Int-1) was examined using cobalt backfills. Profound differences in the dendritic arbors of the deafferented Int-1 neurons were noted. Most obviously, the medial dendrites had sprouted across the midline and had grown into the contralateral auditory neuropil (Fig. 8.1b). Remarkably, this growth restored specific functionality. Int-1 neurons that had been deafferented soon after the cricket had hatched could now be driven in the adult by ultrasound stimulation applied to the contralateral ear (Hoy et al. 1985).

This initial work was quickly followed by a number of different studies that expanded our understanding of this phenomenon. Homologous and additional neurons in other cricket species were capable of the same type of response. For example, three different types of auditory neurons were examined in *Gryllus bimaculatus*, and all three, AN-1, AN-2, and ON-1, were found to sprout dendrites across the midline into the contralateral auditory neuropil in response to unilateral deafferentation (Schildberger et al. 1986). For all three types of cells, the tuning curves that depict threshold and intensity characteristics were quite similar after recovery as compared to intact cells. For example, AN-1 and ON-1, which both respond to the 5 kHz chirps of other crickets, maintained this type of response, while AN-2, which is mainly tuned to higher frequencies so as to detect bat ultrasound, maintained its characteristic response after deafferentation. Attempts have been made to correlate behavioral recovery with anatomical changes in auditory dendrites (Schmitz et al. 1988) and to explore the sprouting of intact auditory afferents in response to unilateral deafferentation (Schmitz 1989).

The relative simplicity of this system allows for the exploration of a number of questions that could reveal fundamental principles of dendritic and axon growth and plasticity. For example, what are the direct effects of deafferentation on dendrites and what molecules induce this response? Do deafferented cells maintain a specific volume even though the dendritic arbor has been rearranged? How do competitive interactions influence this anatomical plasticity? Are there indirect effects of deafferentation, and what are they?

8.3 Plasticity and Competition

When nervous systems are injured, axons and dendrites are damaged and then often lost due to Wallerian degeneration (Vargas and Barres 2007). Secondarily, many dendrites are deafferented when their presynaptic partners are lost. The responses of deafferented dendrites vary depending on the system and developmental stage. The typical response during development appears to be dendritic loss. In many systems, deafferentation leads to dendritic withdrawal and even cell death (Parks 1979; Trune 1982a, b; Nordeen et al. 1983; Born and Rubel 1985). The positive growth response of the auditory system of the cricket in response to injury represents one of a handful of exceptions. Though many of these exceptions are in insects (Lakes et al. 1990; Buschges et al. 1992; Wolf and Büschges 1997), there are some examples in mammals as well (Hickmott and Steen 2005; Tailby et al. 2005). Understanding the mechanisms behind these exceptions is beneficial as this knowledge might give us insight into the limited plasticity of other systems.

If deafferentation occurs in the juvenile cricket and is maintained throughout larval life (chronic), the extent of dendritic reorganization of auditory dendrites is quite large when observed in the adult (Fig. 8.2a, b). Medial dendrites sprout across the midline (arrowheads, Fig. 8.2b), while lateral dendrites are reduced in complexity as compared to controls (arrows, Fig. 8.2a, b). While the reduction in complexity of the medial dendrites on the deafferented side is roughly balanced by the new growth of dendrites across the midline into the contralateral neuropil, the reduction in lateral dendrites is extensive (Fig. 8.2c). In fact the overall

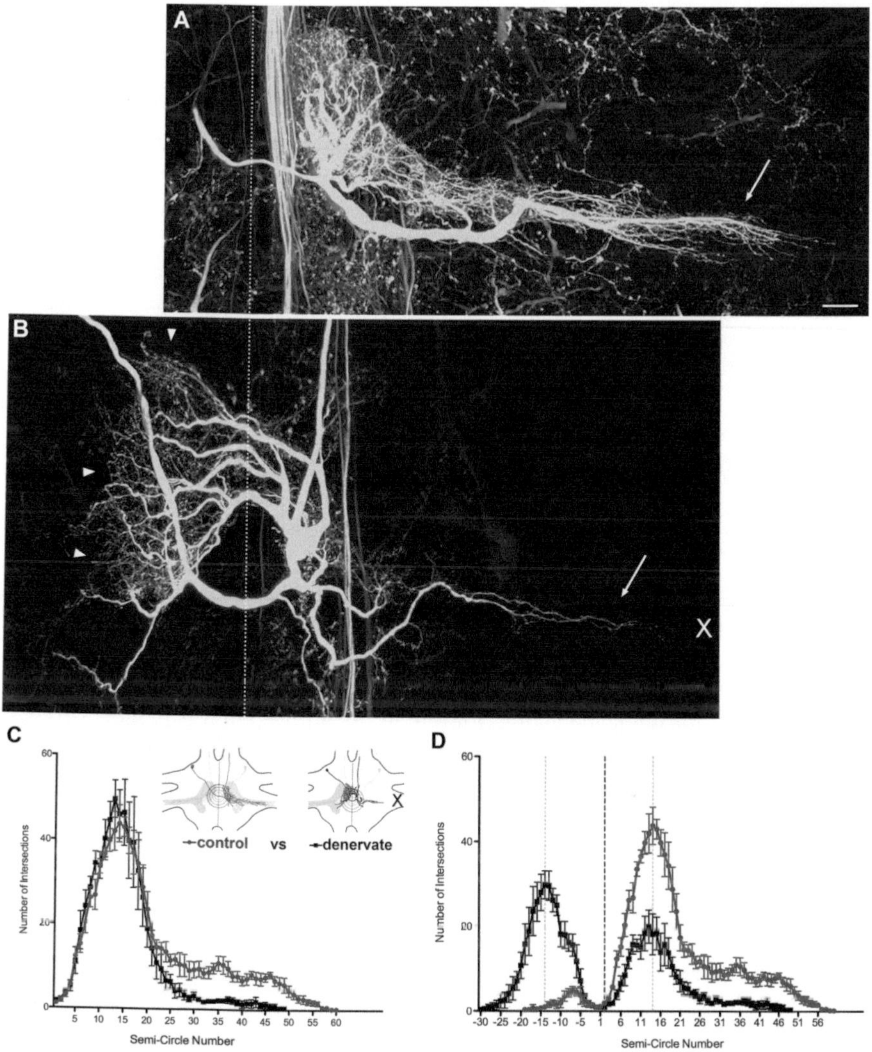

Fig. 8.2 Qualitative and quantitative changes in chronically deafferented dendritic arbors. Alignment of an individual (**a**) control and (**b**) chronically deafferented AN neuron about the midline reveals qualitative changes of both the medial (*arrowheads*) and the lateral (*arrows*) dendrites. X = deafferented side. (**c**) Sholl analysis quantifies the differences in control (*gray*) and deafferented (*black*) AN dendrites by counting the number of dendrites that intersect each ring. Insct shows a schematic of the Sholl analysis completed on control (*left*) and deafferented (*right*) AN neurons. (**d**) Intersections per semicircle (*left* and *right*) quantify the complexity of control (*gray*) and deafferented (*black*) dendrites on each side of the midline. *Dark dashed line* = midline; lighter dashed lines indicate symmetrical peaks in complexity (Figure adapted from Horch et al. 2011)

complexity, as assessed with a Sholl analysis, is reduced after deafferentation (Horch et al. 2011). Interestingly, when medial dendrites sprout across the midline, their contralateral branch elaboration is located in an equivalent position with regards to the midline (Fig. 8.2d).

Once the auditory neurons reorganize in response to deafferentation, is the overall volume of their dendritic arbor maintained? Observations in a variety of species have led to the "conservation of membrane" hypothesis, which proposes that neurons are predetermined to be a particular size (Schneider 1973; Devor 1975). For example, axotomy induces reductions in axonal arbor size that are roughly balanced by reciprocal increases in dendrites, resulting in an arbor that is reorganized but approximately equal in volume to the original (Shankland et al. 1982; Roederer and Cohen 1983; Hall and Cohen 1988). While volume was not directly measured in the cricket, Sholl analysis indicates an overall decrease in dendritic complexity (Fig. 8.2 and Horch et al. 2011). While Sholl analysis is an imperfect proxy for volume, these results predict that measurements of total volume would be reduced after deafferentation. Additional experiments in which arbor volume is directly measured would clarify the answer to this important question, and the auditory system of the cricket is a good system in which to examine this. The development of a transgenic cricket in which GFP was expressed in a small number of neurons or only in auditory neurons would be beneficial to these studies.

The plasticity of this system was further explored by asking how central auditory cells respond if the damaged auditory afferents are allowed to regenerate. This can easily be achieved by crushing rather than cutting nerve 5. Remarkably, the central auditory neuron maintains both the deafferentation-induced aberrant dendritic growth and the novel synapses even when the original, appropriate afferents regenerate and reestablish connections (Pallas and Hoy 1986). These previously deafferented neurons are then left receiving auditory information from both ears, a highly unusual situation. Given the role of competition for postsynaptic space in other systems (Purves 1976; Purves and Hume 1981; Sanes and Lichtman 1999), one might have expected the aberrant connections to the contralateral ear to be lost once the ipsilateral afferents regenerated and reformed synapses.

In addition to the direct effects of deafferentation on postsynaptic dendrites, unilateral loss of auditory afferents leads to indirect effects on other neurons in the ganglion. This is likely due to the bilateral contribution of the ONs, which provide reciprocal inhibition designed to sharpen directional tuning. In addition, the intact auditory afferents, which normally forms synapses specifically with ipsilateral dendrites, must contend with a large number of "invading" dendrites from the contralateral side after deafferentation. One would predict that this could have structural ramifications for contralateral dendrites and axons. Therefore, it is reasonable to wonder how the physiological and anatomical changes induced by deafferentation affect the form and function of the intact AN dendrites on the contralateral side. In fact, the intact cell shows no changes in strength or latency of physiological responses upon adult deafferentation (Brodfuehrer and Hoy 1988). Thus, even after its territory is invaded by contralateral dendrites, which form new

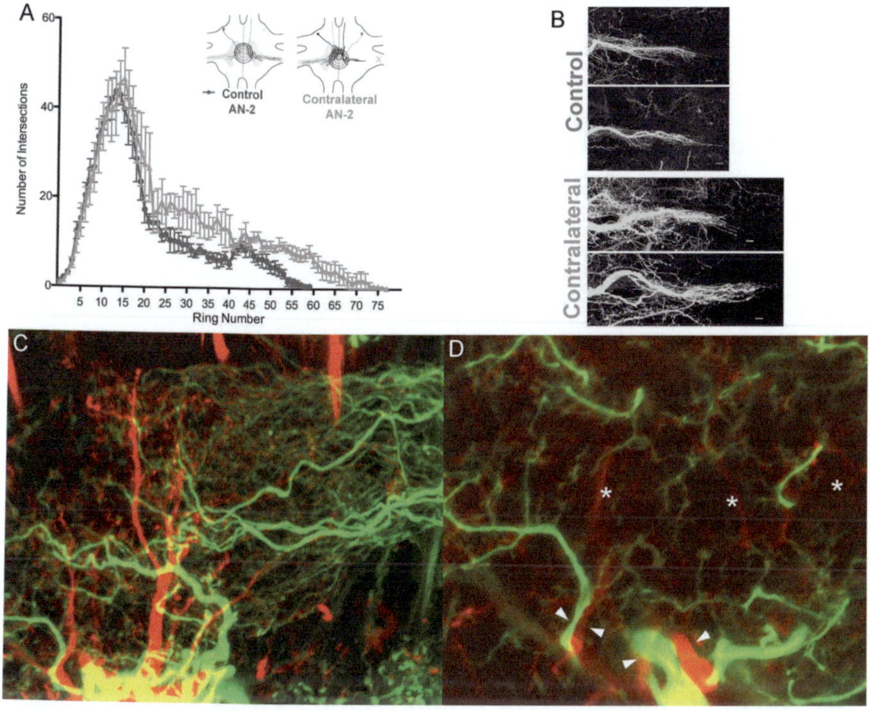

Fig. 8.3 Effects of deafferentation on contralateral axons and dendrites. (**a**) Sholl analysis quantifying the number of dendritic intersections per ring for a control AN (*black circles*) as compared to the intact AN contralateral to deafferentation (*gray triangles*) shows lateral sprouting of intact AN dendrites. Plots are average ± SEMs. Inset shows a schematic of concentric rings centered at the midline. (**b**) Confocal images of the lateral dendrites of both controls and intact ANs, aligned at the midline and cropped to illustrate the increased lateral sprouting in intact ANs contralateral to deafferentation. (**c**) No tiling was seen between intact AN dendrites (*red*) and deafferented AN dendrites (*green*) when deafferentation forced dendrites into the same neuropil. (**d**) A single optical section of this region shows that dendrites sometimes grow in close apposition (*arrowheads*), though there are some regions where one set of dendrites predominate (*asterisks*). Scale bars in **b** = 10 μm, scale bars in **c** and **d** = 5 μm (Adapted from Horch et al. 2011)

synapses with the remaining auditory afferents, no obvious physiological changes are evident. This lack of physiological change after acute deafferentation is in contrast to quantified anatomical changes to these cells after chronic deafferentation. Specifically, the lateral dendritic arbor of the intact AN-2 projects farther away from the deafferented side and becomes more complex than in non-deafferented tissue (Fig. 8.3a, b). It will be important to determine whether this increased growth and branching in the lateral dendrites only occurs after chronic deafferentation or whether it also occurs upon acute deafferentation in the adult. The functional significance of this sprouting is currently unknown. Likewise, it is important to understand whether chronic deafferentation induces physiological changes in the intact AN dendrites in the adult.

While the growth of lateral dendrites of intact AN-2 neurons is surprising, one might predict that the medial dendrites of the intact ANs would be altered when "invaded" by dendrites from the deafferented side. When forced together, do these two arbors stake out different territorial regions or "tiles," as has been demonstrated in several other systems (Grueber et al. 2001, 2002; Sugimura et al. 2003; Emoto et al. 2004; Soba et al. 2007; Ferguson et al. 2009)? The answer is no. There does not appear to be clear reorganization or "tiling" when these two different arbors are forced together (Fig. 8.3c, d). Though there are some regions where one set of dendrites or the other predominate (asterisks, Fig. 8.3d), there is no evidence of "tiling." In contrast, there are clearly areas where the two sets of dendrites grow in tight apposition (arrowheads, Fig. 8.3d), raising the possibility that there is some recognition, coordination, or fasciculation between the dendrites of these two arbors.

The indirect effects of unilateral deafferentation are also seen in the intact auditory afferents on the contralateral side. Though a few axons cross the midline in controls, there is an increase in the extent and volume of intact axons crossing from the contralateral side as well as an increase in skeletal length (Horch et al. 2011; Pfister et al. 2013). Axons also sprout out of the "claw"-shaped auditory neuropil, though many of these axons do not cross the midline (Horch et al. 2011).

8.4 Sexual Dimorphism of Compensatory Growth

Much of the early work exploring this injury-induced compensatory growth either did not specify the sex of animals used or selected only one sex for experiments. Only very recent work has compared the growth patterns of deafferented dendrites between males and females. Quantification of several characteristics of dendrites over time after deafferentation in the adult reveals striking sexually dimorphic growth patterns (Pfister et al. 2013). In females, dendrites grow rapidly across the midline but stop growing approximately 3–5 days after deafferentation. In contrast, male dendritic growth rates are linear, with dendrites growing, on average, twice as long as female dendrites by 20 days. Skeletal length, a measurement that takes into account branching, follows the same sexually dimorphic growth patterns (Pfister et al. 2013).

Dendritic branching also showed a sexual dimorphism, though axonal branches did not. To distinguish different levels of axonal or dendritic branches, each axon or dendrite physically crossing the midline was defined as a "midline-primary" (M1) axon or dendrite. Subsequent branches arising from M1 branches were defined as M2, M3, and M4. The use of the "M" prior to branch number is an acknowledgment of the uncertainty of whether these M1 dendrites and axons branched before crossing the midline. Patterns in M1, M2, M3, and M4 branch formation over time were analyzed in male and female data sets separately. The nonparametric Spearman Rank correlation was used to determine whether there were significant trends in the number of M1, M2, M3, and M4 branches forming over time. If Spearman Rank indicated significant trends, data were fit with linear and nonlinear models, as described in Pfister et al. (2013).

For all data, except where there were too few data for males, a *t*-test or Kruskal-Wallis test was used to compare the end point of measured variables for males versus females. In addition, differences in male and female data were determined with an ANCOVA if the nonlinear model was not appropriate for either of the sexes and if a significantly non-zero linear slope existed for at least one of the sexes. When there were no linear trends in the data, male and female data sets were compared with a simple *t*-test.

Male M1, M2, M3, and M4 AN dendritic branch formation numbers were best fit by an increasing linear regression model (Fig. 8.4; Table 8.1). Although the mean number of branches increased linearly, male M2 (Fig. 8.4b), M3 (Fig. 8.4c), and M4 (Fig. 8.4d) branch means did not increase at 20 days, suggesting that branch formation in males may have slowed or stopped 14 days after deafferentation. Female AN M1 (Fig. 8.4a), M2 (Fig. 8.4b), and M3 (Fig. 8.4c) dendritic branch

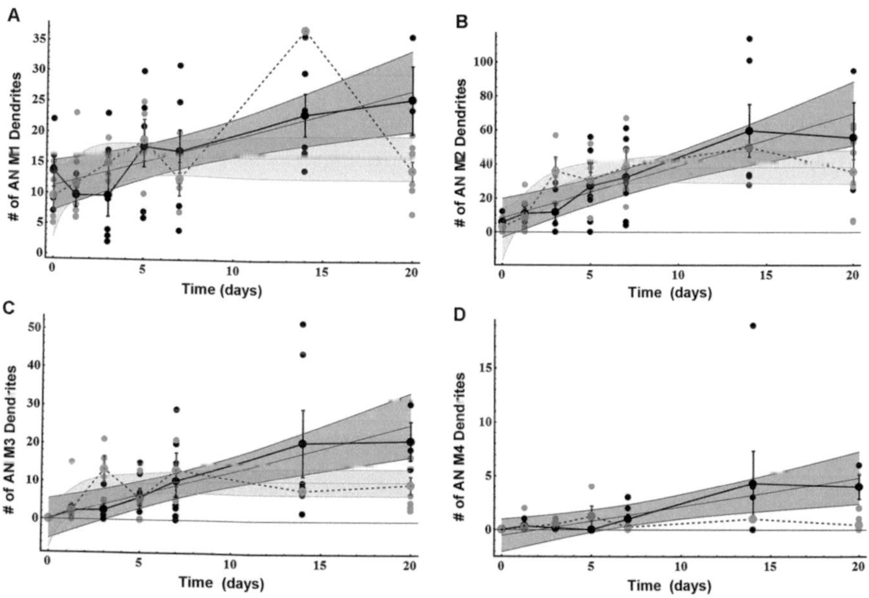

Fig. 8.4 The number of AN branches increased nonlinearly in females and linearly in males. (**a**) M1, (**b**) M2, and (**c**) M3 branch formation was nonlinear in females and linear in males. Female data appears in gray; male data appears in black. Nonlinear growth was characterized by initial rapid formation of branches followed by slowed formation of branches during the later deafferentation time points. (**d**) M4 branch formation increased linearly in males but did not change over time in females. Larger data points represent the mean. Error bars represent standard error of the mean and the shaded areas represent the 95 % confidence interval. Female AN control ($n=4$), 30 h ($n=9$), 3 day ($n=4$), 5 day ($n=4$), 7 day ($n=7$), 14 day ($n=1$), and 20 day ($n=9$). Male AN control ($n=5$), 30 h ($n=4$), 3 day ($n=6$), 5 day ($n=8$), 7 day ($n=7$), 14 day ($n=6$), and 20 day ($n=3$). Note that some points appearing on the graphs may represent more than one data point with equal values. For regression statistics see Table 8.1

Table 8.1 Summary of variables and statistics for AN branching analysis

Variable		AN M1 dendrites (#)		AN M2 dendrites (#)		AN M3 dendrites (#)		AN M4 dendrites (#)	
Sex		Male	Female	Male	Female	Male	Female	Male	Female
Spearman rank	Sample size	37	35	37	35	37	35	37	35
	r	0.50	0.29	0.66	0.61	0.69	0.62	0.57	0.25
	One-tailed P (df)	0.0008 (35)	0.04 (33)	<0.0001 (35)	<0.0001 (33)	<0.0001 (35)	<0.0001 (33)	0.0001 (35)	0.07 (33)
Compare linear to nonlinear	ΔAICc	2.3	1.0	1.0	8.0	2.0	4.8	2.4	0.4
	Probability linear model is correct	0.76	0.38	0.63	0.02	0.73	0.08	0.77	0.55
	Preferred model	Linear	Nonlinear	Linear	Nonlinear	Linear	Nonlinear	Linear	No trend
r^2 and fitted parameters for preferred model	r^2	0.26	0.13	0.40	0.42	0.36	0.28	0.23	na
	k (per day)	na	0.50	na	0.40	na	0.50	na	na
	m (μm per day)	0.8	na	3.1	na	1.3	na	0.3	na
	lo (μm)	na	9.0	na	−1.2	na	−1.2	na	na
	b (μm)	11.2	na	8.4	na	0.2	na	−0.5	na
	l_∞ *(μm)*	na	16.1	na	38.4	na	10.3	na	na
	p slope=0 (d.f.)	0.001 (1,35)	na	<0.0001 (1,35)	na	<0.0001 (1,35)	na	0.003 (1,35)	na
t-test or Kruskal-Wallis (K-W) comparing end point	p(df) M?F	0.02(10) M>F (t-test)		0.26(10) M=F (t-test)		0.04(10) M>F (t-test)		0.003(10) M>F (K-W)	

To determine if significant trends existed for each variable over time, data were analyzed with the nonparametric Spearman rank correlation. If Spearman rank showed a significant correlation, linear and nonlinear models were fit to the data (Eqs. 1 and 2 in Pfister et al. (2013)) and the preferred model was determined using the Akaike Information Criterion corrected model (AICc). Preferred model parameters for each variable are shown, with units in parentheses. For linear cases, the probability and degrees of freedom that the regression line slope is zero are given. Finally, a *t*-test or Kruskal-Wallis was used to compare end points after testing assumptions of normality and equal variance formation was nonlinear, with branch numbers rapidly increasing until 5 days after deafferentation when branch numbers plateaued (Table 8.1). Female M4 branching did not change over time (Fig. 8.4d). For both sexes, M2 dendritic branches were the most numerous while M4 dendritic branching was generally sparse. At 20 days after deafferentation, males formed on average more M1, M3, and M4 branches than females (Table 8.1, Kruskal-Wallis comparing end point $p < 0.05$).

A similar analysis was completed for axonal branching. Axonal branching showed no sexual dimorphism; thus, all male and female axonal branching data were pooled for analysis. M1 and M4 branch numbers did not change over time while M2 and M3 branch numbers increased linearly (data not shown). There was no within-animal correlation between the amount of AN and N5 branching in any case ($p > 0.05$; data not shown). A similar lack of correlation was found when examining dendritic and axonal length (Pfister et al. 2013).

When adults are deafferented, functional recovery has been detected between about 4–6 days postdeafferentation (Brodfuehrer and Hoy 1988), though the sex of subjects was not reported in this study. It is interesting to note that the timing of this functional recovery matches the point at which growth rates plateau in females. In addition, by 5–7 days, male dendritic growth has essentially caught up with females. One might predict, based on these data, that females would recover more quickly than males. Furthermore, it seems likely that male tuning curves may continue to improve over time as compared to females, given the steady growth of dendrites in males. It would be interesting to look for sexually dimorphic differences in tuning curve recovery over time after deafferentation and be able to correlate functional recovery with the extent of dendritic growth and branching (c.f. Schmitz et al. 1988).

What might be responsible for this sexual dimorphism? While the answer is unknown, hormones are obvious candidates. Hormones have been shown to influence both dendritic and axonal structures in vertebrates as well as invertebrates (DeVoogd and Nottebohm 1981; Cooke and Woolley 2005; Williams and Truman 2005). The hormones that drive metamorphosis and act as sex hormones in insects are ecdysteroids. They appear to influence the branching and growth of dendrites and axons in other systems (Bowen et al. 1984; Truman and Reiss 1988; Weeks and Ernst-Utzschneider 1989; Prugh et al. 1992; Truman and Reiss 1995; Kraft et al. 1998; Matheson and Levine 1999; Cayre et al. 2000; Williams 2005). As such, they are good candidates for influencing dendritic responses to injury in a sexually dimorphic manner.

Differences in ecdysone levels between males and females could be responsible for the differences in growth rates between the sexes. Higher ecdysone titers have been found in *Gryllus firmus* females as compared to males. Life history choices strongly influence the levels of ecdysone in females (Chap. 15 and Zera et al. 2007). The ecdysone levels in male and female *G. bimaculatus* have never been compared; however, it seems likely that similar differences would exist in this species (A. Zera personal communication).

It is generally assumed that AN growth and compensatory synaptogenesis occurs in both sexes so that the auditory system can recover functionality. It also seems likely that both males and females need to recover the ability to hear 5 kHz cricket song as well as bat ultrasound, which is typically 15 kHz and higher. Though the quantified sexually dimorphic growth may have included both AN-1 and AN-2 dendrites, AN-2 dendrites were the vast majority of those measured. Based on these results, one might predict that females recover ultrasound detection more rapidly than males, though what evolutionary forces may have driven this are unclear. Perhaps females have a higher risk of predation as they fly in search of males (Raghuram et al. 2015). It is also intriguing to speculate on whether male dendritic growth, which is on average twice as long by 20 days, makes males more sensitive to bat ultrasound in some behaviorally relevant way.

8.5 Molecular Basis of Anatomical Plasticity

The detailed characterization and quantification of the growth and plasticity of dendrites and axons after deafferentation is important because it informs a search for the molecular candidates that contribute to this unusual plasticity. Understanding where and when transcriptional changes occur helps investigators target particular time points and anatomical locations for investigation. For example, the uncorrelated nature of dendritic and axonal growth within each animal (Pfister et al. 2013) implies that independent molecular cues are used. This complicates the search for molecules that might be involved. It is likely that a complex milieu of factors is generated upon deafferentation, which separately instruct or guide axonal and dendritic growth. Indeed, we have some evidence that multiple factors are differentially regulated upon deafferentation (Horch et al. 2009).

Thus far, we have taken two different approaches to identifying the molecular bases for the growth and plasticity seen. First, we have taken an "unbiased" approach, where we simply ask which genes are differentially expressed after deafferentation. The second "usual suspects" approach focuses on well-conserved guidance molecules, such as semaphorins, for investigation. Both will be described below.

We have used two different "unbiased" techniques to survey the changes that occur after deafferentation, suppression subtractive hybridization (SSH) and differential display (DD). SSH works by amplifying cDNA fragments that differ between a control and an experimental transcriptome (Diatchenko et al. 1996). Our strategy compared control and 3-day deafferented transcriptomes that were generated from prothoracic ganglia from *Gryllus bimaculatus* of the antepenultimate instar stage. We used both a forward and backward subtraction strategy in order to identify both up- and downregulated genes. Our results identified 49 candidates that were downregulated by at least twofold 3 days after deafferentation, but no candidates met the criteria for upregulation at this time point (Horch et al. 2009). The time point of 3 days was selected because it is a period of intense growth (Pfister et al. 2013) and is likely the initial point when compensatory synapses are just beginning to be formed (Brodfuehrer and Hoy 1988). A majority of the candidates identified (27) had E-values greater than 0.001 when analyzed in 2009 using the BLAST web database at the National Center for Biotechnology Information (NCBI) database and were thus defined as having no significant hits (see Table 2 in Horch et al. 2009). The identification of so many potentially novel candidates is exciting, though the implications are difficult to predict. For example, these could represent noncoding RNAs of unknown regulatory function (Genikhovich et al. 2006), or they might represent particularly poorly conserved portions of known genes.

In 2013, a de novo assembly of a *G. bimaculatus* transcriptome was completed (Zeng et al. 2013). This resource gives us additional information about our candidates. Reanalyzing our 2009 SSH results with the aid of ASGARD (Assembled Searchable Giant Arthropod Read Database; Zeng and Extavour 2012), we are now able to report updated results. Of the 27 unidentified candidates, nine match ASGARD entries that, given their larger length, match protein coding genes in *Drosophila* (Table 8.2). Several of these predicted proteins have unknown functions. However, the presence of a Rho GTPase-activating protein (GAP; candidate C2C5) is intriguing, given the role of GTPases in neurite outgrowth (Luo 2000). Ten of the previously unidentified candidates match sequences in ASGARD but are predicted to be noncoding. And finally, the last nine of these candidates were not represented in ASGARD or the NCBI database. While three of these are predicted to be noncoding, the remaining six appear to have open reading frames (defined as a predicted stretch of at least 50 continuous amino acids), though BLAST searches yield no predicted proteins (Table 8.2).

The candidates that do match entries in the NCBI database are shown in Table 8.3, and the information for the matching accessions in NCBI have been updated here since originally published in 2009 (Horch et al. 2009). Candidates are grouped for convenience into categories including enzymes, stress response factors, protein degradation, and protein synthesis candidates. Several of the candidates identified could act to influence protein levels by altering protein transcription rates or changing the rate of protein degradation by influencing proteasome system function. Two candidates identified in our screen, *pushover* and *ubiquitin-specific protease 14*, encode proteins that are involved in proteasome-dependent protein degradation. Evidence indicates that the proteasome system may play an important role in synapse formation and synaptic plasticity (Bingol and Schuman 2005). Pushover is a large, highly conserved ubiquitin ligase that may be calcium sensitive since it has a calmodulin binding domain (Xu et al. 1998; Hegde and DiAntonio 2002; Sutton and Schuman 2005). Others have shown that E3 ligases influence midline patterning by regulating the expression of molecules such as the Slit receptor and Netrin signaling pathway components (Campbell and Holt 2001; Myat et al. 2002; Ing et al. 2007). Intriguingly, the E3 ubiquitin ligase UBE3B was strongly upregulated after noise trauma in the chick (Lomax et al. 2000). Regulation of the function of the ubiquitin/proteasome system (UPS) via transcriptional changes in ubiquitin ligases may be an important regulator of injury response in the auditory system. Likewise, translational control of proteins can occur rapidly and can have a profound impact on cellular function. For example, growth cone

Table 8.2 Suppression subtraction hybridization candidates whose E-values do not meet criteria ($E > 0.001$) in the NCBI database

Representative clone (# seq)	*Gb* accession #	Size (bp)	ASGARD match	ASGARD description
C1D12 (8)	GE653120	565	Isotig10110	Mitotic 15 (FBgn0004643)/cell cycle
C1G3	GE653129	465	Isotig04843, isotig04842	CG4729 (FBgn0036623)/ acyltransferase
C2A8	GE653121	233	Contig13157	Pmi (FBgn0044419)/ unknown
C2B6	GH160417	521	Isotig00090	CG33253 (FBgn0030992)/ phagocytosis
C2C5	GE653133	220	Isotig07718	RhoGAP1A (FBgn0025836)/GTPase activator
C2D5	GE653123	372	Isotig09632	CG32441 (FBgn0052441)/unknown
C2D9	GE653153	291	Isotig06524	No prediction/zinc finger domain
C2F6	GE653140	330	Contig23674	AIF (FBgn0031392)/apoptosis inducing factor
C3F6	GE653161	250	Isotig07965	Lethal (2) 09851 (FBgn0022288)/unknown
C1C6	GE653142	322	Isotig13861	Noncoding
C1D1	GE653166	195	GFCP6CO01EFJTI	Noncoding
C1D9	GE653158	433	Isotig09430	Noncoding
C1G5	GE653132	441	Isotig14117	Noncoding
C2D1 (2)	GE653154	586	Isotig13003	Noncoding
C3B2	GE653139	199	GFCP6CO01CXBXX, GFCP6CO02HI234	Noncoding
C3B6 (2)	GE653144	535	GE8SX9M01CLDJU	Noncoding
C3E8	GE653163	340	Isotig10518	Noncoding
C3F1	GE653164	275	Isotig10658	Noncoding
C2E3	GE653124	319	NSH	Noncoding
C3A9	GE653159	277	NSH	Noncoding
C3H1	GE653146	278	NSH	Noncoding
C1H1	GE653122	155	NSH	ORF/NSH
C1C12 (3)	GE653150	209	NSH	ORF/NSH
C2C2	GE653157	188	NSH	ORF/NSH
C3C4 (5)	GE653130	308	NSH	ORF/NSH
C3D6	GE653165	372	NSH	ORF/NSH
C3H7	GE653148	317	NSH	ORF/NSH

For each candidate, a representative clone is listed along with accession number, size in base pairs, any ASGARD match, and the ASGARD description if relevant. "NSH" indicates that no hit or match was found in ASGARD. If an open reading frame of more than 50 amino acids was predicted, it is noted as ORF/NSF. Otherwise, it is predicted to be noncoding (Adapted from Horch et al. (2009))

Table 8.3 Suppression subtraction hybridization candidates whose E-values met criteria ($E < 0.001$) are grouped by category (*left*)

Category	Representative clone (# seq)	*bimaculatus* accession #	Size (bp)	Best match description	E-value	Best match accession #	% Identity % similarity	Species of best match
Enzyme	C2A4	GE653135	334	Alcohol dehydrogenase	1.00E-31	XM_011148733.1	62/87	*Harpegnathos saltator*
	C1B3 (14)	GE653162	633	Alpha-amylase	1.00E-33	KJ146681.2	78/89	*Odoiporus longicollis*
	C1H8	GE653138	128	Alpha-amylase	1.00E-08	AF208002	56/75	*Diabrotica virgifera virgifera*
	C2A12 (8)	GE653128	332	Alpha-amylase	6.00E-34	XM_008206979.1	56/71	*Nasonia vitripennis*
	C2A7 (2)	GE653152	418	Alpha-amylase	3.00E-38	DQ355516	68/79	*Blattella germanica*
	C3C12 (2)	GE653137	246	Alpha-amylase	8.00E-27	FJ489868.1	71/79	*Anagasta kuehniella*
	C3F10 (9)	GE653155	549	Alpha-amylase	3.00E-61	KC207403.1	74/84	*Blattella germanica*
	C3C10	GE653136	247	Beta-1,4-endoglucanase	5.00E-14	JQ292846.1.1	72/81	*Macrotermes barneyi*
	C1F1 (6)	GE653141	175	*bimac*. EST: lysozyme	3.00E-52	AY226458	63/78	*Gryllus bimaculatus*
	C3F11	GE653156	543	Cytochrome oxidase I (mit)	2.00E-75	GPU88332	91/93	*Gryllus pennsylvanicus*
	C2B8	GE653134	272	Oxidoreductase	5.00E-05	XM_001844134.1	55/78	*Culex quinquefasciatus*
	C1A6 (10)	GE653118	461	Salivary lysozyme	9.00E-35	AY226458.1	63/76	*Anopheles stephensi*
	C1A5 (2)	GE653117	442	Trehalase	.002	NM_059489.6	48/67	*Caenorhabditis elegans*
	C2H7 (2)	GE653127	368	Trehalase	1.00E-26	NM_001112671.1	48/67	*Apis mellifera*
	C3H9	GE653149	284	Trehalase	1.00E-13	KJ789159.1	54/77	*Blattella germanica*
Protein degradation	C2F2	GE653126	505	Pushover	1.00E-45	AF096896	54/74	*Drosophila melanogaster*
	C1H4	GE653145	290	Ubiquitin-specific protease	5.00E-27	EF092057.1	71/91	*Maconellicoccus hirsutus*

(continued)

Table 8.3 (continued)

Category	Representative clone (# seq)	*bimaculatus* accession #	Size (bp)	Best match description	E-value	Best match accession #	% Identity % similarity	Species of best match
Protein synthesis	C3D1	GH160416	222	Ribosomal protein S8	2.00E-22	EF638994.1	87/92	*Triatoma infestans*
	C1G9	GE653143	311	Serine carboxypeptidase	3.00E-32	NM_001159303.1	66/86	*Apis Mellifera*
	C3C5	GE653131	335	Translation initiation factor 3	4.00E-39	XM_001842255.1	71/89	*Culex quinquefasciatus*
Stress response	C1A9	GE653119	284	Pr-5-like protein	7.00E-23	AY863030.1	77/82	*Lysiphlebus testaceipes*
	C3F3 (4)	GE653160	413	Regulcalcin	1.00E-38	XM-002432396.1	58/72	*Pediculus humanus corporis*

Representative clones are given, and the number of matching clones is included in parentheses. Accession numbers, size, best match at time of submission (July 2015), and E-value are given. The accession number of the best match, the % identity and similarity, and the species of best match are listed (Adapted from Horch et al. (2009))

guidance is dependent on rapid and local protein synthesis via the activation of translation initiation factors (Campbell and Holt 2001) similar to the translation initiation factor 3 identified in our screen (Table 8.3). All of these results support the hypothesis that protein levels may be changing in functionally important ways even though no alteration in transcription is detectable.

Differential display (DD) was also used in an "unbiased" way to ask which transcriptional changes occurred in this tissue. DD relies on PCR and oligo-dT primers paired with random upstream primers to amplify a large percentage of cDNAs from any tissue (Liang and Pardee 1992). We deafferented a mix of penultimate and antepenultimate larval nymphs and harvested control and deafferented tissue 24 h and 7 days later. Comparing the DD results of amplified control to either the 24 h or to the 7-day deafferent products yielded a roughly equal number of upregulated and downregulated amplified candidates (subsample of the 7-day comparison shown in Fig. 8.5).

SSH and DD have complementary strengths and, in fact, tend to identify different, nonoverlapping, mRNA populations (Wan and Erlander 1997). Though differential display has the potential to have a higher false-positive rate than SSH (Miele et al. 1998), there appears to be no reason to assume that there would be a directional bias to this rate. This implies that either our differential display experiment was more sensitive in identifying rare transcripts or that there might be a large amount of upregulation at 7 days that had not yet occurred 3 days after deafferentation.

Thus far, we have sequenced 18 of the candidate amplicons identified in our DD screen (Table 8.4). Seven candidates have ORFs and match entries either in the NCBI or ASGARD database, though two of these have no known function. Four match nothing in either database, while seven are predicted to be noncoding. One of these, candidate DD9, was downregulated and is identical to clone C3B6, which was independently identified using SSH. Given that DD amplicons are amplified using oligo-dT primers, the sequence is mainly from the 3′ end of the transcripts and, if short, may include only the 3′ untranslated region and no ORF. Obtaining additional sequence from these candidates would be informative.

Several candidates would be interesting to characterize further. For example, DD31 may be a nuclear hormone receptor of some type. The differential regulation of a hormone receptor is particularly interesting to think about in light of the sexual dimorphism discussed above. DD22 shows similarity to the synaptobrevin family member neuronal synaptobrevin (n-syb). Q-PCR experiments confirmed that n-syb was upregulated after deafferentation in the cricket prothoracic ganglion (Horch et al. 2011). Synaptobrevins are known to be involved in the growth of neurites during development (Martinez-Arca et al. 2000; Shirasu et al. 2000), and they have also been shown to be differentially regulated after neuronal injury in mammals (Jacobsson et al. 1998). While the role of n-syb in the compensatory plasticity of the cricket auditory system is unknown, it is a good candidate for future functional studies.

In addition to these "unbiased" approaches, we also attempted to clone molecules known to regulate axonal and dendritic guidance in other system and which

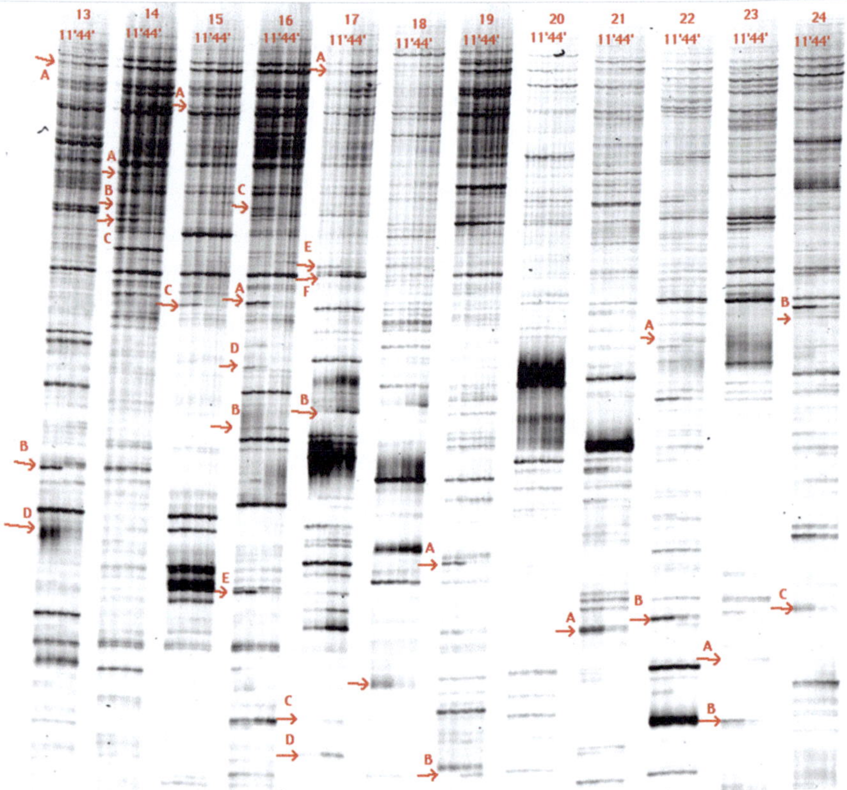

Fig. 8.5 Sample of differential display results. Prothoracic ganglia from late larval instars were deafferented. Seven days later, control prothoracic ganglia (1, 1′) and deafferented ganglia (4, 4′) were dissected and processed for DD. Roughly equal numbers of males and females were combined and used for RNA extraction and subsequent amplification. Each group of lanes (13, 14, etc.) were amplified using a different random upstream primer, and control and 3-day deafferented lanes were run in duplicate (1,1′ and 4,4′). *Red* letters and *arrows* indicate potential candidates that are either upregulated (c.f. column 13, *arrow* **a**) or downregulated (c.f. column 17, *arrow* **b**) after deafferentation

might play a role in the compensatory growth and plasticity in the cricket. Semaphorins (Semas) are a family of guidance molecules that were first identified as factors that influence axonal growth and branching in a number of different species (Tessier-Lavigne and Goodman 1996). More recently, Semas have been shown to influence the guidance and morphology of developing dendrites as well (Polleux et al. 2000; Jan and Jan 2003; Fenstermaker et al. 2004; Kim and Chiba 2004; Morita 2006; Yamashita et al. 2007; Gontheir et al. 2009; Nakamura et al. 2009; Tran et al. 2009). In addition, Semas are upregulated after injury in mammals (De Winter et al. 2002; Harel and Strittmatter 2006; Pasterkamp and Giger 2009), indicating that they may be well-suited to influence compensatory responses in the cricket auditory system after injury. Sema2a is a secreted protein that is expressed

Table 8.4 Differential display candidates up or downregulated 24 h or 7 days post deafferentation

Candidate	*Gb* accession no.	Size (bp)	Regulation	Best match description	Match ID number
DD1	GT564726	669	Up (24h)	CG12264 cysteine desulfurase	*Culex quinquefasciatus* (XM_001845110.1) ASGARD: contig15400, 15399
DD3	GT564727	234	Up (24 h)	Noncoding	ASGARD: isotig19318
DD4	GT564728	303	Down (24 h)	Noncoding	ASGARD: GFJY65E02GEVR7
DD5	GT564729	286	Down	ORF/unknown function	ASGARD: FZTBZRY02HG02Y *G. bimac* (AK277621.1)
DD8	GT564730	255	Down	ORF/unknown function	ASGARD: GE8SX9M01CAH0Y
DD9	GT564731	342	Down	Noncoding	Matches C3B6 from SSH (Horch et al. 2009)
DD11	GT564732	291	Up	NSH	–
DD12	GT564733	220	Up	NucB1 (DNA-binding function)	*Gryllus bimaculatus* (AK280893.1) ASGARD: isotig 01170, 10068, 01169, 01167
DD13	GT564734	451	Up	NSH	–
DD14	GT564735	328	Down	Enhancer of rudimentary (transcriptional regulator)	*Gryllus bimaculatus* (AK255500.1) ASGARD: isotig06378
DD15	GT564736	234	Down	Noncoding	ASGARD: GFJY65E02HTR24
DD16	GT564737	295	Up	Noncoding	ASGARD: FQTBZRY02GDFH3
DD17	GT564738	531	Up	NSH	
DD19	GT564739	303	Up	NSH	
DD22	GT564740	215	Up	Neuronal synaptobrevin (exocytosis)	*Gryllus bimaculatus* (BAK23254) ASGARD: isotig11254
DD23	GT564741	728	Up	Noncoding	ASGARD: isotig13713
DD27	GT564742	384	Up	Noncoding	ASGARD: GFJY65E01EUTQ3
DD31	GT564743	563	Up	Nuclear hormone receptor-like 39	*Gryllus firmus* (GU479987.1) ASGARD: isotig13512

Most candidates were identified in a screen completed 7 days postdeafferentation, but the first three, DD1, DD3, and DD4, were identified in a 24 h screen. Amplified candidates are listed along with their NCBI accession numbers and their size. Based on the DD gels, candidates are categorized as upregulated (*up*) or downregulated (*down*) 7 days after deafferentation. The description of the best match and the species and accession number or the ASGARD number of that match are given (queried July, 2015). NSH means there were no E-value "hits" lower than 0.001 in either NCBI or ASGARD. If an open reading frame of more than 50 amino acids was predicted, it is noted as ORF/unknown function. Otherwise, it is predicted to be noncoding both in the developing and adult crickets (Maynard et al. 2007). Sema1a is a transmembrane protein that has not yet been well characterized in the cricket. Ongoing work in our lab is assessing the changes in expression of *sema1a* and *2a* over time after deafferentation. Future studies will explore the correlative and causative role of Semas in the injury-induced compensatory plasticity of the auditory system of the cricket.

8.6 Conclusion

The possible conceptual gains that can be gleaned from further study of the molecular basis of compensatory plasticity in the auditory system of the cricket are exciting. Elucidation of the mechanisms at play in this cricket system will offer insight into the general mechanisms that can trigger neuronal growth in adult nervous systems. In addition, further exploration of sexually dimorphic responses to injury in this and other systems will likely reveal insight into the basic principles of how neurons recover from injury. As the rest of this volume highlights, there are a growing number of molecular tools available that can be applied to this problem. The anticipated completion of the genome, and the continued refinement of techniques such as transgenesis and RNAi knockdown, will provide increasingly sophisticated experimental approaches for the study of this phenomenon.

Acknowledgment This project was supported by a grant from the National Institute of General Medical Sciences (8 P20 GM103423-12) from the National Institutes of Health.

References

Bingol B, Schuman EM (2005) Synaptic protein degradation by the ubiquitin proteasome system. Curr Opin Neurobiol 15:536–541

Born D, Rubel E (1985) Afferent influences on brain stem auditory nuclei of the chicken: neuron number and size following cochlea removal. J Comp Neurol 231:435–445

Bowen MF, Bollenbacher WE, Gilbert LI (1984) In vitro studies on the role of the brain and prothoracic glands in the pupal diapause of Manduca Sexta. J Exp Biol 108:9–24

Brodfuehrer PD, Hoy RR (1988) Effect of auditory deafferentation on the synaptic connectivity of a pair of identified interneurons in adult field crickets. J Neurobiol 19:17–38

Buschges A, Ramirez JM, Pearson KG (1992) Reorganization of sensory regulation of locust flight after partial deafferentation. J Neurobiol 23:31–43

Campbell D, Holt C (2001) Chemotropic responses of retinal growth cones mediated by rapid local protein synthesis. Neuron 32:1013–1026

Cayre M, Strambi C, Strambi A, Charpin P, Ternaux JP (2000) Dual effect of ecdysone on adult cricket mushroom bodies. Eur J Neurosci 12:633–642

Cooke BM, Woolley CS (2005) Gonadal hormone modulation of dendrites in the mammalian CNS. J Neurobiol 64:34–46

De Winter F, Oudega M, Lankhorst A, Hamers F, Blits B, Ruitenberg M, Pasterkamp R, Gispen W, Verhaagen J (2002) Injury-induced class 3 semaphorin expression in the rat spinal cord. Exp Neurol 175:61–75

DeVoogd T, Nottebohm F (1981) Gonadal hormones induce dendritic growth in the adult avian brain. Science 214:202–204

Devor M (1975) Neuroplasticity in the rearrangement of olfactory tract fibers after neonatal transection in hamster. J Comp Neurol 166:49–72

Diatchenko L, Lau Y, Campbell A, Chenchik A, Moqadam F, Huang B, Lukyanov S, Lukyanov K, Gurskaya N, Sverdlov E, Siebert P (1996) Suppression subtractive hybridization: a method for generating differentially regulated or tissue-specific cDNA probes and libraries. Proc Natl Acad Sci U S A 93:6025–6030

Easter S, Purves D, Rakic P, Spitzer N (1985) The changing view of neural specificity. Science 230:507–511

Emoto K, He Y, Ye B, Grueber WB, Adler PN, Jan LY, Jan YN (2004) Control of dendritic branching and tiling by the Tricornered-kinase/Furry signaling pathway in Drosophila sensory neurons. Cell 119:245–256

Fenstermaker V, Chen Y, Ghosh A, Yuste R (2004) Regulation of dendritic length and branching by semaphorin 3A. J Neurobiol 58:403–412

Ferguson K, Long H, Cameron S, Chang W-T, Rao Y (2009) The conserved Ig superfamily member Turtle mediates axonal tiling in Drosophila. J Neurosci 29:14151–14159

Genikhovich G, Kurn U, Hemmrich G, Bosch T (2006) Discovery of genes expressed in Hydra embryogenesis. Dev Biol 289:466–481

Gontheir B, Koncina E, Satkauskas S, Perraut M, Roussel G, Aunis D, Kapfhammer J, Bagnard D (2009) A PKC-dependent recruitment of MMP-2 controls semaphorin-3A growth-promoting effect in cortical dendrites. PLoS ONE 4, e5099

Groh C, Meinertzhagen IA (2010) Brain plasticity in Diptera and Hymenoptera. Front Biosci (Scholar Ed) 2:268

Grueber WB, Graubard K, Truman JW (2001) Tiling of the body wall by multidendritic sensory neurons in Manduca sexta. J Comp Neurol 440:271–283

Grueber WB, Jan LY, Jan YN (2002) Tiling of the Drosophila epidermis by multidendritic sensory neurons. Development 129:2867–2878

Hall GF, Cohen MJ (1988) The pattern of dendritic sprouting and retraction induced by axotomy of lamprey central neurons. J Neurosci 8:3584–3597

Harel NY, Strittmatter SM (2006) Can regenerating axons recapitulate developmental guidance during recovery from spinal cord injury? Nat Rev Neurosci 7:603–616

Hegde AN, DiAntonio A (2002) Ubiquitin and the synapse. Nat Rev Neurosci 3:854–861

Hickmott PW, Steen PA (2005) Large-scale changes in dendritic structure during reorganization of adult somatosensory cortex. Nat Neurosci 8:140–142

Horch HW, McCarthy SS, Johansen SL, Harris JM (2009) Differential gene expression during compensatory sprouting of dendrites in the auditory system of the cricket Gryllus bimaculatus. Insect Mol Biol 18:483–496

Horch HW, Sheldon E, Cutting CC, Williams CR, Riker DM, Peckler HR, Sangal RB (2011) Bilateral consequences of chronic unilateral deafferentation in the auditory system of the cricket gryllus bimaculatus. Dev Neurosci 33:21–37

Hoy RR, Nolen TG, Casaday GC (1985) Dendritic sprouting and compensatory synaptogenesis in an identified interneuron following auditory deprivation in a cricket. Proc Natl Acad Sci U S A 82:7772–7776

Ing B, Shteiman-Kotler A, Castelli M, Henry P, Pak Y, Stewart B, Boulianne G, Rotin D (2007) Regulation of commissureless by the ubiquitin ligase DNedd4 is required for neuromuscular synaptogenesis in Drosophila melanogaster. Mol Cell Biol 27:481–496

Jacobsson G, Piehl F, Meister B (1998) VAMP-1 and VAMP-2 gene expression in rat spinal motoneurones: differential regulation after neuronal injury. Eur J Neurosci 10:301–316

Jan YN, Jan L (2003) The control of dendrite development. Neuron 40:229–242

Kim S, Chiba A (2004) Dendritic guidance. Trends Neurosci 27:194–202
Kraft R, Levine RB, Restifo LL (1998) The steroid hormone 20-hydroxyecdysone enhances neurite growth of Drosophila mushroom body neurons isolated during metamorphosis. J Neurosci 18:8886–8899
Lakes R, Kalmring K, Engelhard KH (1990) Changes in the auditory system of locusts (Locusta migratoria and Schistocerca gregaria) after deafferentation. J Comp Physiol A 166:553–563
Liang P, Pardee AB (1992) Differential display of eukaryotic messenger RNA by means of the polymerase chain reaction. Science 257:967–971
Lomax M, Huang L, Cho Y, Gong T-W, Altschuler R (2000) Differential display and gene arrays to examine auditory plasticity. Hear Res 147:293–302
Luo L (2000) Rho GTPases in neuronal morphogenesis. Nat Rev Neurosci 1:173–180
Martinez-Arca S, Alberts P, Zahraoui A, Louvard D (2000) Role of tetanus neurotoxin insensitive vesicle-associated membrane protein (TI-VAMP) in vesicular transport mediating neurite outgrowth. J Cell Biol 149:889–899
Matheson SF, Levine RB (1999) Steroid hormone enhancement of neurite outgrowth in identified insect motor neurons involves specific effects on growth cone form and function. J Neurobiol 38:27–45
Matsumoto Y (2004) Context-dependent olfactory learning in an insect. Learn Mem 11:288–293
Matsumoto Y, Mizunami M (2000) Olfactory learning in the cricket Gryllus bimaculatus. J Exp Biol 203:2581–2588
Matsumoto Y, Mizunami M (2002) Lifetime olfactory memory in the cricket Gryllus bimaculatus. J Comp Physiol A 188:295–299
Matsumoto Y, Noji S, Mizunami M (2003) Time course of protein synthesis-dependent phase of olfactory memory in the cricket Gryllus bimaculatus. Zool Sci 20:409–416
Matsumoto Y, Unoki S, Aonuma H, Mizunami M (2006) Critical role of nitric oxide-cGMP cascade in the formation of cAMP-dependent long-term memory. Learn Mem 13:35–44
Maynard KR, McCarthy SS, Sheldon E, Horch HW (2007) Developmental and adult expression of semaphorin 2a in the cricket Gryllus bimaculatus. J Comp Neurol 503:169–181
McLean M, Edwards JS (1976) Target discrimination in regenerating insect sensory nerve. J Embryol Exp Morphol:36:19–39
Meinertzhagen IA (2001) Plasticity in the insect nervous system. In: Evans PD (ed) Advances in insect physiology, vol 28. Academic, New York, pp 84–167
Miele G, MacRae L, McBride D, Manson J, Clinton M (1998) Elimination of false positives generated through PCR re-amplification of differential display cDNA. BioTechniques 25:138–144
Morita A (2006) Regulation of dendritic branching and spine maturation by semaphorin3A-Fyn signaling. J Neurosci 26:2971–2980
Murphey RK (1986) The myth of the inflexible invertebrate: competition and synaptic remodelling in the development of invertebrate nervous systems. J Neurobiol 17:585–591
Murphey RK, Mendenhall B, Palka J, Edwards JS (1975) Deafferentation slows the growth of specific dendrites of identified giant interneurons. J Comp Neurol 159:407–418
Murphey RK, Johnson SE, Walthall WW (1981) The effects of transplantation and regeneration of sensory neurons on a somatotopic map in the cricket central nervous system. Dev Biol 88:247–258
Murphey RK, Bacon JP, Sakaguchi DS, Johnson SE (1983a) Transplantation of cricket sensory neurons to ectopic locations: arborizations and synaptic connections. J Neurosci 3:659–672
Murphey RK, Johnson SE, Sakaguchi DS (1983b) Anatomy and physiology of supernumerary cercal afferents in crickets: implications for pattern formation. J Neurosci 3:312–325
Myat A, Henry P, McCabe V, Flintoft L, Rotin D, Tear G (2002) Drosophila Nedd4, a ubiquitin ligase, is recruited by commissureless to control cell surface levels of the roundabout receptor. Neuron 35:447–459
Nakamura F, Ugahin K, Yamashita N, Okada T, Uchida Y, Taniguchi M, Ohshima T, Goshima Y (2009) Increased proximal bifurcation of CA1 pyramidal apical dendrites in sema3A mutant mice. J Comp Neurol 516:360–375

Nordeen KW, Killackey HP, Kitzes LM (1983) Ascending projections to the inferior colliculus following unilateral cochlear ablation in the neonatal gerbil, Meriones unguiculatus. J Comp Neurol 214:144–153

Palka J (1984) Precision and plasticity in the insect nervous system. Trends Neurosci 7:455–456

Palka J, Edwards JS (1974) The cerci and abdominal giant fibres of the house cricket, Acheta domesticus. II. Regeneration and effects of chronic deprivation. Proc R Soc B Biol Sci 185:105–121

Pallas SL, Hoy RR (1986) Regeneration of normal afferent input does not eliminate aberrant synaptic connections of an identified auditory interneuron in the cricket, Teleogryllus oceanicus. J Comp Neurol 248:348–359

Parks T (1979) Afferent influences on the development of the brain stem auditory nuclei of the chicken: otocyst ablation. J Comp Neurol 183:665–678

Pasterkamp RJ, Giger RJ (2009) Semaphorin function in neural plasticity and disease. Curr Opin Neurobiol 19:263–274

Pfister A, Johnson A, Ellers O, Horch HW (2013) Quantification of dendritic and axonal growth after injury to the auditory system of the adult cricket Gryllus bimaculatus. Front Physiol 3:367

Polleux F, Morrow T, Ghosh A (2000) Semaphorin 3A is a chemoattractant for cortical apical dendrites. Nature 404:567–573

Prugh J, Croce Della K, Levine RB (1992) Effects of the steroid hormone, 20-hydroxyecdysone, on the growth of neurites by identified insect motoneurons in vitro. Dev Biol 154:331–347

Purves D (1976) Competitive and non-competitive re-innervation of mammalian sympathetic neurones by native and foreign fibres. J Physiol 261:453–475

Purves D, Hume RI (1981) The relation of postsynaptic geometry to the number of presynaptic axons that innervate autonomic ganglion cells. J Neurosci 1:441–452

Raghuram H, Deb R, Nandi D, Balakrishnan R (2015) Silent katydid females are at higher risk of bat predation than acoustically signalling katydid males. Proc R Soc B Biol Sci 282:20142319

Roederer E, Cohen MJ (1983) Regeneration of an identified central neuron in the cricket. J Neurosci 3:1835–1847

Sanes JR, Lichtman JW (1999) Development of the vertebrate neuromuscular junction. Annu Rev Neurosci 22:389–442

Schildberger K, Wohlers DW, Schmitz B, Kleindienst HU, Huber F (1986) Morphological and physiological changes in central auditory neurons following unilateral foreleg amputation in larval crickets. J Comp Physiol A 158:291–300

Schmitz B (1989) Neuroplasticity and phonotaxis in monaural adult female crickets (Gryllus bimaculatus de Geer). J Comp Physiol A: Neuroethol 164:343–358

Schmitz B, Kleindienst HU, Schildberger K, Huber F (1988) Acoustic orientation in adult, female crickets (Gryllus- Bimaculatus Degeer) after unilateral foreleg amputation in the larva. J Comp Physiol A-Sens Neural Behav Physiol 162:715–728

Schneider GE (1973) Early lesions of superior colliculus: factors affecting the formation of abnormal retinal projections. Brain Behav Evol 8:73–109

Scotto-Lomassese S, Strambi C, Strambi A, Aouane A, Augier R, Rougon G, Cayre M (2003) Suppression of adult neurogenesis impairs olfactory learning and memory in an adult insect. J Neurosci 23:9289–9296

Selverston AI, Kleindienst HU, Huber F (1985) Synaptic connectivity between cricket auditory interneurons as studied by selective photoinactivation. J Neurosci 5:1283–1292

Shankland M, Bentley D, Goodman CS (1982) Afferent innervation shapes the dendritic branching pattern of the Medial Giant Interneuron in grasshopper embryos raised in culture. Dev Biol 92:507–520

Shirasu M, Kimura K, Kataoka M, Takahashi M, Okajima S, Kawaguchi S, Hirasawa Y, Ide C, Mizoguchi A (2000) VAMP-2 promotes neurite elongation and SNAP-25A increases neurite sprouting in PC12 cells. Neuroscience 37:265–275

Soba P, Zhu S, Emoto K, Younger S, al E (2007) Drosophila sensory neurons require Dscam for dendritic self-avoidance and proper dendritic field organization. Neuron

Sugimura K, Yamamoto M, Niwa R, Satoh D, Goto S, Taniguchi M, Hayashi S, Uemura T (2003) Distinct developmental modes and lesion-induced reactions of dendrites of two clases of Drosophila sensory neurons. J Neurosci 23:3752–3760

Sutton MA, Schuman EM (2005) Local translational control in dendrites and its role in long-term synaptic plasticity. J Neurobiol 64:116–131

Tailby C, Wright LL, Metha AB, Calford MB (2005) Activity dependent maintenance and growth of dendrites in adult cortex. Proc Natl Acad Sci U S A 102:4631–4636

Tessier-Lavigne M, Goodman CS (1996) The molecular biology of axon guidance. Science 274:1123–1133

Tran TS, Rubio ME, Clem RL, Johnson D, Case L, Tessier-Lavigne M, Huganir RL, Ginty DD, Kolodkin AL (2009) Secreted semaphorins control spine distribution and morphogenesis in the postnatal CNS. :1–7

Truman JW, Reiss SE (1988) Hormonal regulation of the shape of identified motoneurons in the moth Manduca sexta. J Neurosci 8:765–775

Truman JW, Reiss SE (1995) Neuromuscular metamorphosis in the moth Manduca sexta: hormonal regulation of synapses loss and remodeling. J Neurosci 15:4815–4826

Trune DR (1982a) Influence of neonatal cochlear removal on the development of mouse cochlear nucleus: I. Number, size, and density of its neurons. J Comp Neurol 209:409–424

Trune DR (1982b) Influence of neonatal cochlear removal on the development of mouse cochlear nucleus: II. Dendritic morphometry of its neurons. J Comp Neurol 209:425–434

Vargas ME, Barres BA (2007) Why is Wallerian degeneration in the CNS so slow? Annu Rev Neurosci 30:153–179

Wan JS, Erlander MG (1997) Cloning differentially expressed genes by using differential display and subtractive hybridization. Methods Mol Biol 85:45–68

Weeks JC, Ernst-Utzschneider K (1989) Respecification of larval proleg motoneurons during metamorphosis of the tobacco hornworm, Manduca sexta: segmental dependence and hormonal regulation. J Neurobiol 20:569–592

Williams D (2005) Remodeling dendrites during insect metamorphosis – Williams – 2005 – Journal of Neurobiology – Wiley Online Library. J Neurobiol

Williams DW, Truman JW (2005) Remodeling dendrites during insect metamorphosis. J Neurobiol 64:24–33

Wohlers DW, Huber F (1982) Processing of sound signals by 6 types of neurons in the prothoracic ganglion of the cricket, Gryllus-campestris. J Comp Physiol 146:161–173

Wolf H, Büschges A (1997) Plasticity of synaptic connections in sensory-motor pathways of the adult locust flight system. J Neurophysiol 78:1276–1284

Xu X-ZS, Wes P, Chen H, Li H-S, Yu M, Morgan S, Liu Y, Montell C (1998) Retinal targets for calmodulin include proteins implicated in synaptic transmission. J Biol Chem 47:31297–31307

Yamashita N, Morita A, Uchida Y, Nakamura F, Usui H, Ohshima T, Taniguchi M, Honnorat J, Thomasset N, Takei K, Takahashi T, Kolattukudy P, Goshima Y (2007) Regulation of spine development by semaphorin3A through cyclin-dependent kinase 5 phosphorylation of collapsin response mediator protein 1. J Neurosci 27:12546–12554

Zeng V, Extavour CG (2012) ASGARD: an open-access database of annotated transcriptomes for emerging model arthropod species. Database (Oxford) 2012:bas048

Zeng V, Ewen-Campen B, Horch HW, Roth S, Mito T, Extavour CG (2013) Developmental gene discovery in a hemimetabolous insect: de novo assembly and annotation of a transcriptome for the cricket Gryllus bimaculatus Dearden PK, ed. PLoS ONE 8, e61479

Zera AJ, Zhao Z, Kaliseck K (2007) Hormones in the field: evolutionary endocrinology of Juvenile Hormone and Ecdysteroids in field populations of the wing-dimorphic cricket Gryllus firmus. Physiol Biochem Zool 80:592–606

Chapter 9
Learning and Memory

Makoto Mizunami and Yukihisa Matsumoto

Abstract Crickets have excellent capabilities for olfactory and visual learning and thus are useful organisms in which to study the mechanisms of learning and memory. Our studies on crickets have revealed detailed information about signaling cascades underlying long-term memory (LTM) formation, namely, that the serial activation of the NO-cGMP system, cyclic nucleotide-gated (CNG) channel, the calcium/calmodulin system, and cAMP-protein kinase A (PKA) underlies LTM formation. Our studies also suggest that octopaminergic (OA-ergic) and dopaminergic (DA-ergic) neurons convey information about appetitive or aversive unconditioned stimuli (US), respectively, in conditioning of odors, visual patterns, and color cues. Our studies also suggest that activation of OA-ergic and DA-ergic neurons is needed for retrieval of appetitive and aversive memory, respectively, in olfactory learning and visual learning. Many of these findings differ from those reported for the fruit fly *Drosophila*, suggesting unexpected diversity in the mechanisms of learning and memory in different species of insects. Studies of the functional significance and underlying evolutionary history of such diversity should emerge as important areas of research. Recently, new techniques such as RNA interference and transgenesis have been successfully applied to crickets, which should help deepen the study of the cellular and molecular mechanisms of learning and memory in crickets.

Keywords *Gryllus bimaculatus* • Olfactory learning • Visual learning • NO signaling • Octopamine • Dopamine • Classical conditioning • Long-term memory

---

M. Mizunami (✉)
Faculty of Science, Hokkaido University, Sapporo 060-0810, Japan
e-mail: mizunami@sci.hokudai.ac.jp

Y. Matsumoto
College of Liberal Arts and Science, Tokyo Medical and Dental University, 2-8-30 Kounodai, Ichikawa 272-0827, Japan
e-mail: yukihisa.las@tmd.ac.jp

H.W. Horch et al. (eds.), *The Cricket as a Model Organism*,
DOI 10.1007/978-4-431-56478-2_9

9.1 Introduction

Insects are useful organisms for the study of the neural mechanisms of learning and memory. This is because they exhibit a rich variety of learning, and because their brains, which we refer to as microbrains (Mizunami et al. 1999, 2004), are accessible to detailed experimental analysis. Previously, most studies on the mechanisms of learning and memory in insects have been performed on only two species of insects, namely, the fruit fly *Drosophila melanogaster* (Davis 2005, 2011) and the honeybee *Apis mellifera* (Menzel and Giurfa 2006). Recently, we demonstrated that the cricket *Gryllus bimaculatus* and the cockroach *Periplaneta americana* (Mizunami et al. 1998b; Watanabe et al. 2011) are also useful for exploring the mechanisms of learning and memory. In this chapter we review recent progress from our studies on olfactory and visual learning in crickets.

9.2 Olfactory Learning in Crickets

We used a "classical conditioning and operant testing procedure" in our conditioning experiments (Matsumoto and Mizunami 2002a; Mizunami and Matsumoto 2010; see Chap. 17). For olfactory conditioning, a filter paper soaked with an odor (conditioned stimulus, CS) was brought near the antennae of the cricket, and then a drop of water or sodium chloride solution (appetitive or aversive unconditioned stimulus, US) was applied to the mouth. In the operant odor preference test, crickets were individually placed in a test chamber and allowed to visit two odor sources on the floor (e.g., banana and apple odor sources). The time that the crickets explored each odor source with the mouth or palpi was measured for evaluating relative odor preference of the crickets. Similar procedures were employed for conditioning of visual patterns (Unoki et al. 2006) and color cues (Nakatani et al. 2009).

The first form of learning we studied in crickets was olfactory learning. We found that one conditioning trial was sufficient to establish a memory lasting for several hours (midterm memory, MTM) in appetitive olfactory conditioning (Fig. 9.1a). Two conditioning trials (with an intertrial interval (ITI) of 5 min) induced memory that lasted for at least 1 day (Unoki et al. 2005), which was characterized as protein-synthesis-dependent long-term memory (LTM) (Matsumoto et al. 2003). In aversive olfactory conditioning, two trials were sufficient to establish 30-min retention, but six trials (with a 5-min ITI) were needed to establish 1-day retention (Unoki et al. 2005). Based on the results of subsequent studies, we concluded that memory after aversive learning is less durable than after appetitive learning when learning odors, visual patterns (Unoki et al. 2006), or color cues (Nakatani et al. 2009).

Subsequently, we showed that crickets are capable of (1) retaining memory for life (Matsumoto and Mizunami 2002b), (2) simultaneously memorizing seven odor

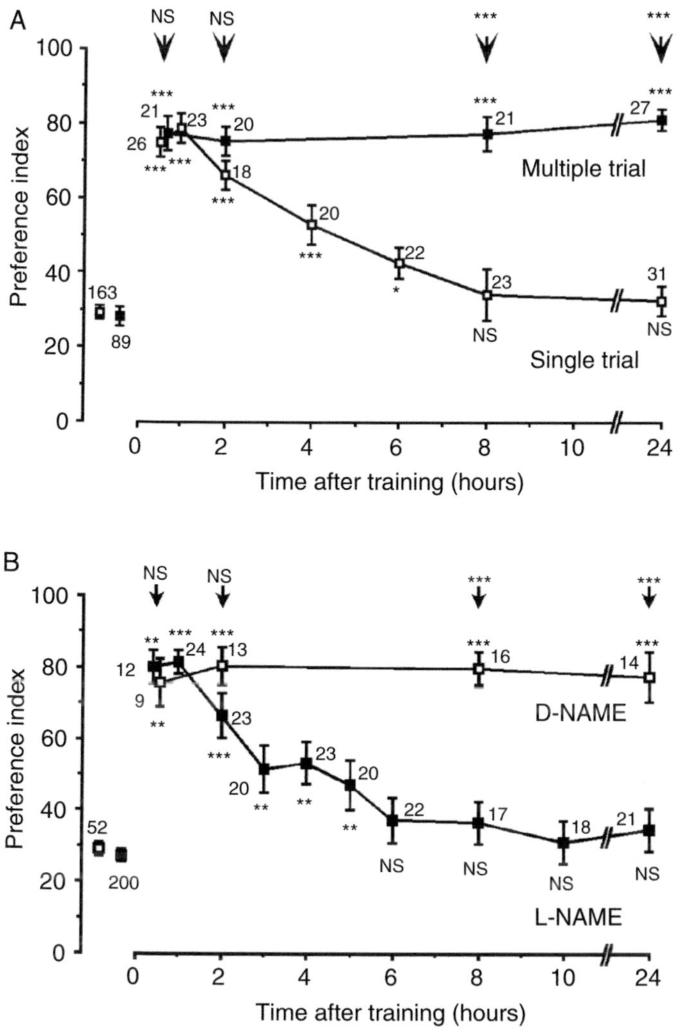

Fig. 9.1 (**a**) Memory retention after single- and multiple-trial appetitive olfactory conditioning. Seven groups of animals were subjected to single-trial conditioning (*open squares*) and another four groups were subjected to four-trial conditioning, with an ITI of 5 min (*black squares*). (**b**) Effects of L-NAME, an inhibitor of NO synthase, or D-NAME, a noneffective isomer, on LTM formation. Prior to the four-trial conditioning, animals in ten groups were each injected with 3 μl saline containing 400 μM L-NAME (*black squares*), and animals in another four control groups were each injected with 3 μl saline containing 400 μM D-NAME (*open squares*). Odor preference tests were given to animals before and at various times after conditioning. Preference indices (PIs) for the rewarded odor are shown as means ± SEM. PIs before conditioning are shown as pooled data for each category of animal groups. Statistical comparisons of odor preferences were made before and after conditioning for each group (Wilcoxon's test) and between single- and multiple-trial groups at each time after conditioning (Mann-Whitney test). The results are shown at each data point and above the *arrow*, respectively (*$p < 0.05$; **$p < 0.01$; ***$p < 0.001$; NS $p > 0.05$). The number of animals is shown at each data point (Modified from Matsumoto et al. 2006)

pairs (Matsumoto and Mizunami 2006), and (3) performing context-dependent discrimination learning, i.e., selecting one of a pair of odors and avoiding the other in one context and the opposite pairing in another context (Matsumoto and Mizunami 2004). Moreover, we found that crickets exhibit second-order conditioning, i.e., crickets that had been subjected to pairing of a stimulus (CS1) and a US and then subjected to pairing of another stimulus (CS2) and the CS1 exhibited conditioned responses to CS2, although they had never experienced pairing of the CS2 with the US (Mizunami et al. 2009). Therefore, although our current research focuses on the mechanisms of elemental associative learning between CS and US, crickets may emerge as organisms to study the mechanisms of sophisticated forms of associative learning.

9.3 Role of the NO-cGMP System in Formation of LTM

Nitric oxide (NO) is a membrane-permeable molecule that functions in intercellular signaling. It is produced by NO synthase (NOS), diffuses into neighboring cells, and stimulates soluble guanylyl cyclase (sGC) to produce cyclic GMP (cGMP). Studies on honeybees have suggested that the NO-cGMP signaling system and cAMP system act in parallel and complementarily for the formation of LTM (Müller 2000). Our studies on crickets, however, brought us to a different conclusion (Matsumoto et al. 2006, 2009). Multiple (two or more) appetitive olfactory conditioning trials led to LTM that lasted for at least 1 day in crickets. On the other hand, memory induced by single-trial appetitive conditioning decayed within several hours (Fig. 9.1a). Injection of inhibitors of the enzyme catalyzing the formation of NO, cGMP, or cAMP into the hemolymph prior to multiple-trial conditioning blocked formation of LTM, as is shown in the example in Fig. 9.1b. In contrast, injection of an NO donor, a cGMP analog, or a cAMP analog prior to single-trial conditioning induced LTM, suggesting participation of the NO-cGMP system and cAMP system in LTM formation. Induction of LTM by injection of an NO donor or a cGMP analog paired with single-trial conditioning was blocked by inhibition of the cAMP system. However, induction of LTM by a cAMP analog was unaffected by inhibition of the NO-cGMP system. The results suggest that the cAMP pathway is a downstream target of the NO-cGMP pathway for LTM formation. We also obtained evidence suggesting that cyclic nucleotide-gated (CNG) channels and calcium-calmodulin intervene between the NO-cGMP system and the cAMP system. We have thus proposed that serial activation of the NO-cGMP system, CNG channel, and calcium-calmodulin and cAMP systems underlies formation of LTM in crickets (Fig. 9.2).

Further, we have found that RNA interference (RNAi) of the NOS gene impairs LTM formation in crickets (Takahashi et al. 2009). Crickets injected with double-stranded RNA (dsRNA) into the hemolymph 2 days before conditioning exhibited impairment of 1-day memory retention, although 30-min retention was intact. In

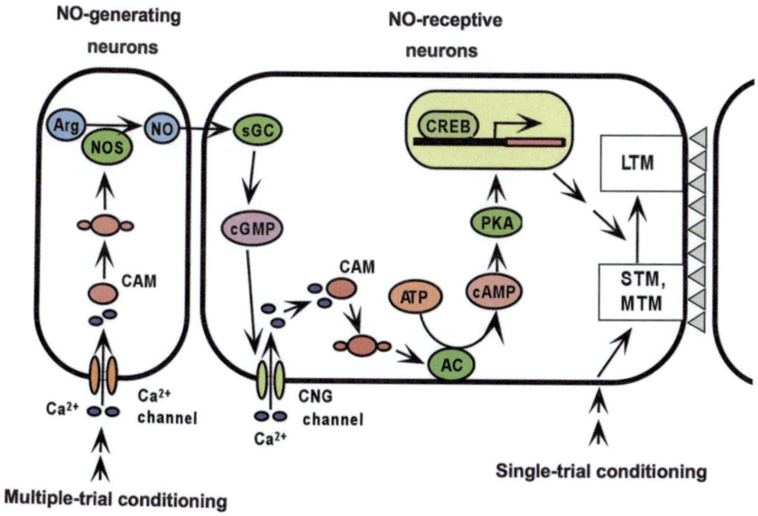

Fig. 9.2 A model of signaling cascades underlying LTM formation in crickets, proposing a serial arrangement of the NO-cGMP system and the cAMP system for LTM formation. Single-trial conditioning induces synaptic plasticity of limited durability, which is thought to underlie short-term memory and midterm memory. Multiple-trial conditioning activates the NO-cGMP system, and this in turn activates adenylyl cyclase (*AC*) and then PKA, via the cyclic nucleotide-gated (*CNG*) channel and calcium calmodulin (*CAM*) system. Activation of PKA is assumed to activate a transcription factor, cAMP-responsive element-binding protein (*CREB*), which leads to protein synthesis that is necessary to achieve long-term plasticity of synaptic connection upon other neurons. *Arg* arginine, *NOS* NO synthase, *sGC* soluble guanylyl cyclase (Modified from Matsumoto et al. 2009)

situ hybridization demonstrated a high level of expression of NOS mRNA in one class of Kenyon cells (intrinsic neurons) of the mushroom body, in addition to some neurons around the antennal lobe (primary olfactory center) and the optic lobe (visual center). The mushroom body is a multisensory association center of the insect brain (Mizunami et al. 1998a, b) and participates in olfactory learning in honeybees (Menzel and Giurfa 2006), fruit flies (Davis 2011), and cockroaches (Watanabe et al. 2011). RNAi will likely become a useful method for study of the mechanisms of learning and memory in crickets.

Interestingly, despite the accumulation of information on the molecular and neuronal mechanisms of LTM formation in *Drosophila*, there have been no reports suggesting participation of NO in LTM formation in this species (Davis 2005, 2011). Moreover, we also obtained evidence showing that the cAMP system does not participate in the formation of short-term memory (STM) in olfactory learning in crickets (Matsumoto et al. 2006), in contrast to the well-established fact that the cAMP system plays critical roles in STM formation in olfactory learning in *Drosophila* (Davis 2005). We thus suggest that there is a diversity in the molecular mechanisms of learning and memory in different insects.

9.4 Roles of OA-ergic and DA-ergic Neurons in Olfactory Memory Formation

We studied the roles of octopaminergic (OA-ergic) neurons and dopaminergic (DA-ergic) neurons in appetitive and aversive olfactory conditioning in crickets (Unoki et al. 2005). In mammals, midbrain DA-ergic neurons convey appetitive and aversive signals in various forms of learning (Schultz 2006). In insects, earlier studies suggested that OA- and DA-ergic neurons play roles in appetitive and aversive olfactory conditioning, respectively, in honeybees (Hammer and Menzel 1998; Farooqui et al. 2003) and the fruit fly *Drosophila* (Schwaerzel et al. 2003), although recent studies on *Drosophila* have suggested that DA-ergic neurons participate in both appetitive and aversive learning, as will be discussed later.

We found that crickets injected with an OA receptor antagonist (epinastine or mianserin) into the hemolymph before conditioning exhibited an impairment of appetitive conditioning to an odor with water reward. In contrast, these animals exhibited no impairment of aversive conditioning to an odor with sodium chloride punishment. The latter finding indicates that OA receptor antagonists do not impair sensory function, motor function or the motivation necessary for learning. We thus conclude that OA-ergic neurons are specifically involved in conveying water reward. We also found that injection of a DA receptor antagonist (fluphenazine, chlorpromazine, or spiperone) impaired aversive learning with sodium chloride punishment but not appetitive learning with a water reward. We thus conclude that DA-ergic neurons are specifically involved in conveying sodium chloride punishment. Overall, we can conclude that OA- and DA-ergic neurons convey information about appetitive and aversive US, respectively, in olfactory conditioning in crickets.

9.5 Roles of OA-ergic and DA-ergic Neurons in Formation of Memory for Visual Patterns and Color Cues

We next studied the roles of OA-ergic and DA-ergic neurons in appetitive and aversive conditioning of a visual pattern (Unoki et al. 2006) and a color cue (Nakatani et al. 2009). Crickets injected with an OA receptor antagonist (epinastine or mianserin) into the hemolymph before visual pattern conditioning exhibited an impairment of appetitive learning, whereas aversive learning of a visual pattern was unaffected (Unoki et al. 2006). In contrast, a DA receptor antagonist (fluphenazine, chlorpromazine, or spiperone) impaired aversive learning but not appetitive learning. Similarly, crickets injected with an OA receptor antagonist into the hemolymph before color conditioning exhibited an impairment of appetitive learning without any effect on aversive color learning (Nakatani et al. 2009). In contrast, injection of a DA receptor antagonist into the hemolymph impaired aversive color learning but had no effect on appetitive color learning. These results indicate that the roles of

OA-ergic and DA-ergic neurons in conveying information about appetitive and aversive US, respectively, are ubiquitous in learning of odor, visual pattern, and color stimuli. OA-ergic and DA-ergic neurons may serve as general reward and punishment systems, respectively, for learning in crickets.

Recent studies on *Drosophila*, on the other hand, have suggested that different classes of DA-ergic neurons mediate reward and punishment in olfactory conditioning (Liu et al. 2012; Burke et al. 2012), a finding fundamentally different from that in crickets. This strengthens our suggestion that there is fundamental diversity in the mechanisms of learning and memory in insects.

9.6 Participation of OA-ergic Neurons and DA-ergic Neurons in Appetitive and Aversive Memory Retrieval

We then studied the roles of OA-ergic and DA-ergic neurons in appetitive and aversive memory retrieval (Mizunami et al. 2009). Crickets were subjected to appetitive or aversive olfactory conditioning. Then they were injected with OA or DA receptor antagonists before a retention test. Injection of an OA receptor antagonist completely impaired appetitive olfactory memory retrieval but had no effect on aversive olfactory memory retrieval (Fig. 9.3a). In contrast, injection of a DA receptor antagonist completely impaired aversive memory retrieval but had no effect on appetitive memory retrieval (Fig. 9.3b). Moreover, we observed that injection of an OA and DA receptor antagonist before the retention test impaired appetitive and aversive memory retrieval, respectively, in visual pattern learning. Therefore, we concluded that participation of OA- and DA-ergic neurons in the retrieval of appetitive memory and aversive memory, respectively, is ubiquitous in learning of odors and visual patterns. This differs from reports on *Drosophila* that impairment of OA- or DA-ergic transmission had no effect on memory retrieval (Schwaerzel et al. 2003; Liu et al. 2012; Burka et al. 2012).

9.7 Proposal of a New Model of Classical Conditioning in Insects

Findings in crickets described above were not consistent with conventional neural models of classical conditioning in *Drosophila*. Figure 9.4a illustrates a model proposed by Schwaerzel et al. (2003) to account for the roles of extrinsic and intrinsic neurons of the mushroom body in appetitive or aversive olfactory conditioning in *Drosophila*. The model assumes, at first, that the "CS" neurons (Kenyon cells of the mushroom body) that convey information of the CS make synaptic connections with dendrites of "CR" neurons (efferent (output) neurons in the lobes of the mushroom body). A conditioned response (CR) that mimics an

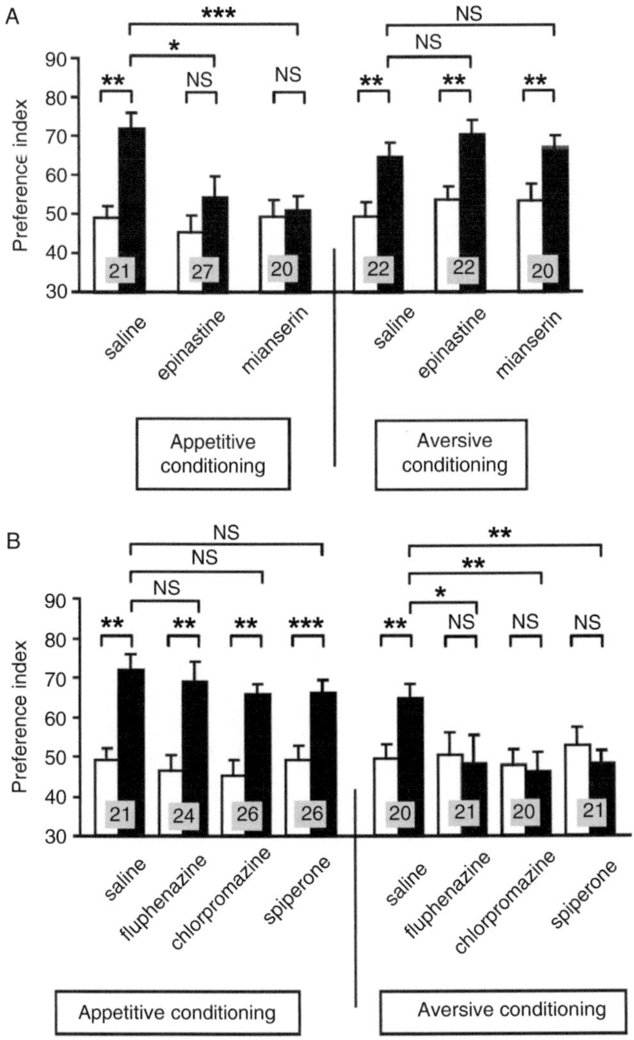

Fig. 9.3 OA and DA receptor antagonists impair appetitive and aversive olfactory memory retrieval, respectively. Effects of OA (**a**) and DA (**b**) receptor antagonists on olfactory memory retrieval. Fourteen groups of crickets were subjected to appetitive (*left*) or aversive (*right*) olfactory conditioning trials. The next day, each group was injected with 3 μl of saline or saline containing 1 μM epinastine, 1 μM mianserin, 500 μM fluphenazine, 500 μM chlorpromazine, or 500 μM spiperone before the final test. Preference indices for the rewarded odor (in the case of appetitive conditioning) or unpunished control odor (in the case of aversive conditioning) before (*white bars*) and 1 day after (*black bars*) conditioning are shown with means + SEM. The results of statistical comparison before and after conditioning (Wilcoxon's test) and between experimental and saline-injected control groups (Mann-Whitney test) are shown as *asterisks* ($*p < 0.05$; $**p < 0.01$; $***p < 0.001$, NS $p > 0.05$). The number of crickets is shown at each data point (Modified from Mizunami et al. 2009)

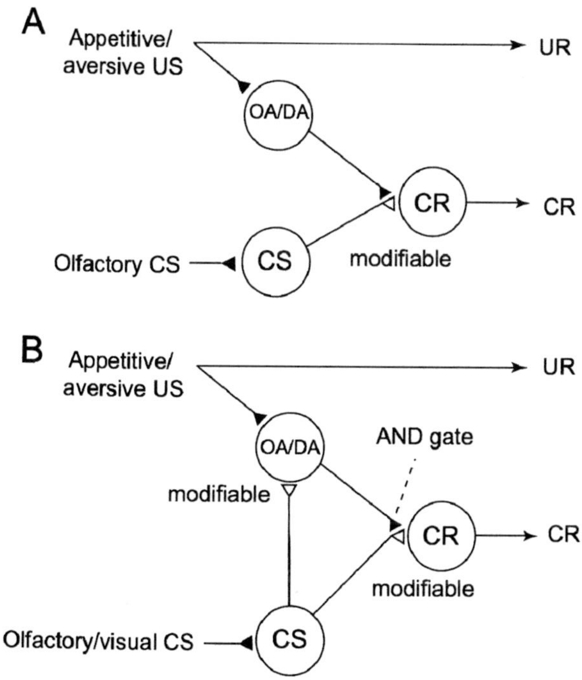

Fig. 9.4 Conventional and new models of classical conditioning in insects. (**a**) A model proposed by Schwaerzel et al. (2003) to account for the roles of intrinsic and extrinsic neurons of the mushroom body in olfactory conditioning in *Drosophila*. OA-ergic and DA-ergic neurons ("OA/DA" neurons) project to the lobes of the mushroom body and convey information about appetitive and aversive US, respectively. "CS" neurons, which are Kenyon cells of the mushroom body and convey information about the CS, make synaptic connections with "CR" neurons, which are efferent neurons of the lobes. "CR" neurons are assumed to induce a conditioned response (CR), the efficacy of the connection being strengthened by conditioning. "OA/DA" neurons make synaptic connections with axon terminals of "CS" neurons. (**b**) Our new model of classical conditioning. In the model, it is assumed that coincident activation of "OA/DA" neurons and "CS" neurons is needed to activate "CR" neurons to lead to a CR (AND gate). It is also assumed that conditioning strengthens the efficacy of synaptic transmission from "CS" neurons to "OA/DA" neurons (Modified from Mizunami et al. 2009)

unconditioned response (UR) can activate these output neurons, but these synaptic connections are weak or silent before conditioning. Secondly, it is assumed that OA- and DA-ergic efferent neurons projecting to the lobes ("OA/DA" neurons), which convey information about appetitive and aversive US, respectively, make synaptic connections with axon terminals of "CS" neurons. Thirdly, it is assumed that the efficacy of synaptic transmission from "CS" neurons to "CR" neurons, which induces a CR, is strengthened by coincident activation of "CS" neurons and "OA/DA" neurons in conditioning. This model matches our finding that activation of OA- or DA-ergic neurons is needed for memory acquisition. However, it does not account for our finding that activation of these neurons is needed for memory retrieval.

We have, therefore, proposed a new model (Fig. 9.4b, Mizunami et al. 2009), which minimally modifies the conventional model. In our model, it is assumed, at first, that activation of "OA/DA" neurons is needed to "gate" the synaptic pathway from "CS" neurons to "CR" neurons after conditioning. Secondly, it is assumed that synaptic connections from "CS" neurons to "OA/DA" neurons, which encode appetitive/aversive US, are strengthened by coincident activation of "CS" neurons and "OA/DA" neurons by pairing of a CS with a US. In short, this model assumes formation of two kinds of memory traces by conditioning. Results of our pharmacological analysis coupled with a second-order conditioning procedure confirmed predictions from the model (Mizunami et al. 2009). Moreover, this model provides a framework to explain neural mechanisms of sensory preconditioning, a higher-order learning phenomenon (Matsumoto et al. 2013).

9.8 Diversity in the Mechanisms of Learning and Memory in Insects

Our studies on crickets suggest that there are some fundamental differences in the basic mechanisms of learning and memory between the cricket and the fruit fly. Such differences are summarized in Table 9.1. It could be argued that some of these differences might be due to differences in experimental approach (i.e., pharmacology in crickets, genetic manipulation in flies), but it is difficult to believe that methodological differences could account for all of the distinctions noted. One of our research goals is to confirm such diversity and to evaluate its functional significance and underlying evolutionary history. In conclusion, studies on crickets, as well as other species of insects such as fruit flies, honeybees, moths, and cockroaches, promise to yield a better understanding of the diversity and evolution of brain mechanisms underlying learning and memory in insects.

Table 9.1 Proposed differences in the mechanisms of learning and memory in crickets and fruit flies

	Cricket *Gryllus bimaculatus*	Fruit fly *Drosophila melanogaster*
Conditioning	OA or DA neurons participate in appetitive or aversive conditioning, respectively	DA neurons participate in both appetitive and aversive conditioning
Memory retrieval	OA or DA neurons participate in appetitive or aversive memory retrieval, respectively	OA or DA neurons do not participate in memory retrieval
STM formation	The cAMP system does not participate in STM formation	The cAMP system participates in STM formation
LTM formation	The NO-cGMP system and cAMP system participate in LTM formation	The cAMP system, but not NO-cGMP system, participates in LTM formation

STM short-term memory, *LTM* long-term memory. For references, see text

References

Burke CJ, Huetteroth W, Owald D et al (2012) Layered reward signaling through octopamine and dopamine in *Drosophila*. Nature 492:433–438

Davis RL (2005) Olfactory memory formation in *Drosophila*: from molecular to systems neuroscience. Annu Rev Neurosci 28:275–302

Davis RL (2011) Traces of *Drosophila* memory. Neuron 70:8–19

Farooqui T, Robinson K, Vaessin H et al (2003) Modulation of early olfactory processing by an octopaminergic reinforcement pathway in the honeybee. J Neurosci 23:5370–5380

Hammer MR, Menzel R (1998) Multiple sites of associative odor learning as revealed by local brain microinjections of octopamine in honeybees. Learn Mem 5:146–156

Liu C, Plaçais PY, Yamagata N et al (2012) A subset of dopamine neurons signals reward for odour memory in *Drosophila*. Nature 488:512–516

Matsumoto Y, Mizunami M (2002a) Temporal determinants of olfactory long-term retention in the cricket *Gryllus bimaculatus*. J Exp Biol 205:1429–1437

Matsumoto Y, Mizunami M (2002b) Lifetime olfactory memory in the cricket *Gryllus bimaculatus*. J Comp Physiol A 188:295–299

Matsumoto Y, Mizunami M (2004) Context-dependent olfactory learning in an insect. Learn Mem 11:288–293

Matsumoto Y, Mizunami M (2006) Olfactory memory capacity of the cricket *Gryllus bimaculatus*. Biol Lett 2:608–610

Matsumoto Y, Noji S, Mizunami M (2003) Time course of protein synthesis-dependent phase of olfactory memory in the cricket *Gryllus bimaculatus*. Zool Sci 20:409–416

Matsumoto Y, Unoki S, Aonuma H et al (2006) Critical role of nitric oxide-cGMP cascade in the formation of cAMP-dependent long-term memory. Learn Mem 13:35–44

Matsumoto Y, Hatano A, Unoki S et al (2009) Stimulation of the cAMP system by the nitric oxide-cGMP system underlying the formation of long-term memory in an insect. Neurosci Lett 467:81–85

Matsumoto Y, Hirashima D, Mizunami M (2013) Analysis and modeling of neural processes underlying sensory preconditioning. Neurobiol Learn Mem 101:103–113

Menzel R, Giurfa M (2006) Dimensions of cognition in an insect, the honeybee. Behav Cogn Neurosci Rev 5:24–40

Mizunami M, Matsumoto Y (2010) Roles of aminergic neurons in formation and recall of associative memory in crickets. Front Behav Neurosci 4:172

Mizunami M, Okada R, Li Y, Strausfeld NJ (1998a) Mushroom bodies of the cockroach: the activity and identities of neurons recorded in freely moving animals. J Comp Neurol 402:501–519

Mizunami M, Weibrecht JM, Strausfeld NJ (1998b) Mushroom bodies of the cockroach: their participation in place memory. J Comp Neurol 402:520–537

Mizunami M, Yokohari F, Takahata M (1999) Exploration into the adaptive design of the arthropod "microbrain". Zool Sci 16:703–709

Mizunami M, Yokohari F, Takahata M (2004) Further exploration into the adaptive design of the arthropod "microbrain": I. Sensory and memory-processing systems. Zool Sci 21:1141–1151

Mizunami M, Unoki S, Mori Y et al (2009) Roles of octopaminergic and dopaminergic neurons in appetitive and aversive memory recall in an insect. BMC Biol 7:46

Müller U (2000) Prolonged activation of cAMP-dependent protein kinase during conditioning induces long-term memory in honeybees. Neuron 27:159–168

Nakatani Y, Matsumoto Y, Mori Y et al (2009) Why the carrot is more effective than the stick: different dynamics of punishment memory and reward memory and its possible biological basis. Neurobiol Learn Mem 92:370–380

Schultz W (2006) Behavioral theories and the neurophysiology of reward. Annu Rev Psychol 57:87–115

Schwaerzel M, Monastirioti M, Scholz H et al (2003) Dopamine and octopamine differentiate between aversive and appetitive olfactory memories in *Drosophila*. J Neurosci 23:10495–10502

Takahashi T, Hamada A, Miyawaki K et al (2009) Systemic RNA interference for the study of learning and memory in an insect. J Neurosci Methods 179:9–15

Unoki S, Matsumoto Y, Mizunami M (2005) Participation of octopaminergic reward system and dopaminergic punishment system in insect olfactory learning revealed by pharmacological study. Eur J Neurosci 22:1409–1416

Unoki S, Matsumoto Y, Mizunami M (2006) Roles of octopaminergic and dopaminergic neurons in mediating reward and punishment signals in insect visual learning. Eur J Neurosci 24:2031–2038

Watanabe H, Matsumoto SC, Nishino H et al (2011) Critical roles of mecamylamine-sensitive mushroom body neurons in insect olfactory learning. Neurobiol Learn Mem 95:1–13

Chapter 10
Neurons and Networks Underlying Singing Behaviour

Stefan Schöneich and Berthold Hedwig

Abstract In cricket brains a neuropil area in the anterior ventral protocerebrum next to the pedunculus and the α-lobe is involved in the control of singing behaviour. Command interneurons for singing have dendrites in this neuropil, whereas their axon descends towards the ventral nerve cord. A bilateral calling song command neuron has been identified which drives the singing central pattern generator with tonic spike activity. The control of courtship and rivalry song via the brain is not yet resolved at a cellular level. Electrical and pharmacological brain stimulation reliably elicit normal and fictive singing in crickets. The central pattern generating network for singing seems to extend from the metathoracic ganglion complex to the first unfused abdominal ganglion A3. Crickets immediately stop singing and do not recover, once the connectives anterior to A3 are cut. Opener interneurons have been identified in A3, which modify and reset the singing motor pattern. The response properties of opener and closer interneurons upon hyperpolarising current injection indicate that post-inhibitory rebound mechanisms may be central to motor pattern generation underlying singing. Recordings of flight interneurons and singing interneurons prove that both motor patterns are controlled by separate neuronal networks.

Keywords Brain stimulation • Command interneurons • Calling song • Fictive singing • Central pattern generator • Abdominal ganglia

10.1 Introduction

Crickets show a conspicuous acoustic behaviour for intraspecific communication. Males sing with rhythmic opening and closing movements of their raised front wings and thereby produce a vigorous rivalry song on encounter with another male, a loud calling song to attract females or the more gentle courtship song to initiate

S. Schöneich
Institute for Biology, Leipzig University, Talstraße 33, 04103 Leipzig, Germany

B. Hedwig (✉)
Department of Zoology, University of Cambridge, Cambridge CB2 3EJ, UK
e-mail: bh202@cam.ac.uk

H.W. Horch et al. (eds.), *The Cricket as a Model Organism*,
DOI 10.1007/978-4-431-56478-2_10

mating (Alexander 1962). For more than 50 years, two principle questions about the neural basis of their singing behaviour have been central to cricket neurobiology (Huber 1960). How are the different song patterns controlled by the brain and what is the organisation of the neural networks for singing motor pattern generation? Since the review by Kutsch and Huber (1989), considerable progress has been made at the cellular level to answer both questions. Intracellular recordings of brain neurons allowed identifying command neurons that control the calling song. Furthermore, electrical and pharmacological brain stimulations have been established as a reliable method to elicit fictive singing in the dissected nervous system. This allowed localising the singing network in the ventral nerve cord and characterising component neurons of the central pattern generator (CPG) for singing.

10.2 Control of Singing Behaviour by the Brain

Brain stimulation experiments by Huber (1955, 1960) and electrical stimulation of fibres in the neck connectives (Otto 1971; Bentley 1977) demonstrated that descending brain neurons control the singing behaviour in *G. campestris* and *T. oceanicus*. Tonic activation of these neurons was sufficient to elicit rivalry, calling and/or courtship song, respectively. The brain does not generate the rhythmic motor pattern for singing, but rather drives the singing motor network in the ventral nerve cord by tonic descending activity. Based on the experimental evidence, these brain neurons seemed to be specific labelled lines for the activation of motor activity, i.e. command neurons for singing. Electrical and mechanical brain stimulation experiments pointed towards the anterior protocerebrum as a principle region involved in the control of singing (Huber 1960). Intracellular recordings in the brain of tethered male *G. bimaculatus* finally identified a bilateral pair of descending brain interneurons that can be characterised as command neurons for the control of singing (Hedwig 2000).

10.2.1 Descending Command Neurons for Calling Song

The command neuron's cell body is located at the dorsal side of the brain, the neurite projects anteriorly and dendrites arborise in the anterior protocerebrum ventral to the pedunculus (Fig. 10.1a). The axon crosses the midline ventral to the central body complex and gives off some collaterals that end in the left and right tritocerebrum before the axon finally descends in the contralateral connective in a medial position. The arborisation pattern of the neuron in the ventral nerve cord has not yet been described.

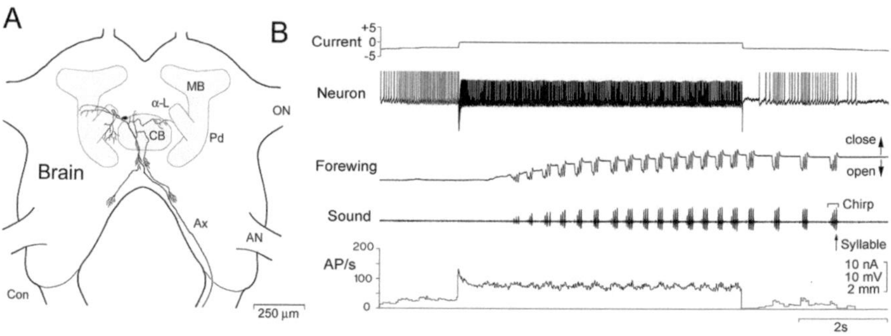

Fig. 10.1 Command neuron for calling song in the cricket *Gryllus bimaculatus*. (**a**) Arborisation pattern of the neuron in the brain, dendrites are located in the anterior protocerebrum, and the axon descends in the contralateral connective. (**b**) Depolarising current injection increases the discharge rate of the neuron and elicits calling song as indicated by opening-closing forewing movements and recording of the sound pattern. *CB* central body, *MB* mushroom body, *Pd* pedunculus, *ON* optical nerve, *AN* antennal nerve, *α-L* alpha-lobe, *Ax* axon of the neuron, *Con* connective (Modified from Hedwig 2000)

When the dendrite of this neuron is penetrated by a microelectrode, transient spike activity of this cell elicits rudimentary singing movements of the front wings. Systematic manipulation of the neuron's activity by intracellular current injection demonstrates that its spike activity is sufficient for calling song motor activity to occur. When the activity in one of the two mirror-image neurons is enhanced to 60–80 AP/s, the cricket will gradually raise its front wings and start sonorous singing of the calling song (Fig. 10.1b). Singing behaviour is maintained for the duration of enhanced interneuron activity. With the end of current injection, the interneuron spike activity as well as the chirp rate of singing decreases until singing stops. During singing the neuron's spike activity exhibits a weak modulation in the chirp rhythm, the significance of which is not yet evident. Spiking activity of the neuron is also necessary for singing to occur. In crickets which started to sing spontaneously, or after stable singing had been evoked by sequences of current injection into the command neuron, singing can be stopped by hyperpolarising current injection that blocks spiking of the neuron. Thus, in *G. bimaculatus*, this descending brain neuron is sufficient and necessary to control the calling song behaviour (Hedwig 2000) and therefore represents an example of a classical command neuron (Kupfermann and Weiss 1978).

Transitions from calling song to either courtship song or rivalry song were reported during electrical connective stimulation in *G. campestris* (Otto 1971) and single fibre stimulation in *T. oceanicus* (Bentley 1977). Therefore, the activity level of a single descending interneuron may be sufficient to control different song patterns in *T. oceanicus*, whereas in *G. campestris* different fibres might be involved. In *G. bimaculatus*, preliminary data indicate separate command neurons for the control of calling, rivalry and courtship song (Hedwig unpublished); however, so far the evidence for the descending control of the different song patterns is not yet conclusive.

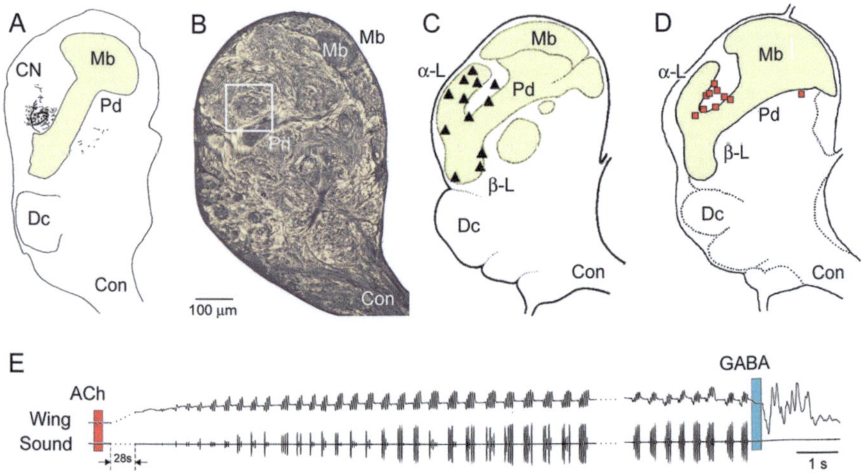

Fig. 10.2 (**a**) Brain neuropil area involved in the control of calling song as indicated by the dendritic arborisation pattern of the command neuron. (**b**) Histological parasagittal section of the brain with the projection area of the calling song command neuron identified as circular neuropil in front of the pedunculus, indicated by *white square*. (**c**) Electrical stimulation sites in the anterior brain and (**d**) microinjection sites of cholinergic agonists that elicited calling song. (**e**) Sequence of calling song after injection of *ACh* into the brain and subsequent termination of the behaviour by *GABA* injection. β-*L* beta-lobe, *DC* deutocerebrum, *CN* command neuron; other abbreviations as in Fig. 10.1 (Modified from Hedwig 2006)

10.2.2 Brain Structures for Singing and Targeted Pharmacological Brain Stimulation

Labelling the calling song command neuron with fluorescent tracers and subsequent histological analysis of the brain revealed the region between the pedunculus and the α-lobe as housing the main dendritic branches of the command neuron. Comparing the stimulation sites that elicited calling song (Huber 1960) with the arborisation site of the calling song command neuron (Hedwig 2000) reveals a close match between the command neuron's dendritic arborisations and effective stimulation sites. In parasagittal sections the corresponding neuropil area stands out as a circular region within the ventral protocerebrum (Fig. 10.2a–d).

Otto (1978) demonstrated that each of the three song patterns can be released by local injection of acetylcholine (ACh) into the brain. The identification of a specific brain neuropil involved in the control of singing allowed to systematically target this area in pharmacological brain stimulation experiments (Fig. 10.2e). Testing of neuroactive substances revealed that also cholinergic agonists such as nicotine or muscarine are effective in releasing singing (Wenzel and Hedwig 1999). Microinjections of nicotine elicited short bouts of singing behaviour with a short latency, whereas muscarine gradually activated singing activity within 60 s that then lasted for several minutes. A very effective way to release enduring singing activity is the injection of eserine, an ACh-esterase inhibitor that causes a gradual accumulation of ACh in the neuropil around the injection site. Correspondingly, eserine injection will activate singing only after several minutes, but then it can last for several hours. Once singing has been elicited by cholinergic drugs, it can be stopped by local brain injection of the inhibitory transmitter GABA. Thus, the pathway for singing is not only under a cholinergic but also under GABAergic control, which may be required for fast and effective termination of singing. The pharmacological experiments do not allow to conclude weather the command neurons are directly activated by cholinergic substances or driven by presynaptic neurons with cholinergic receptors. Under natural conditions singing behaviour seems to depend on a circadian input as the males preferentially produce the calling song at species-specific times of the day (Wiedenmann and Loher 1984; Nakatani et al. 1994; Fergus and Shaw 2013). In contrast rivalry and courtship song may require more specific sensory inputs as these song types are only generated after physical encounter with conspecific males or females, respectively (Adamo and Hoy 1994, 1995).

10.3 Motor Pattern Generation Underlying Singing

Sound production in crickets is based on rhythmic opening and closing movements of the front wings. As the antagonistic wing-opener and wing-closer muscles and their motoneuron machinery are located in the anterior thoracic segments (Kutsch and Huber 1989; Hennig 1989), it was initially assumed that the CPG for singing might be located in the corresponding ganglia of the thoracic nerve cord. Recent studies, however, revealed a spatial separation between the mesothoracic ganglion that produces the motor output and the neural network that generates the singing motor pattern, which apparently spans from the metathoracic to the first unfused abdominal ganglion (Hennig and Otto 1996; Schöneich and Hedwig 2011). The localisation of the singing CPG eventually allowed to analyse this network at the cellular level (Schöneich and Hedwig 2012).

10.3.1 Locating the Central Pattern Generator for Singing

Early behavioural observations in *G. campestris* and *A. domesticus* with lesions in the CNS lead to the assumption that the neural network generating the singing motor pattern may be housed in the mesothoracic ganglion, where it receives descending commands from the brain and sensory information from the terminal genital apparatus (Huber 1955, 1960; Kutsch and Otto 1972). More systematic micro-lesioning of the thoracic nervous system in combination with recordings of wing movements and muscle activity indicated that the metathoracic ganglion complex (T3-A2, comprising the third thoracic and the two subsequent abdominal neuromers) houses the singing CPG (Hennig and Otto 1996). However, the

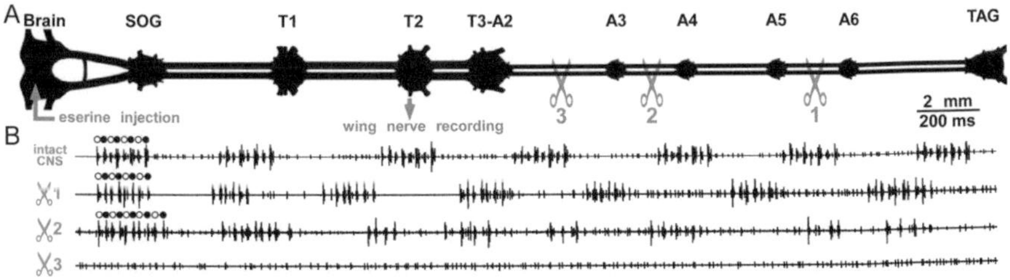

Fig. 10.3 (**a**) Cricket CNS with the location of pharmacological brain stimulation and the recording site for the fictive singing motor activity indicated. (**b**) Sectioning of abdominal connectives and their effect on fictive singing; numbered scissor symbols indicate different cuts. Singing motor activity fails immediately if connectives anterior to A3 are sectioned (Modified from Schöneich and Hedwig 2011)

observation that local cooling of the thoracic ganglia has only a minor effect on song pattern generation (Pires and Hoy 1992) and the structure of descending thoracic interneurons which are activated during singing (Hennig 1990) pointed towards a contribution of the abdominal ganglia to singing pattern generation. To clarify the situation, further connective lesioning experiments in the abdominal nerve cord were required. In a *G. bimaculatus* preparation where fictive singing was elicited by pharmacological brain stimulation, the connectives between the abdominal ganglia were systematically sectioned, while singing motor activity was monitored by extracellular recording the activity of singing motoneurons in the mesothoracic wing nerve N3A (Fig. 10.3). In the nerve recording, rhythmically alternating spike activity of wing-opener and wing-closer motoneurons reliably reflects the syllable pattern of the calling song chirps. When connectives were transected anywhere between A4 and the terminal ganglion, TAG singing continued although the motor pattern became more variable as more ganglia were disconnected. After the A3-A4 connective was disrupted, singing motor activity continued for another minute or so and then gradually waned. All animals stopped singing immediately when the A2-A3 connectives were cut, and they never recovered over the next hour of observation. These experiments in fictive singing crickets demonstrate that the abdominal ganglion A3 is essential for the motor pattern generation of the calling song and indicate that other abdominal ganglia may contribute as well.

10.3.2 Identification and Functional Characterisation of Singing CPG Interneurons

The interneurons of the central singing network can be classified as opener and closer interneurons depending on the wing movement and corresponding muscle/motoneuron activity they precede with their rhythmic spike activity. Accordingly, opener interneurons depolarise and spike in the wing-opening phase of each

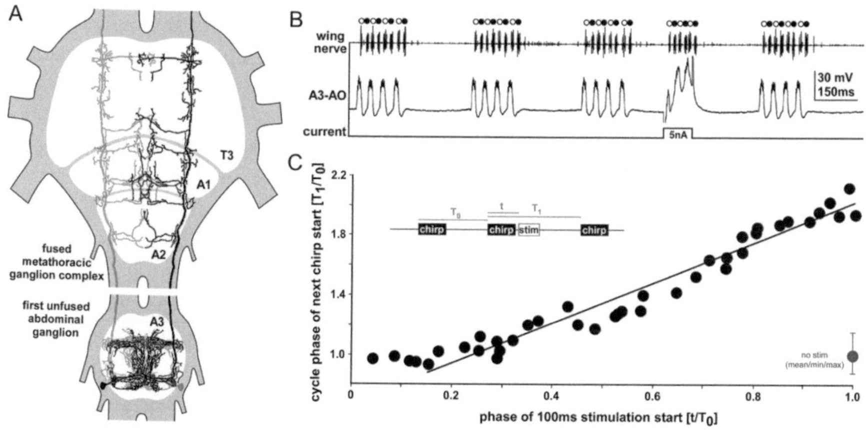

Fig. 10.4 (**a**) Structure of the bilateral neuron pair A3–AO with soma and dendrites in A3 and axonal projection in the fused metathoracic ganglion complex. (**b**) Activity pattern of A3–AO during fictive singing; depolarising current injection elicits additional singing motor activity that resets the ongoing chirp pattern. (**c**) Phase-response diagram; the shift in the chirp rhythm depends linearly on stimulus phase (Modified from Schöneich and Hedwig 2012)

syllable and are inhibited in the wing-closing phase, whereas closer interneuron activity always starts with an inhibition in the wing-opening phase. With evidence that the singing network extends into the abdominal nerve cord, experiments to analyse the singing CPG and to identify its components do now focus on the metathoracic and anterior abdominal ganglia. Interneurons with rhythmic membrane potential oscillations in phase with the syllable pattern of the calling song and an axon descending from the metathoracic into the abdominal ganglia chain were first identified by Hennig (1990) in *T. commodus*. Recently, interneurons have been identified in *G. bimaculatus* that clearly qualify as elements of the calling song CPG (Schöneich and Hedwig 2012).

The singing interneurons exist as bilateral mirror-image pairs and exhibit characteristic arborisations in the dorsal midline neuropils of the fused metathoracic and/or the first unfused abdominal ganglion. Their spike activity precedes the corresponding motoneuron spike bursts in the wing nerve by 5–15 ms. Component neurons of the central pattern generator, however, should also alter or reset the ongoing singing rhythm whenever their activity is disturbed by intracellular current injection (Selverston 2010). Neurons that comply with these criteria were found in both the metathoracic T3 and the first unfused abdominal ganglion A3 (Schöneich and Hedwig 2012). One of the identified CPG neurons is an ascending opener interneuron with its cell body and extensive dendritic arborisations in the abdominal ganglion A3 (A3–AO; Fig. 10.4a). It's ascending axon projects in all three thoracic ganglia, giving off distinct collaterals in the metathoracic ganglion complex as well as the meso- and prothoracic ganglion.

The spike activity of A3–AO has a strong impact on singing motor activity. Stimulation of the neuron by constant depolarising current injection in its dendrite leads to continuous depolarisation-hyperpolarisation cycles of its membrane potential for the duration of the current pulse. Each depolarisation phase produces a short burst of action potentials, which reliably elicits rhythmically alternating bursts of wing-opener and wing-closer motoneurons reflecting the syllable pattern of the calling song motor activity in the ipsi- and contralateral wing nerves. Depolarising current pulses of just 100 ms duration elicited strictly three depolarisation-hyperpolarisation cycles and the motor pattern of a three-syllable chirp (Fig. 10.4b), whereas 500 ms current pulses caused unnaturally long chirps with 14–15 syllables. Evoking activity in just one of the two A3–AO interneurons is sufficient to continuously drive the coordinated motor pattern of the calling song. Dye-coupling between the bilateral mirror-image pair of A3–AO neurons indicates that they may be electrically coupled via gap junctions.

Eliciting additional chirps by current injection in the A3–AO dendrite during chirp intervals reliably resets the chirp pattern, as the interval between the last syllable of the elicited chirp and the first syllable of the subsequently generated chirp remains constant (Fig. 10.4c). Depolarising current injection within a chirp, however, does not reset the singing rhythm. Interestingly, membrane potential oscillations that elicit additional syllables can also be evoked by the release from hyperpolarising current injection. The rhythmic activity may arise from post-inhibitory rebound depolarisation properties of the membrane in combination with feedback inhibition by yet unidentified closer interneurons. A post-inhibitory rebound in the A3–AO opener interneuron is also supported by the observation that for the last syllable of each chirp, the inhibition that terminates the large amplitude depolarisation of its dendrite is always followed by a small subthreshold depolarisation.

The importance of inhibition and post-inhibitory rebound depolarisation is more obvious in closer interneurons, where each syllable of fictive singing starts with an inhibition in phase with the opener motor activity followed by a depolarising membrane potential and spike activity in phase with the closer motor activity (Fig. 10.5). In some closer interneurons, volleys of inhibitory postsynaptic potentials (IPSPs) can be recorded several seconds before singing activity starts, and over a sequence of singing motor activity, the membrane potential gradually drops by few millivolt. Hyperpolarising current injection into closer interneurons reliably evokes a subsequent rebound depolarisation that triggers spiking activity. This clearly demonstrates that post-inhibitory rebound depolarisation occurs in these neurons and may be a basic neural mechanism underlying singing motor pattern generation.

10.3.3 Functional Organisation of the Singing Pattern Generating Network

The remarkable consistency of the singing pattern in intact animals (Verburgt et al. 2011) compared to the motor activity of pharmacologically induced fictive singing (Schöneich and Hedwig 2012) demonstrates that the underlying CPG operates

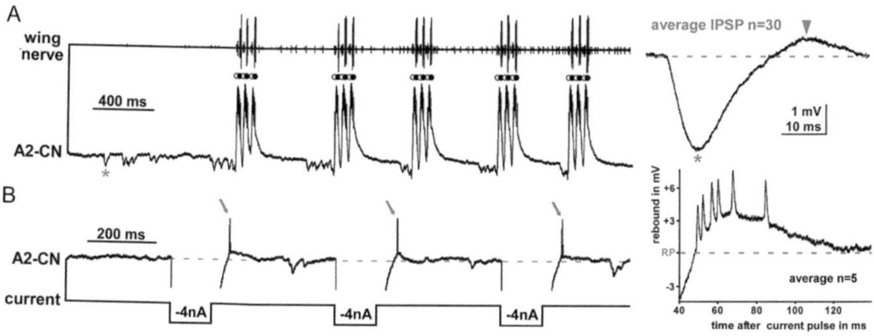

Fig. 10.5 (**a**) Intracellular recording of a non-identified closer interneuron in A2 during fictive singing. An average of 30 individual IPSPs (*asterisks*) shows subsequent post-inhibitory rebound depolarisation (*arrow head*). *Dashed lines* indicate the resting membrane potential. (**b**) Rebound depolarisation triggers spiking after the dendrite is released from 4 nA hyperpolarising current injection (Modified from Schöneich and Hedwig 2012)

virtually independent of sensory feedback. During sonorous singing, mechanosensory feedback merely adjusts the angular position of the moving front wings to ensure the required engaging force (Möss 1971; Elliott and Koch 1983).

Whereas in the *G. bimaculatus* calling song, the interval between subsequent opener activation may gradually increase within the chirps, there is a strict latency coupling between opener and closer motoneuron activity (Kutsch 1969; Kutsch and Huber 1989; Schöneich and Hedwig 2011). Based on motoneuron recordings, Bentley (1969) speculated that the wing-closer activity might be triggered by post-inhibitory rebound depolarisation in order to explain the tight temporal coupling between opener and closer motor activity. Thorough statistical analysis, however, did not support a direct motoneuron coupling (Elepfandt 1980). As antagonistic singing motoneurons may be recruited as agonists during flight motor activity (Hennig 1989), a direct synaptic motoneuron coupling would not be functional. However, as the closer interneurons are inhibited in phase with the depolarisation of the opener interneurons, post-inhibitory rebound in closer interneurons could secure the latency-fixed coupling to the opener activity.

Intracellular recordings of singing interneurons in *G. bimaculatus* (Schöneich and Hedwig 2012) indicate that inhibitory connections between opener and closer interneurons with post-inhibitory rebound properties may form a rather simple neural network to generate the motor pattern underlying rhythmic sound production. Computer modelling already demonstrated that small circuits of reciprocally inhibitory neurons exhibiting weak post-inhibitory rebound produce stable rhythmic output whenever the network is tonically activated by a command neuron (Perkel and Mulloney 1974; Satterlie 1985). Further studies are, however, needed to identify the remaining elements of the singing pattern generator in *G. bimaculatus* and to investigate their detailed circuitry and membrane properties. This should also reveal the neural mechanisms that control the chirp rhythm, which may include activity-dependent changes of membrane conductance, presynaptic inhibition of the command neuron input and/or periodical recovery of strongly depressing synapses.

10.4 Singing and Flight: Two Behaviours of Rhythmic Wing Movements

Crickets use their wings for different rhythmic behaviours. Whereas singing consists of rhythmical opening-closing movements of the front wings, flight comprises continuous up-down movements of front wings and hindwings. The motor patterns for singing and flight share the alternating activation of some antagonistic wing muscles and motoneurons, and in some crickets both behaviours are based on a similar cycle period. Early studies on the flight and stridulatory motor patterns in *G. campestris* (Huber 1962; Kutsch 1969) therefore suggested that both behaviours may be driven by a common central oscillator, while the chirps of the calling song were described as short flight sequences.

A comparison of the synaptic and spike activation patterns of singing and flight interneurons in *T. commodus* (Hennig 1990), however, revealed that flight interneurons are not activated during singing motor activity and that singing interneurons were inhibited during flight (Fig. 10.6). The study also demonstrated that there is no strict coupling between the singing and flight motor pattern when these two rhythmic behaviours, which are normally mutually exclusive, are elicited at the same time. Although the most conclusive experiments would require intracellular recordings of the corresponding CPG interneurons, these experiments convincingly demonstrate that the flight and singing patterns are generated by functionally separate networks that do not share a common pool of interneurons.

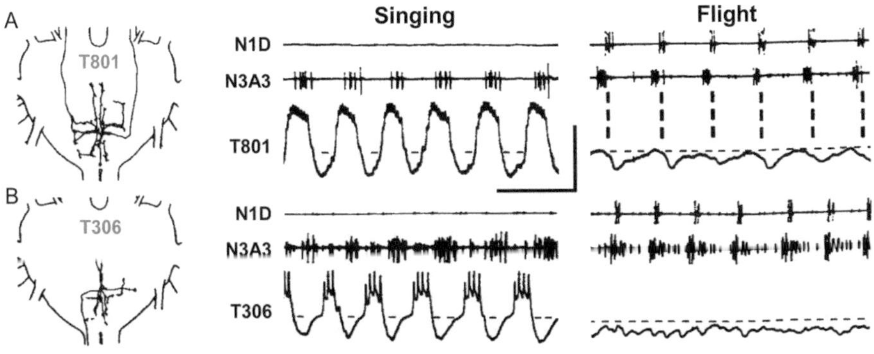

Fig. 10.6 Activity patterns of singing interneurons in *T. commodus* during generation of the singing and flight motor pattern. Although rhythmically active during singing, closer (**a**) and opener (**b**) interneurons are not activated or even inhibited during the generation of the flight motor pattern. Calibration: vertical: 20 mV (*top*) and 25 mV (*bottom*); horizontal 100 ms (Modified from Hennig 1990)

10.5 Future Perspectives

Acoustic communication in crickets is based on three different song patterns. The command neurons for calling song have been identified, but the control of courtship song and rivalry song and the generation of the underlying motor patterns are yet to be understood. Further systematic recordings of central neurons will be required.

The dendritic arborisation pattern of the calling song command neurons identifies a neuropil area in front of the pedunculus to be involved in the control of singing behaviour. As calling song generation is under circadian control, we may expect direct functional links between the circadian clock in the medulla and this area of the ventral protocerebrum (Yukizane et al. 2002).

At the level of calling song pattern generation, closer interneurons have been recorded in the metathoracic ganglion complex and the first unfused abdominal ganglion A3 but have not been anatomically identified yet. Furthermore, the singing CPG might have homologous interneurons in the more posterior abdominal ganglia which may contribute to motor pattern generation. Such an organisation would shed new light on the evolution of the singing CPG in crickets.

The enormous variety of song patterns makes crickets an ideal model system to study evolution and species-specific alterations in pattern generating systems. In Hawaiian Laupala crickets, sexual selection of song patterns caused a rapid evolution of sister species, identical in morphological traits but clearly different in respect to their songs (Mendelson and Shaw 2005). With the basic elements of the neural networks for singing revealed, comparative studies should now allow us to understand which of the network properties have been conserved and at which level changes occurred that allowed for the species-specific motor patterns.

Moreover crickets are on the verge of becoming a genetic model system (Hamada et al. 2009; Watanabe et al. 2012). Having genetic tools available to generate knockouts or to introduce suited reporter genes for cell lines and neuronal activity in the central nervous system will provide an enormous step forward. As compared to other species, in crickets these tools could be easily combined with intracellular recording techniques and a wealth of existing data on identified neurons.

References

Adamo SA, Hoy RR (1994) Mating behaviour of the field cricket *Gryllus bimaculatus* and its dependence on social and environmental cues. Anim Behav 47:857–868

Adamo SA, Hoy RR (1995) Agonistic behaviour in male and female field crickets, *Gryllus bimaculatus*, and how behavioural context influences its expression. Anim Behav 49:1491–1501

Alexander RD (1962) Evolutionary change in cricket acoustical communication. Evolution 16 (4):443–467

Bentley DR (1969) Intracellular activity in cricket neurons during generation of song patterns. J Comp Physiol A 62(3):267–283

Bentley D (1977) Control of cricket song patterns by descending interneurons. J Comp Physiol A 16(1):19–38
Elepfandt A (1980) Morphology and output coupling of wing muscle motoneurons in the field cricket (Gryllidae, Orthoptera). Zool Jahrb Physiol 84:26–45
Elliott CJH, Koch UT (1983) Sensory feedback stabilizing reliable stridulation in the field cricket *Gryllus campestris* L. Anim Behav 31:887–901
Fergus DJ, Shaw KL (2013) Circadian rhythms and period expression in the Hawaiian cricket genus Laupala. Behav Genet 43(3):241–253
Hamada A, Miyawaki K, Honda-Sumi E, Tomioka K, Mito T, Ohuchi H, Noji S (2009) Loss-of-function analyses of the fragile X-related and dopamine receptor genes by RNA interference in the cricket *Gryllus bimaculatus*. Dev Dyn 238(8):2025–2033
Hedwig B (2000) Control of cricket stridulation by a command neuron: efficacy depends on the behavioral state. J Neurophysiol 83(2):712–722
Hedwig B (2006) Pulses, patterns and paths: neurobiology of acoustic behaviour in crickets. J Comp Physiol A 192:677–689
Hennig RM (1989) Neuromuscular activity during stridulation in the cricket *Teleogryllus commodus*. J Comp Physiol A 165:837–846
Hennig RM (1990) Neuronal control of the forewings in two different behaviours: stridulation and flight in the cricket, *Teleogryllus commodus*. J Comp Physiol A 167(5):617–627
Hennig RM, Otto D (1996) Distributed control of song pattern generation in crickets revealed by lesions to the thoracic ganglia. Zoology 99:268–276
Huber F (1955) Sitz und Bedeutung nervöser Zentren für Instinkthandlungen beim Männchen von *Gryllus campestris* L. Z Tierpsychol 12(1):12–48
Huber F (1960) Untersuchungen über die Funktion des Zentralnervensystems und insbesondere des Gehirnes bei der Fortbewegung und der Lauterzeugung der Grillen. J Comp Physiol A 44 (1):60–132
Huber F (1962) Central nervous control of sound production in crickets and some speculations on its evolution. Evolution 16(4):429–442
Kupfermann I, Weiss KR (1978) The command neuron concept. Behav Brain Sci 1:3–10
Kutsch W (1969) Neuromuscular activity in three cricket species during various behavioural patterns. J Comp Physiol A 63(4):335–378
Kutsch W, Huber F (1989) Neural basis of song production. In: Huber F, Moore TE, Loher W (eds) Cricket behaviour and neurobiology. Cornell University Press, Ithaca, pp 262–309
Kutsch W, Otto D (1972) Evidence for spontaneous song production independent of head ganglia in *Gryllus campestris* L. J Comp Physiol A 81(1):115–119
Mendelson TC, Shaw KL (2005) Rapid speciation in an arthropod. Nature 433:375–376
Möss D (1971) Sense organs in the wing region of the field cricket (*Gryllus campestris* L.) and their role in the control of stridulation and wing position. J Comp Physiol A 73(1):53–83
Nakatani I, Adachi T, Murayama O (1994) Selection of light or darkness, locomotor, and stridulatory activities in the cricket, *Gryllus bimaculatus* DeGeer (Orthoptera: *Gryllidae*). J Insect Physiol 40:1007–1015
Otto D (1971) Central nervous control of sound production in crickets. J Comp Physiol A 74 (3):227–271
Otto D (1978) Changes of parameters in cricket song (*Gryllus campestris* L.) after injection of drugs into the brain. Verh Dtsch Zool Ges 245
Perkel DH, Mulloney B (1974) Motor pattern production in reciprocally inhibitory neurons exhibiting postinhibitory rebound. Science 185(4146):181–183
Pires A, Hoy RR (1992) Temperature coupling in cricket acoustic communication II. Localization of temperature effects on song production and recognition networks in *Gryllus firmus*. J Comp Physiol A 171:79–92
Satterlie RA (1985) Reciprocal inhibition and post-inhibitory rebound produce reverberation in a locomotor pattern generator. Science 229:402–404

Schöneich S, Hedwig B (2011) Neural basis of singing in crickets: central pattern generation in abdominal ganglia. Naturwissenschaften 98:1069–1073

Schöneich S, Hedwig B (2012) Cellular basis for singing motor pattern generation in the field cricket (*Gryllus bimaculatus* DeGeer). Brain Behav 2(6):707–725

Selverston AI (2010) Invertebrate central pattern generator circuits. Philos Trans R Soc B 365 (1551):2329–2345

Verburgt L, Ferreira M, Ferguson JWH (2011) Male field cricket song reflects age, allowing females to prefer young males. Anim Behav 81:19–29

Watanabe T, Ochiai H, Sakuma T, Horch HW, Hamaguchi N, Nakamura T, Bando T, Ohuchi H, Yamamoto T, Noji S, Mito T (2012) Non-transgenic genome modifications in a hemimetabolous insect using zinc-finger and TAL effector nucleases. Nat Commun 3:1–8

Wenzel B, Hedwig B (1999) Neurochemical control of cricket stridulation revealed by pharmacological microinjections into the brain. J Exp Biol 202:2203–2216

Wiedenmann G, Loher W (1984) Circadian control of singing in crickets: two different pacemakers for early-evening and before-dawn activity. J Insect Physiol 30:145–151

Yukizane M, Kaneko A, Tomioka K (2002) Electrophysiological and morphological characterization of the medulla bilateral neurons that connect bilateral optic lobes in the cricket, *Gryllus bimaculatus*. J Insect Physiol 48:631–641

Chapter 11
The Cricket Auditory Pathway: Neural Processing of Acoustic Signals

Gerald S. Pollack and Berthold Hedwig

Abstract The auditory afferents of crickets project from the hearing organs in the front legs toward the auditory neuropil in the prothoracic ganglion. They respond best to either the carrier frequency of the communication signals or to ultrasound, such as occurs in bat echolocation calls. Local interneurons (ON1) and two ascending interneurons (AN1 and AN2) establish the first stage of auditory processing. The recurrent inhibitory connections of ON1 support bilateral auditory contrast enhancement. The ascending neurons forward either spike patterns related to the calling song (AN1) or to ultrasound (AN2) to the brain. The temporal integration by the interneurons (ON1, AN1) of sensory activity in response to cricket-like sound patterns leads to an optimum information transfer at the pulse repetition rate of the song patterns. The temporal structure of song patterns may be further processed in a delay-coincidence-detection circuit of local neurons in the brain, which show selective responses to the pulse rate of the male calling song. Spike bursts in ON1 and AN2 seem to be a salient response in the processing of ultrasound. They can even trigger avoidance steering motor activity whereas the neural link between pattern recognition and phonotaxis is not yet characterized.

Keywords Afferent tuning • Central processing • Pattern recognition • Ultrasound avoidance

11.1 Introduction: Signals and Behavior

The auditory system of crickets serves two specific functions: intraspecific communication and predator detection. Male crickets sing to attract females, to entice them to mate, and they sing in agonistic interactions with rivals.

G.S. Pollack (✉)
Department of Biological Sciences, University of Toronto Scarborough, Toronto, ON, Canada
e-mail: gerald.pollack@utoronto.ca

B. Hedwig
Department of Zoology, University of Cambridge, Cambridge CB2 3EJ, UK
e-mail: bh202@cam.ac.uk

H.W. Horch et al. (eds.), *The Cricket as a Model Organism*,
DOI 10.1007/978-4-431-56478-2_11

Crickets can also hear sound frequencies that are higher than those used for communication but fall within the range of ultrasonic echolocation calls of hunting bats. The frequency and temporal structures of the signals, and of the neuronal responses that they evoke, play important roles in determining behavioral responses. In this chapter, we first briefly describe the relationships between stimulus structure and behavior and then examine how stimulus temporal features are analyzed in the nervous system. We describe the coding properties of auditory afferents, the network of thoracic interneurons, and finally the response properties of local auditory brain neurons which may be involved in pattern recognition and ultrasound processing.

11.1.1 Communication Signals

Cricket songs consist of series of discrete sound pulses or syllables that are produced by stridulatory wing movements with species-specific temporal patterns. Although the songs of different species may differ in sound frequency, the behavioral impact of this is only to determine how audible a signal is, that is, how well the sound frequency matches the frequency selectivity of the ear. As long as a signal falls within a low-frequency range of about 3–6 kHz and is loud enough to be heard, it can be effective in eliciting behavioral responses.

The main cue for recognizing the songs of conspecific singers is thus the song's temporal pattern, or rhythm, which may be different for calling, rivalry, and courtship songs. Tests with computer-generated song models have shown that temporal features, such as the durations of sound pulses, the intervals between them, and their grouping into higher-order structures such as chirps or trills, all may contribute to phonotactic behavior and may serve as species-identifying characteristics (Pollack and Hoy 1979; Weber and Thorson 1989).

11.1.2 Bat Echolocation Calls

Like cricket songs, bat calls consist of a series of discrete sounds. However, whereas the dominant sound frequency of most cricket songs is in the range 3–6 kHz, bat calls are ultrasonic, with frequencies ranging from ca. 20 to >100 kHz. A second difference from cricket songs is that the temporal structure of bat calls is variable, both between species and within the call sequence emitted during a single bat-insect encounter, where call duration decreases from a few ms to <1 ms and emission rate increases from about 10 to >100 Hz as the bat approaches its prey (Hoy 1992).

11.2 Neural Processing: Overview

The spectral and temporal structure of an acoustic signal is first encoded by the spike trains of auditory receptor neurons in the front leg hearing organ. Each ear contains ca. 60–70 receptor neurons. The majority of these are most sensitive to low sound frequencies similar to those used for communication, but a subset is tuned to the higher sound frequencies that occur in bat calls (Imaizumi and Pollack 1999).

Receptor neurons terminate in the prothoracic ganglion, where they provide synaptic input to a number of bilaterally paired interneurons that, based on their anatomical and physiological properties, are uniquely identified cells that can be recognized in all individuals, both within and across species (Fig. 11.1, Wohlers and Huber 1982; Atkins and Pollack 1987a, b). Three of these have been shown to be instrumental in contributing to behavioral responses and their synaptic connections have been elucidated. These have been given a number of different acronyms; the labeling of neurons is still an unsolved and unsatisfactory issue. In this chapter, we refer to these as ascending neuron 1 (AN1), ascending neuron 2 (AN2), and omega neuron 1 (ON1). The interneurons receive synaptic input from receptor neurons in the prothoracic ganglion. Both ascending neurons project to the brain, whereas the omega neuron is a local prothoracic neuron.

ON1 receives excitatory synaptic input mainly from receptor neurons of one ear and provides inhibitory input to AN1, AN2, and ON1 that receive excitatory input from receptors of the other ear, thus increasing binaural contrast (Selverston et al.

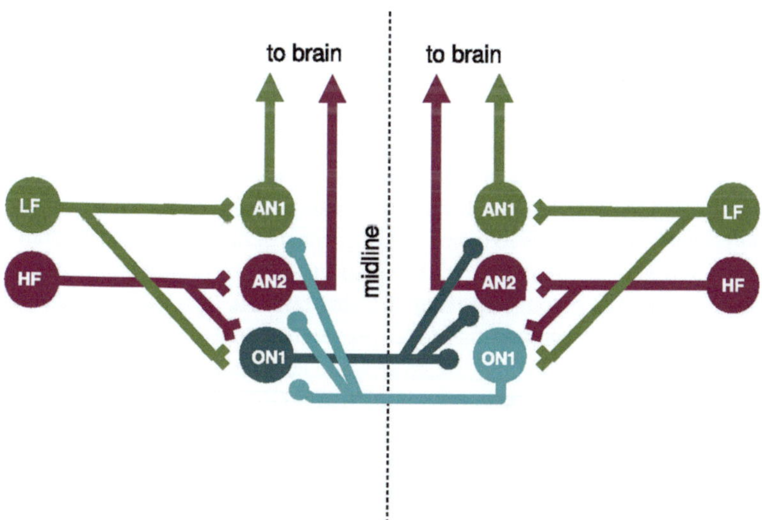

Fig. 11.1 First stage of auditory processing in the prothoracic ganglion. *LF*, *HF* receptor neurons tuned to low (LF) and to high (HF) sound frequencies, *AN1*, *AN2*, and *ON1* identified interneurons, as described in the text

1985). ON1 appears to be the sole source of contralateral inhibition, serving this role for both cricket-like and bat-like stimuli (Faulkes and Pollack 2000; Marsat and Pollack 2007). ON1's role in behavior has not yet been clearly demonstrated. When ON1 responses were attenuated by injection of hyperpolarizing current while walking female crickets performed open-loop acoustic orientation toward a male calling song, the tendency to turn toward the side of the manipulated ON1 was reduced, although not reversed, as occurred when AN1 was hyperpolarized (see below). Moreover, this effect was seen in only about half the animals tested (Schildberger and Hörner 1988). In these experiments, however, the sound sources were situated at 50° from the longitudinal axis, when physical cues for sound localization are large and unambiguous. It is possible that the effect of manipulating ON1 activity would be more pronounced for sound sources positioned closer to the midline where the directional cues for sound localization are small. In other experiments, ON1-mediated contralateral inhibition was shown to affect not only the response strength, i.e., firing rate, of its contralateral target neurons but also the accuracy with which they were able to encode temporal features of a stimulus (Marsat and Pollack 2005).

AN1 is most sensitive to sound frequencies close to the communication signals and provides input to a network of neurons in the brain that constitute a temporal-pattern filter (see below). The importance of AN1 for phonotactic behavior was demonstrated by manipulating its activity through intracellular current injection. These experiments were performed in a restrained cricket, walking on a spherical treadmill and orienting toward the sound source. Normally, crickets steer toward the right when a song model is broadcast from the right and toward the left when the song is played from the left. When the response of one AN1 was suppressed by intracellular injection of hyperpolarizing current, the cricket steered toward the side of the undisturbed AN1, no matter from where the sound emanated, demonstrating that the bilateral difference in AN1 activity is important for determining response direction (Schildberger and Hörner 1988).

AN2 responds best to high sound frequencies, which occur in the courtship song of many species and also in bat echolocation calls. There is some evidence that AN2 may play a role in responses to conspecific signals (Schildberger and Hörner 1988; Hutchings and Lewis 1984; but c.f. Libersat et al. 1994), but its clearest role in behavior is to elicit avoidance steering in flying crickets in response to stimulation with ultrasound. Hyperpolarizing AN2 prevents evasive steering responses to ultrasound stimuli, whereas eliciting AN2 spike activity in the absence of acoustic stimulation evokes steering responses (Nolen and Hoy 1984).

The axons of AN1 and AN2 terminate in the brain, where a number of inter-neurons have been identified that, based on their physiological properties, are likely to mediate responses to cricket songs and to bat calls. The auditory signals are processed by brain networks and then carried to motor centers that control the legs and wings through descending neurons.

11.2.1 Temporal Processing

11.2.1.1 Processing of Song at the Thoracic Level

AN1 is the only interneuron that relays information about songlike stimuli to the brain. Like the auditory afferents, it responds to stimuli with a wide range of temporal features, generally extending beyond the restricted range that best evokes behavioral responses.

Studies using an information-theoretic approach, however, show that, in at least one species, *Teleogryllus oceanicus*, the timing of AN1 spikes most accurately represents song-like rates of change of stimulus amplitude (Marsat and Pollack 2005). In these experiments, stimuli consist of long-lasting tones, the amplitude of which varies randomly through time, and the ability of the neuron to track these amplitude modulations is extracted by comparing moment-to-moment changes in firing rate with the corresponding changes in stimulus amplitude (Fig. 11.2a). The results, in units of information transferred by the neuron's spike rate, show that AN1 best represents information about the stimulus for rates of amplitude modulation (AM) that occur in the species' songs. ON1 exhibits similar information coding of stimuli which are presented at the carrier frequency of the cricket song with a cricket-song-like rate of amplitude modulation (Fig. 11.2b).

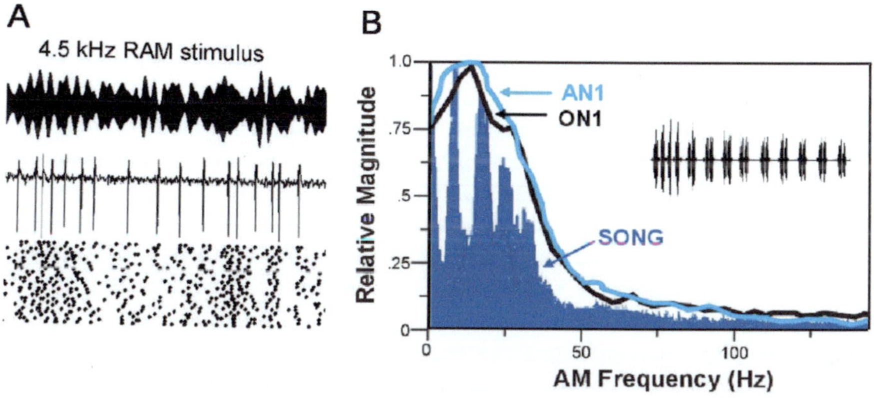

Fig. 11.2 Information-theoretic analysis of temporal coding. (**a**) *Top*: stimulus, consisting of a tone with carrier frequency of cricket song, 4.5 kHz, the amplitude of which is varied randomly through time. *Middle*: example spike train of ON1. *Bottom*: raster plot showing responses to repeated presentations of the stimulus (*rows*); each *dot* represents a single action potential (From Marsat and Pollack (2004)). (**b**) Information curves (normalized) derived from experiments like that shown in (**a**). The *solid blue* area shows the spectrum of the amplitude envelope of *T. oceanicus* calling song (*inset*). The *curves* show that the spike trains of AN1 and ON1 most accurately capture information about the stimulus temporal structure over the restricted range of AM rates that occur in song

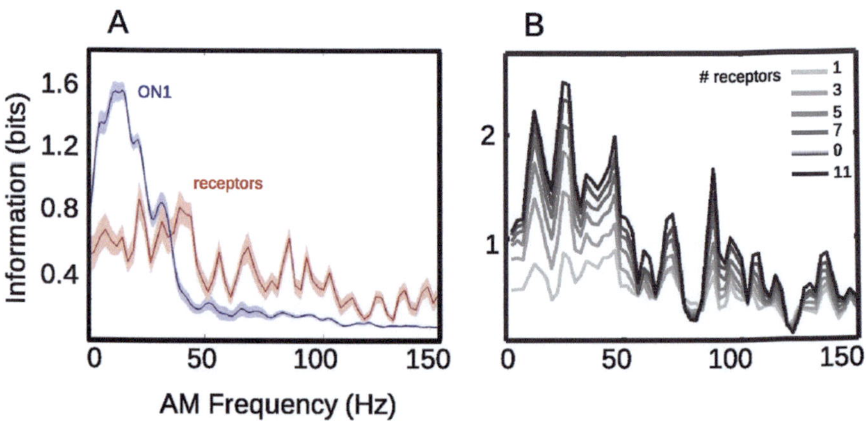

Fig. 11.3 (**a**) Individual auditory receptor neurons encode temporal information poorly, compared to that of interneurons, shown here for ON1. Encoding is measured here in units of information, which reflect the degree to which the rates of AM can be extracted from the neuron's spike train. (**b**) Encoding by populations of receptors (increasingly *darker shades*) approaches that of the interneurons, both in magnitude and the range of AM rates that are best encoded (Modified from Sabourin and Pollack 2010)

Surprisingly, information coding by the low-frequency-tuned receptor neurons that provide input to ON1 and AN1 is relatively poor (Fig. 11.3a). This mismatch between the two levels of processing raises the question of whether intrinsic physiological features of the interneurons, such as the types and dynamics of voltage-sensitive ion channels, might contribute to the selective encoding of cricket-like AM rates. Experiments on ON1 have found no evidence for this. When ON1 is hyperpolarized by intracellular current injection, so that spikes are prevented, encoding of the stimulus amplitude envelope by moment-to-moment variations in membrane potential is restricted to the same, cricket-like range of AM rates that are encoded by the spike train (Sabourin and Pollack 2010). Hyperpolarization would eliminate, or at least partially suppress, the contribution of any depolarization-activated currents, suggesting that these play little role in ON1's temporal response characteristics.

These experiments suggest that the temporal properties of the interneurons' responses have their origins in those of the receptor neurons, despite their poor coding capabilities. This apparent paradox is resolved when one considers that the interneurons can pool inputs from many receptor neurons. When virtual receptor population responses are created by combining the spike trains of individually recorded receptors, the coding performance of the population increases with increasing numbers of receptor neurons and approaches that of the interneurons (Fig. 11.3b). Pooling of afferent information in this manner would be possible only if redundancy of the responses of the contributing receptor neurons was low, i.e., if their spike trains were not overly similar, and this indeed is the case (Sabourin and Pollack 2010).

11.2.1.2 Temporal Processing by Brain Neurons

In contrast to *T. oceanicus*, in *G. campestris* and *G. bimaculatus*, there is no evidence for temporal filtering by AN1; rather, it forwards any auditory patterns presented at the species-specific carrier frequency to the brain where the recognition of species-specific song patterns resides. Based on the response properties of local auditory brain neurons, it was concluded that the band-pass-like tuning that the phonotaxis of female *G. bimaculatus* shows in response to different pulse patterns is due to a serial network of low-pass and high-pass filter neurons, the responses of which result in band-pass tuned neural responses akin to the tuning of phonotaxis (Schildberger 1984).

A different picture emerges from more recent studies on auditory processing in the cricket brain. In recording the activity of brain neurons in tethered crickets, which were allowed to stand and walk on a trackball, Kostarakos and Hedwig (2012) identified a new group of local auditory brain neurons with arborizations that closely match the axonal arborization pattern of AN1. The axonal projections of AN1 as well as the neurites of these brain interneurons form a ringlike arborization in the anterior protocerebrum (Fig. 11.4a), which is implicated in temporal processing of the calling song. Although the local auditory brain neurons are quite similar in structure, their auditory response properties are very different and demonstrate an increasing level of tuning toward temporal patterns. This is evident when the neural response of the local auditory brain neurons (B-LI2 to B-LI-4) to different pulse patterns is compared with the phonotactic responses of females to corresponding patterns. While the temporal tuning of the local auditory brain neurons becomes sharper from B-LI2 to B-LI4, their spike response latency increases from 21 to 37 ms, and their maximum response decreases from 14.3 AP/Chirp to 3.9 AP/Chirp, respectively. This is in line with the concept of sparse coding, shifting the representation of stimulus features from a temporal activity-based code to a neuron-specific place code.

In female *G. bimaculatus*, phonotaxis is tuned toward pulse periods of 34–42 ms. Upon acoustic stimulation the local B-LI2 brain neuron responds phasically with EPSPs and spikes to the individual sound pulses of all patterns with different pulse periods. Its activity even reflects minor differences in the overall sound energy presented. Its response function, however, exhibits no tuning to the different pulse patterns, and, like the activity of the ascending interneuron AN1, it does not match the phonotactic behavior (Fig. 11.4 left). A more complex tuned response pattern is apparent at the level of B-LI3, as the strongest excitatory response occurs to the second pulse of a chirp. Moreover the response of B-LI3 to short and long pulse periods is about 60 % lower than the response to chirps presented at the species-specific pulse period (Fig. 11.4 middle). Tuning to the species-specific temporal pattern is best in B-LI4 neurons, which receive a mixture of excitatory and inhibitory synaptic inputs (Fig. 11.4 right). Inhibition dominates the neuron's response, when chirps with low or high pulse periods are presented; however, at the species-specific pulse period, the excitatory spiking response of the neuron is strongest. This leads to a close match between the tuning of the neural response and the tuning of phonotactic behavior.

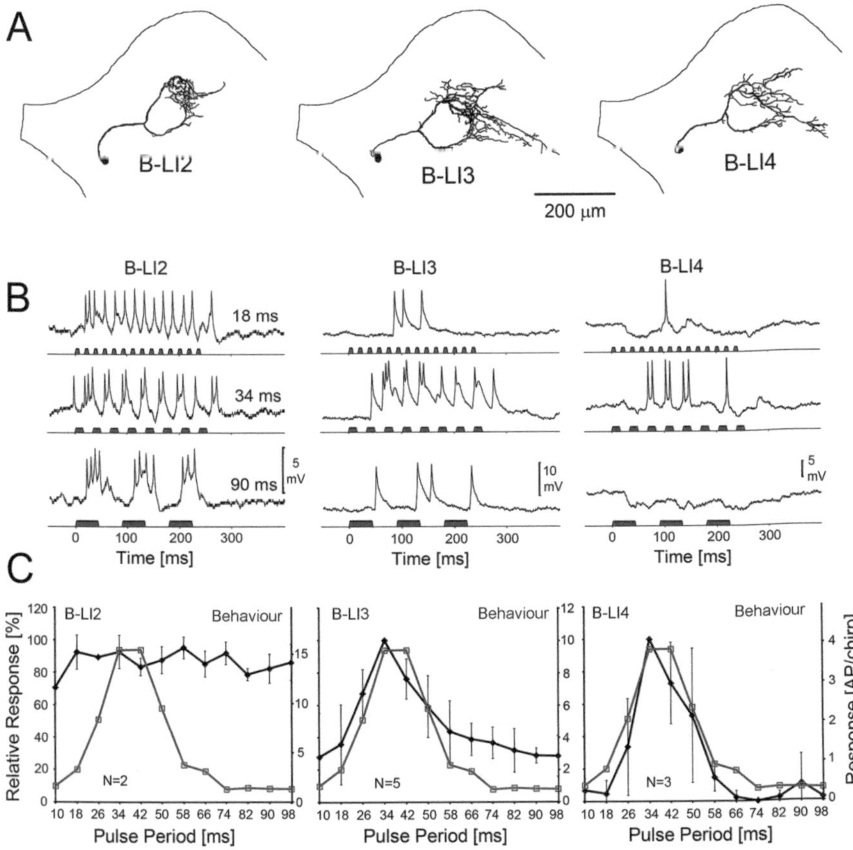

Fig. 11.4 (**a**) Structure of local auditory brain neurons in *G. bimaculatus* with arborization patterns that closely match the ringlike axonal arborization of AN1. (**b**) When exposed to sound patterns with different pulse periods, the spike activity of the brain neurons exhibits different degrees of temporal filtering. B-LI4 receives excitatory and inhibitory synaptic inputs and responds best to the species-specific pulse period of 34 ms. (**c**) Whereas B-LI2 shows no sign of filtering, the responses of B-LI3 and especially B-LI4 closely match the tuning of female phonotactic behavior, which is indicated by a *gray line* (Modified from Kostarakos and Hedwig 2012)

In the cricket brain, temporal processing and tuning to the species-specific pattern seem to occur in a network of local auditory interneurons, closely matching the axonal arborization pattern of AN1. Auditory band-pass properties encountered in other brain interneurons (Schildberger 1984) may simply be the consequence of this. As in frogs (Edwards et al. 2007; Rose et al. 2011), auditory processing appears to be based on fast pulse-by-pulse interactions of excitatory and inhibitory synaptic activity. Recent experiments indicate that a delay-coincidence-detection mechanism, as earlier proposed by Weber and Thorson (1989), may be the basis of species-specific temporal filtering in the ringlike local auditory interneurons. The required delay loop needs to span the duration of the pulse period, which is

35–40 ms in the case of *G. bimaculatus*. In a small insect brain, this cannot be achieved by delay lines based on extended axonal outgrowth but appears to be realized via postinhibitory rebound activity in a non-spiking interneuron. Coincidence detection between the ascending activity of AN1 and the delayed activity occurs subsequently in a local interneuron, which exhibits a high selectivity for acoustic stimulation with paired sound pulses. The response toward the second pulse is strongly enhanced when pulse duration and the interval between the pulses correspond to the species-specific parameters.

How many tuned filters for pulse interval processing might crickets have? As the local auditory brain neurons are only tuned to the species-specific song pattern, it appears that *G. bimaculatus* employs just one filter at the level of pulse intervals but may use additional filtering at the chirp level (Doherty 1985; Meckenhäuser et al. 2013). The situation may be more complex in *T. oceanicus*, with a calling song consisting of chirps and trills with different temporal settings. This processing of communication signals occurs separately from the ability of crickets to respond to sequences of ultrasound pulses.

11.2.2 Processing of Ultrasound

Like many nocturnally flying insects, crickets respond to ultrasound stimuli, similar to the echolocation calls of hunting bats, by steering away from the sound source. This response is triggered by high-frequency firing of AN2 (Nolen and Hoy 1984). AN2's tendency to fire at a high rate is revealed in experiments using ultrasound tones with dynamically varying amplitude envelopes. These stimuli evoke two patterns of firing: bursts, i.e., groups of spikes separated by brief ($<$6 ms) interspike intervals, and non-burst, or isolated spikes, which are separated by longer, and more variable, interspike intervals. Bursts are triggered by conspicuous increases in stimulus amplitude following brief quiet periods, i.e., by the sorts of stimuli that comprise echolocation calls. Analysis of AN2 responses using signal-detection theory shows that bursts are highly reliable at detecting such stimulus features (Marsat and Pollack 2006).

Properly directed steering responses require that the source location be encoded as a bilateral difference in AN2 responses. The difference in bursting between the left and right AN2 responses is pronounced for a lateral sound source, whereas there is no apparent difference in the firing rate of non-burst spikes. Thus, bursts in bilateral AN2 activity not only detect salient stimulus features reliably, they also encode stimulus direction.

The role of bursts in evoking behavioral responses was examined by comparing the timing of behavioral responses and of AN2 bursts throughout the course of an amplitude-modulated stimulus. In *Teleogryllus oceanicus* discrete steering movements directed away from the sound source occur most reliably following AN2 bursts, demonstrating the functional importance of AN2's firing pattern (Fig. 11.5a).

Like AN2, ON1 also produces both bursts and isolated spikes, but only for ultrasound stimuli. Stimulation with the identical amplitude modulation envelope,

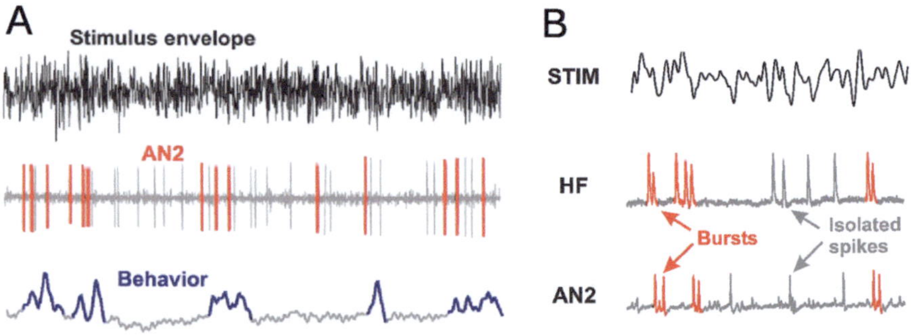

Fig. 11.5 (**a**) AN2 spike bursts elicit behavioral responses. *Top*: amplitude envelope of 30 kHz stimulus with randomly varying amplitude. *Middle*: example AN2 spike train. Bursts are highlighted in *red*. *Bottom*: steering movements of the abdomen (Modified from Marsat and Pollack (2006)). (**b**) Bursts in AN2 are elicited by bursts in HF receptors (Modified from Sabourin and Pollack 2009)

but with cricket-like rather than bat-like sound frequency, does not elicit bursting. With respect to their bursting behavior, ON1 and AN2 are essentially identical; indeed, paired recordings show that both neurons tend to produce bursts at the same time. Coordinated bursting of the two neurons is functionally important, insuring that ON1 bursts are likely to occur with the proper timing to most effectively inhibit bursts in the contralateral AN2, thus helping to generate the large bilateral difference in AN2 bursts mentioned earlier (Marsat and Pollack 2007).

Bursting in AN2 and ON1 for high-frequency stimuli appears to originate at the level of high-frequency-tuned receptor neurons. These receptor neurons, but not cricket-song-tuned afferents, burst in response to conspicuous increases in stimulus amplitude, and their bursts of spike activity are temporally correlated with bursts in the interneurons (Fig. 11.5b). Moreover, bursts tend to occur simultaneously in different high-frequency-tuned receptors, resulting in strong, burst-inducing synaptic input to the interneurons (Sabourin and Pollack 2009).

In the brain, local and descending interneurons have been identified in *G. bimaculatus* and *T. oceanicus*, which respond to high-frequency acoustic stimuli (Boyan and Williams 1981; Brodfuehrer and Hoy 1989, 1990). From the physiological and morphological data available so far, however, no clear picture emerges as to where and how descending motor commands for ultrasound avoidance steering may emerge.

11.3 From Auditory Processing to Motor Responses

Cricket communication signals as well as sonar pulse patterns of predatory bats are processed in the brain. During negative and positive phonotaxis, flying as well as walking crickets show short latency reflex-like steering responses to sound stimuli (Pollack and Hoy 1981; Hedwig and Poulet 2004). Moreover, pattern recognition transiently modulates the amplitude of steering responses toward non-attractive test pulses (Poulet and Hedwig 2005). In the brain, auditory information is forwarded to the lateral accessory lobes (Brodfuehrer and Hoy 1990; Zorovic and Hedwig 2011), which are implicated in the higher control of insect motor activity. Currently, however, we have no proper understanding of how the networks in the brain that recognize auditory patterns subsequently drive motor responses underlying phonotactic walking or ultrasound avoidance through auditory steering. Overall about 200 interneurons descend from the brain toward the thoracic ganglia (Staudacher 1988). When depolarized by intracellular current injection in resting animals, some of these interneurons elicit motor response-like walking or turning movements (Böhm and Schildberger 1992; Zorovic and Hedwig 2013). Generally, all descending interneurons identified so far exhibit only weak auditory responses, which are not sufficient to explain the rapid, reflex-like steering response observed.

Based on current data, we still cannot rule out that local thoracic networks are involved in generating auditory steering responses. Imaizumi and Pollack (2005) describe a group of low frequency auditory afferents (BC type), with an axonal arborization pattern that does not overlap with the ascending interneurons, but which are theoretically suitable for forwarding auditory information directly to descending auditory interneurons not included in Fig. 11.1 (Wohlers and Huber 1982; Atkins and Pollack 1987a, b). Although these afferents and interneurons have a slightly higher threshold than the ascending neurons, this local thoracic pathway might be involved in auditory-to-motor processing. The thoracic networks could be gated by descending activity, depending on pattern recognition, which does not provide the actual motor commands but allows the local auditory processing to impact on motor activity. As an advantage such a network organization would allow the generation of steering commands based on the activity of the pool of auditory afferents, providing sufficient directional details (Schöneich and Hedwig 2010) as compared to the summed auditory activity forwarded to the brain via a single pair of ascending interneurons. Few attempts to elucidate the mechanisms of auditory behavior have combined microelectrode recordings and phonotactic behavioral techniques (Schildberger and Hörner 1988; Zorovic and Hedwig 2011, 2013). In order to reveal the complete auditory-to-motor loop, recordings of descending brain neurons and/or thoracic interneurons, with a focus on actively behaving crickets, will be required.

References

Atkins G, Pollack GS (1987a) Response properties of prothoracic, interganglionic, sound-activated interneurons in the cricket *Teleogryllus oceanicus*. J Comp Physiol A 161 (5):681–693

Atkins G, Pollack GS (1987b) Correlations between structure, topographic arrangement, and spectral sensitivity of sound-sensitive interneurons in crickets. J Comp Neurol 266(3):398–412

Böhm H, Schildberger K (1992) Brain neurones involved in the control of walking in the cricket *Gryllus bimaculatus*. J Exp Biol 166(1):113–130

Boyan G, Williams J (1981) Descending interneurones in the brain of the cricket. Naturwissenschaften 68(9):486–497
Brodfuehrer PD, Hoy RR (1989) Integration of ultrasound and flight inputs on descending neurons in the cricket brain. J Exp Biol 145(1):157–171
Brodfuehrer PD, Hoy RR (1990) Ultrasound sensitive neurons in the cricket brain. J Comp Physiol A 166(5):651–662
Doherty JA (1985) Trade-off phenomena in calling song recognition and phonotaxis in the cricket, *Gryllus bimaculatus* (Orthoptera, Gryllidae). J Comp Physiol A 156(6):787–801
Edwards CJ, Leary CJ, Rose GJ (2007) Counting on inhibition and rate-dependent excitation in the auditory system. J Neurosci 27(49):11392–13384
Faulkes Z, Pollack GS (2000) Effects of inhibitory timing on contrast enhancement in auditory circuits in crickets (*Teleogryllus oceanicus*). J Neurophysiol 84:1247–1255
Hedwig B, Poulet JFA (2004) Complex auditory behaviour emerges from simple reactive steering. Nature 430(7001):781–785
Hoy RR (1992) The evolution of hearing in insects as an adaptation to predation from bats. In: Webster DB, Fay RR, Popper AN (eds) The evolutionary biology of hearing. Springer, New York, pp 115–129
Hutchings M, Lewis B (1984) The role of two-tone suppression in song coding by ventral cord neurones in the cricket *Teleogryllus oceanicus* (Le Guillou). J Comp Physiol A 154:103–112
Imaizumi K, Pollack GS (1999) Neural coding of sound frequency by cricket auditory receptors. J Neurosci 19(4):1508–1516
Imaizumi K, Pollack GS (2005) Central projections of auditory receptor neurons of crickets. J Comp Neurol 493(3):439–447
Kostarakos K, Hedwig B (2012) Calling song recognition in female crickets: temporal tuning of identified brain neurons matches behavior. J Neurosci 32(28):9601–9612
Libersat F, Murray JA, Hoy RR (1994) Frequency as a releaser in the courtship song of two crickets, *Gryllus bimaculatus* (de Geer) and *Teleogryllus oceanicus*. J Comp Physiol A 174:485–494
Marsat G, Pollack GS (2004) Differential temporal coding of rhythmically diverse acoustic signals by a single interneuron. J Neurophysiol 92:939–948
Marsat G, Pollack GS (2005) Effect of the temporal pattern of contralateral inhibition on sound localization cues. J Neurosci 25:6137–6144
Marsat G, Pollack GS (2006) A behavioral role for feature detection by sensory bursts. J Neurosci 26(41):10542–10547
Marsat G, Pollack GS (2007) Efficient inhibition of bursts by bursts in the auditory system of crickets. J Comp Physiol A 193:625–633
Meckenhäuser G, Hennig RM, Nawrot MP (2013) Critical song features for auditory pattern recognition in crickets. PLOS One. doi:10.1371/journal.pone.0055349
Nolen TA, Hoy RR (1984) Initiation of behavior by single neurons: the role of behavioral context. Science 226:992–994
Pollack GS, Hoy RR (1979) Temporal pattern as a cue for species-specific calling song recognition in crickets. Science 204(4391):429–432
Pollack GS, Hoy R (1981) Phonotaxis to individual rhythmic components of a complex cricket calling song. J Insect Physiol 27:41–45
Poulet JFA, Hedwig B (2005) Auditory orientation in crickets: pattern recognition controls reactive steering. Proc Natl Acad Sci 102(43):15665–15669
Rose GJ, Leary CJ, Edwards CJ (2011) Interval-counting neurons in the anuran auditory midbrain: factors underlying diversity of interval tuning. J Comp Physiol A 197:97–108
Sabourin P, Pollack GS (2009) Behaviorally relevant burst coding in primary sensory neurons. J Neurophysiol 102:1086–1091
Sabourin P, Pollack GS (2010) Temporal coding by populations of auditory receptor neurons. J Neurophysiol 103:1614–1621

Schildberger K (1984) Temporal selectivity of identified auditory neurons in the cricket brain. J Comp Physiol A 155(2):171–185
Schildberger K, Hörner M (1988) The function of auditory neurons in cricket phonotaxis I. Influence of hyperpolarization of identified neurons on sound localization. J Comp Physiol A 163(5):621–631
Schöneich S, Hedwig B (2010) Hyperacute directional hearing and phonotactic steering in the cricket (*Gryllus bimaculatus* deGeer). PLoS One 5(12):e15141
Selverston AI, Kleindienst H-U, Huber F (1985) Synaptic connectivity between cricket auditory interneurons as studied by photoinactivation. J Neurosci 5(5):1283–1292
Staudacher E (1988) Distribution and morphology of descending brain neurons in the cricket *Gryllus bimaculatus*. Cell Tissue Res 294(1):187–202
Weber T, Thorson J (1989) Phonotactic behavior of walking crickets. In: Huber F, Moore TE, Loher W (eds) Cricket behavior and neurobiology. Cornell University Press, New York, pp 310–339
Wohlers DW, Huber F (1982) Processing of sound signals by six types of neurons in the prothoracic ganglion of the cricket, *Gryllus campestris* L. J Comp Physiol A 146(2):161–173
Zorovic M, Hedwig B (2011) Processing of species-specific auditory patterns in the cricket brain by ascending, local, and descending neurons during standing and walking. J Neurophysiol 105 (5):2181–2194
Zorovic M, Hedwig B (2013) Descending brain neurons in the cricket *Gryllus bimaculatus* (de Geer): auditory responses and impact on walking. J Comp Physiol A 199:25–34

Chapter 12
Neuromodulators and the Control of Aggression in Crickets

Paul A. Stevenson and Jan Rillich

Abstract Crickets have emerged as ideal model systems for investigating the mechanisms controlling intraspecific aggressive behaviour. As in many animals, male aggression in crickets is shaped by numerous experiences including physical exertion, past wins and defeats and the acquisition of resources. This chapter reviews work revealing that neuromodulators, primarily octopamine and nitric oxide, mediate such experience-dependent plasticity by modulating the relative behavioural thresholds to fight and to flee. Octopamine, the invertebrate analogue of noradrenaline, promotes the tendency to fight by mediating the effects of flying, winning and shelter occupancy and thus represents the motivational component of aggression. The gaseous neuromodulator nitric oxide, on the other hand, mediates the decision to flee and induces a period of prolonged submissiveness, which is characteristic for social defeat in many animals. Accumulating evidence also suggests a role for serotonin, dopamine and selected peptides in controlling insect aggression. The roles for neuromodulators in insect aggression are in essence similar to those emerging for corresponding signalling molecules in mammals, where their specific behavioural functions are less clear.

Keywords Agonistic behaviour • Octopamine • Serotonin • Nitric oxide • Motivation • Experience-dependent plasticity • Decision-making • Assessment • Social behaviour • *Gryllus bimaculatus*

P.A. Stevenson (✉)
Institute for Biology, Leipzig University, Leipzig, Germany
e-mail: stevenson@rz.uni-leipzig.de

J. Rillich
Institute for Neurobiology, Free University of Berlin, Berlin, Germany

H.W. Horch et al. (eds.), *The Cricket as a Model Organism*,
DOI 10.1007/978-4-431-56478-2_12

12.1 Introduction

12.1.1 Crickets as Model Organisms for the Study of Aggression

Over the past few years, crickets have advanced to the status of a model organism for studying the mechanisms underlying aggressive behaviour (reviews: Stevenson and Rillich 2012; Stevenson and Schildberger 2013; Simpson and Stevenson 2015). But why study crickets? For one, as insects, their miniature brains contain comparatively few, individually identifiable neurons, but they are nonetheless equipped with the capacity to generate sophisticated social interactions (Huber et al. 1989; Guirfa 2012). Fights between crickets are impressive, highly ritualised affairs (Adamo and Hoy 1995; Stevenson et al. 2000, 2005) and accordingly relatively simple to evaluate. On the other hand, their fighting behaviour, as in mammals, is influenced by a wealth of experiences including age and time of day (Dixon and Cade 1986), physical exertion (Hofmann and Stevenson 2000), winning (Rillich and Stevenson 2011), losing (Iwasaki et al. 2006; Stevenson and Rillich 2013), the presence of shelters (Rillich et al. 2011), food (Nosil 2002) or females (Brown et al. 2006; Tachon et al. 1999), courtship and mating (Killian and Allen 2008; Judge et al. 2010), their song (Brown et al. 2007; Rillich et al. 2009; DiRienzo et al. 2012), social isolation and crowding (Adamo and Hoy 1995; Iba et al. 1995; Stevenson and Rillich 2013). They are thus ideal models for investigating mechanisms underlying experience-dependent plasticity of aggressive behaviour, and this will be the main focus of this account. In some quarters, there is a growing tendency to attribute insects with experiencing conscious emotions (see Mendl et al. 2011 for a rational commentary). While this is hard to prove and practically impossible to refute, the experimental data summarised here illustrate that crickets are able to integrate the influences of ongoing and past experiences to generate adaptive aggressive behaviour – without necessitating rational, conscious emotions or reason – simply by exploiting the basic principles of neuromodulation, and this is perhaps the greatest advantage of studying these fascinating insects.

12.1.2 Understanding Aggression

All animals must cope with a simple biological fact: conspecifics are their greatest natural competitors. They compete for the same territories, shelters, food and sexual partners. For this reason, intraspecific aggression is common throughout the animal kingdom, from the lowest multicellular organisms (Brace and Purvey 1978) to our own species (Albert et al. 1993). Aggression can thus be considered as a behavioural strategy for securing some limited resource. However, since aggression is an inherently dangerous activity, it must be exercised with restraint, so that the costs do not exceed the potential benefits. It is, therefore, generally agreed that competing animals must in some way be able to equate potential costs and benefits of aggression, in order to decide whether it would be more opportune to fight or to flee (cf. Archer 1988). The underlying control mechanisms are, however, barely understood. Game theory predicts that animals adopt evolutionarily stable fighting strategies (Maynard Smith and Price 1973), which typically take the form of stereotyped, gradually escalating contests, involving the ritualised exchange of agonistic signals. This is thought to convey increasingly accurate information on the contestants' abilities to secure the disputed resource ("resource holding potential"), from which each individual bases its decision when best to stand and fight or turn and flee (cf. Hurd 2006; Elwood and Arnott 2012). Numerous studies have illustrated that an animal's resource holding potential not only depends on physique (size, strength, weaponry) but also on an animal's "willingness" to invest energy in fighting, i.e. its "aggressive motivation", a factor determined by a wide variety of experiences such as winning, losing and the possession and value of disputed resources (review: Stevenson and Rillich 2012). These largely theoretical considerations (summarised in Fig. 12.1) provide a convenient framework to explain the observed outcomes of animal contests. But what are the proximate mechanisms? How do different experiences and circumstances control aggressive motivation? How is this encoded in the nervous system? How exactly do animals decide whether to fight or flee?

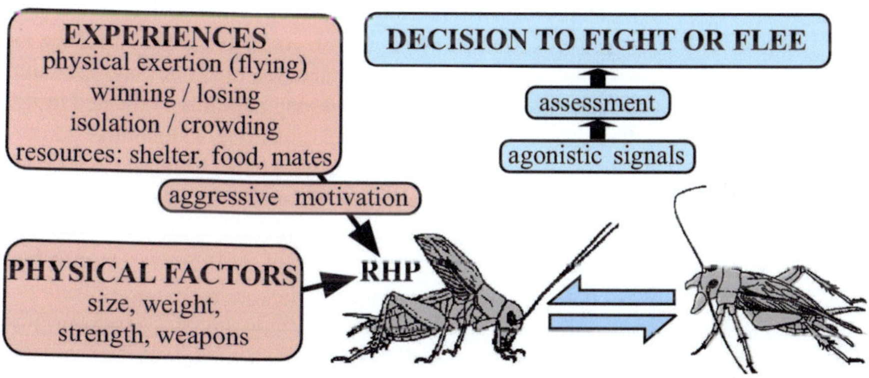

Fig. 12.1 Experience-dependent plasticity of aggression (modified from Stevenson and Rillich 2012). An individual's changes of winning an aggressive encounter is given by its "resource holding potential" (RHP) which depends on physical factors (size, strength, weight, weaponry) as well as on "aggressive motivation" (see Hurd 2006), a factor determined by numerous experiences (physical exertion, winning, losing and presence and value of resources such as territory, food and potential mates). In crickets, flying, winning and shelter occupancy are potentially rewarding experiences that promote aggressiveness via the action of octopamine, which represents the motivational component of aggression (Stevenson et al. 2005; Rillich et al. 2011; Rillich and Stevenson 2011). On confronting a competitor, the agonistic signals exchanged during escalating ritualised fighting are evaluated to assess RHP and to decide when it would be more opportune to persist in fighting or to flee. Crickets conform to the cumulative assessment model (Payne 1998), in that they flee the moment the sum of the opponent's agonistic signals surpasses some critical level (Rillich et al. 2007). The impact of these signals is mediated via activation of the NO/cGMP signalling pathway (Stevenson and Rillich 2015)

12.1.3 Biogenic Amines: Modulators of Aggressive Behaviour

Traditionally nervous systems were thought to generate each behaviour by virtue of the activity of a discrete dedicated circuit of interneurons that control a set of motor neurons and muscles in what David McFarland called the "behavioural final common path" (McFarland and Sibly 1975). It has since become realised (cf. Simpson and Stevenson 2015) that a single given physical circuit can function as a "polymorphic network" (Getting and Dekin 1985) subject to continued functional reconfiguration by the action of numerous neuromodulators and their blood-borne equivalents, neurohormones (Marder 2012). A neuromodulator can be generally defined as any substance released naturally in nervous tissue that alters the efficacy of "classical" synaptic transmission between a pre- and a postsynaptic cell by acting on dedicated metabotropic receptors. Hence, compared to neurotransmitters, the actions of neuromodulators are slower, but longer lasting, and span a far broader variety of effects that depend on the functional types and localities of the targeted receptors.

Neurochemicals with neuromodulator functions include primarily the biogenic amines along with numerous neuropeptides, but also more unconventional signalling molecules such as the gas nitric oxide. Crickets and other insects employ essentially the same neuronal signalling molecules as mammals, and they possess evolutionarily and pharmacologically related receptors (cf. Nagao and Tanimura 1993; Blenau and Baumann 2001; Homberg 2002; Hauser et al. 2006). Regarding neuromodulators, insects employ different though often structurally related neuropeptides and the same or at least analogous and structurally related amines. As in mammals, dedicated insect neurons can synthesise and release the catecholamine, dopamine, indolamine, serotonin (5-hydroxytryptamine) and histamine. Crickets are not known to release noradrenaline and adrenaline, which occur in only trace amounts in insects and other protostomes (Pflüger and Stevenson 2005). Instead, insects convert the catecholamine substrate amino acid L-tyrosine first to the amines tyramine and then to octopamine, which are known only as "trace amines" in the mammalian brain (Evans 1985).

Biogenic amines have long been attributed with influencing the expression of aggressive behaviour. Notably, the adrenergic/noradrenergic system is traditionally viewed as preparing vertebrate animals for fight or flight. However no consistent relationship with aggression has been found, although most recent data points towards promoting effects (Nelson and Trainor 2007; Haden and Scarpa 2007). This paucity in knowledge is at least partly attributable to the fact that research on biogenic amines and aggression in vertebrates has focused almost entirely on serotonin due to its reputed suppressing influence on the expression of aggression in humans and other mammals (Nelson and Trainor 2007).

These generalised actions of amines were thought to be reversed in invertebrates. For example, in crustaceans, serotonin was found to promote aggressiveness, while the invertebrate adrenergic analogue, octopamine, promotes submissiveness (Kravitz and Huber 2003). However, it is now clear that insects do not fit within this scheme. While the role of serotonin in insect aggression is still conjectural (see 1.6 below), work on crickets was the first to show that octopamine promotes aggression in an insect (Stevenson et al. 2000, 2005), and this was later verified for fruit flies (Baier et al. 2002; Hoyer et al. 2008; Zhou et al. 2008) and more recently in ants (Aonuma and Watanabe 2012) and stalk-eyed flies (Bubak et al. 2014). Hence, studies of insect model systems have already made a significant contribution to current understanding of how intraspecific aggression is controlled in the animal kingdom. Before turning to details, the following provides a brief summary of fighting behaviour in crickets.

12.2 Stereotyped Fighting and Its Initiation

12.2.1 Role of the Antennae

When two crickets meet, they are faced with the choice to court, fight or flee, and the outcome is largely controlled by information exchanged during antennal contact (Adamo and Hoy 1995; Hofmann and Schildberger 2001; Rillich and Stevenson 2015; see also Fernandez et al. 2010 on *Drosophila*). Species and sex are perceived by the pheromone signature (Iwasaki and Katagiri 2008), which induces males to court conspecific females. Antennal contact between males, which takes the form of "fencing", is a sufficient and necessary releasing stimulus for initiating cricket aggression and involves both mechanical and olfactory components (Hofmann and Schildberger 2001; Iwasaki and Katagiri 2008; Sakura and Aonuma 2013; Rillich and Stevenson 2015).

Agonistic responses, such as mandible spreading, can be evoked by simply lashing the antennae with a bristle (Alexander 1961; Rillich and Stevenson 2015) or by contact with male pheromones (Iwasaki and Katagiri 2008), which have been specifically identified in fruit flies (Wang and Anderson 2010). The antennal afferent pathways in the cricket brain are known in some detail (Staudacher et al. 2005; Yoritsune and Aonuma 2012) along with descending interneurons that are directly excited by mechanical antennal stimulation (Schöneich et al. 2011). These interneurons descend to thoracic motor centres; some respond to cricket song (Staudacher and Schildberger 1998) and can initiate walking, turning (Böhm and Schildberger 1992; Zorovic and Hedwig 2012) and possibly escape or backward walking as found for homologous neurons in other insects (Comer and Baba 2011; Bidaye et al. 2014).

12.2.2 Levels of Escalating Aggression

Once initiated, aggressive interactions between male crickets follow a stereotyped sequence of levels to which a fight escalates, before one contestant retreats (Hofmann and Stevenson 2000; Fig. 12.2). Antennal contact is always followed by mandible spreading, then mandible engagement and pushing, which culminates in a grappling contest. Most fights involve physical contact without injury

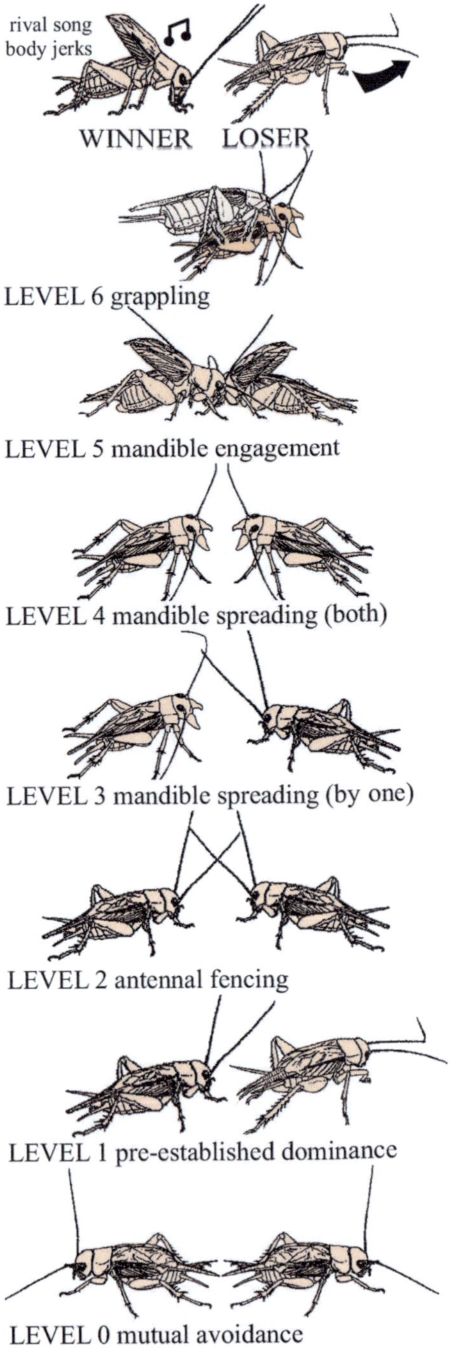

Fig. 12.2 Escalating levels of aggression for adult male crickets. Level 0 mutual avoidance: no aggressive interaction. Level 1 pre-established dominance: one cricket attacks, the other retreats. Level 2 antennal fencing. Level 3 mandible spreading (by one): one cricket displays spread mandibles. Level 4 mandible spreading (both): both crickets display spread mandibles. Level 5 mandible engagement: the mandibles interlock and the animals push against each other. Level 6 grappling: an all out fight, the animals may disengage and reengage to bite other body parts. Fights can be concluded at any level by one opponent, the loser, retreating, the established winner typically produces the rival song and body jerking movements (Modified from Stevenson et al. 2000, in part redrawn from Huber et al. 1989)

and last only several seconds, but fights can go on for minutes and result in the loss of an antenna or leg. The end of a fight is marked by the loser retreating, after which the winner frequently starts to sing a characteristic rivalry song and produce erratic jerking body movements. Winners repeatedly attack losers, while losers actively avoid contact with other males for hours after defeat. There is no firm evidence for individual recognition. Losers will also retreat from unfamiliar opponents (Hofmann and Stevenson 2000), although in *Drosophila*, losers appear to fight differently against familiar and unfamiliar opponents (Yurkovic et al. 2006).

Female crickets rarely interact, but fight vigorously in the presence of a courting male or his courtship song, and winning females have a greater probability of receiving the male's spermatophore (Rillich et al. 2009). In contrast, *Drosophila* males and females adopt different fighting strategies, whereby the males but not the females establish dominance relationships (Nilsen et al. 2004). The sexually dimorphic fighting patterns in fruit flies are specified by sex-specific splicing of the fruitless gene (Vrontou et al. 2006) and controlled by specific subsets of neurons expressing the male form of fruitless proteins (Chan and Kravitz 2007).

12.2.3 Amines and the Initiation of Fighting

Biogenic amines have frequently been shown to have the capacity of initiating selected motor behaviours by directly activating the underlying central pattern generators. The classical example in insects is initiation of flight in locusts (Sombati and Hoyle 1984; Stevenson and Kutsch 1987). Recent studies, however, indicate that a cholinergic, rather than aminergic, mechanism is necessary for initiating flight (Buhl et al. 2008), while amines act as accessory neuromodulators that promote release and fine-tune the motor score (Rillich et al. 2013).

Studies in crickets indicate a similar principle for aggression. The octopamine, dopamine and serotonin content of the cricket central nervous system can be effectively depleted by treatment with reserpine (Stevenson et al. 2000), which blocks the molecular transporter that loads free amines from the cytoplasm into storage vesicles in nerve terminals for subsequent release (Henry and Scherman 1989). Reserpinised crickets are extremely lethargic, and have severely depressed escape responses. However, they are still capable of exhibiting all major elements of aggressive behaviour, albeit they often need to be coaxed to do so by repeated mechanical stimulation of their antennae (Stevenson et al. 2000). Essentially the same response also occurs following semi-selective depletion of octopamine and dopamine using the competitive synthesis inhibitor alpha-methyl-p-tyrosine (AMT). In contrast, crickets with nervous systems depleted of serotonin by treatment with alpha-methyltryptophan (AMTP) exhibit hyperactivity, and enhanced escape responses, but seemingly unchanged aggressive behaviour (Stevenson et al. 2000). On the other hand, the tendency of male crickets to court rather than fight other males after antennectomy (Hofmann and Schildberger 2001) was suggested to result from loss of serotonin in the brain following this operation (Murakami and Itoh 2003). This seems unlikely, however, since antennectomy had no effect on the intensity of serotonin-like immunoreactivity in the cricket brain, and drugs that selectively block serotonergic, octopaminergic or dopaminergic signalling had no effect on the efficacy of antennal stimulation as an aggression-releasing stimulus (Rillich and Stevenson 2015).

Taken together, current evidence thus suggests that biogenic amines are not essential for initiating the basic motor elements of aggressive behaviour. Interestingly, however, prior antennal stimulation with a fresh cut, male antenna is followed by elevated expression of aggression in subsequent encounters with a male, via a mechanism that is dependent on the amine octopamine, but not serotonin or dopamine (Rillich and Stevenson 2015). This priming effect of octopamine is only one example where this amine acts as a neuromodulator to promote the expression of aggression (Rillich and Stevenson 2015). In the following we outline other examples in more detail.

12.3 Experience-Dependent Promotion of Aggression

12.3.1 Octopamine, Physical Exertion and the Flight Effect

Physical exertion, stress, challenge and fighting are frequently accompanied by fluctuations in the brain- or blood-content of neuromodulators, neurohormones and hormones (Wingfield et al. 1990; Bhatia et al. 2011). In vertebrates, stress-induced discharges of adrenalin/noradrenalin are thought to underlie the classical fight or flight response described originally by Walter Bradford Cannon (1915). Causal relationships between such changes and behaviour have, however, rarely been established. Insects exhibit a similar response (Lihoreau et al. 2009; Sokolowski 2010), whereby stressful and a variety of other experiences can lead to an almost tenfold elevation in the haemolymph content of octopamine, the invertebrate analogue of noradrenaline (Davenport and Evans 1984; Evans 1985). In crickets, increases in octopamine levels in the haemolymph or central nervous system are known to occur following a variety of experiences that influence aggressive behaviour (Fig. 12.3), including male antennal contact, copulation, fighting, flying (Adamo et al. 1995), grouping (Iba et al. 1995) and exposure to a mock predator (Adamo and Baker 2011).

Studies in crickets were the first to draw a correlation between activity-dependent promotion of octopaminergic signalling and aggression. Cricket fighting has been a popular pastime for centuries in China, where aficionados recommend "punishing" poor fighters by shaking and launching them in the air several times (Hofmann 1996), a treatment similar to that used to evoke stress-induced release of octopamine (Davenport and Evans 1984). This treatment in fact works surprisingly well, but it proved to be far more effective to simply induce the animals to fly for a minute or so (Hofmann and Stevenson 2000). After flying, crickets become

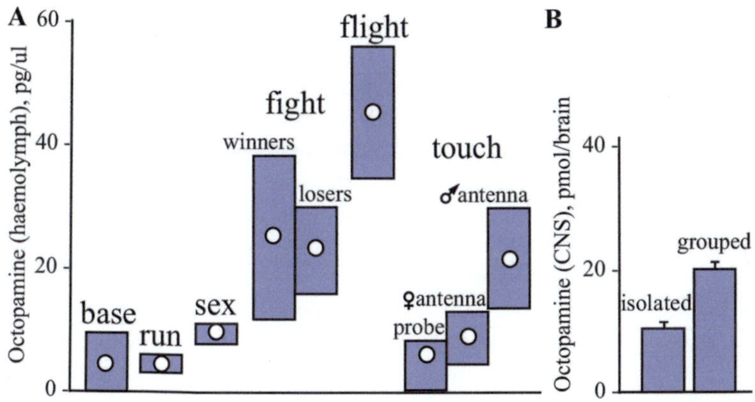

Fig. 12.3 Experience-dependent changes in levels of octopamine. (**a**) Haemolymph content of adult male crickets (pg/ul, *circles* median, *bars* interquartile range) at rest (base) after running (run), copulation (sex), fighting (fight, data for winners and losers), flying (flight) and touching the antennae (touch) with either a probe or male or female antenna (Adopted from data generated by Adamo et al. 1995). (**b**) Brain content (pmol/brain, *bar* mean, whisker standard error of mean) of adult crickets that were kept either in isolation or grouped (Adopted from data generated by Iba et al. 1995)

extremely aggressive, and their fights nearly always escalate to the highest level of aggression (6) and last two to three times longer than usual. These effects occurred without enhancing general excitability as evaluated from the animals' startle responses and clearly depended on the execution of flight motor activity, but not on wind stimulation alone.

Subsequent studies demonstrated that this flight effect is mediated by octopamine. It can be mimicked by treatment with the tissue-permeable octopamine agonist, chlordimeform, and abolished following octopamine/dopamine depletion with AMT or after selective blockade of octopamine receptors (Stevenson et al. 2000, 2005). Flying also modulates cricket courtship behaviour (Dyakonova and Krushinskii 2008) and the responsiveness of identified neurons to sensory stimuli in the same way as chlordimeform (Jung et al. 2011). Whether flying confers any advantage on migrant crickets over residents in securing territory is not known.

While flying leads to a pronounced surge of octopamine in the haemolymph (Adamo et al. 1995; Adamo and Baker 2011), the concentration is too low to pass the insect "blood-brain" barrier and will hence be without effect on the nervous control of aggressive behaviour (cf. Stevenson et al. 2005). However, this surge is largely due to heightened activation of central neurons (cf. Roeder 1999) which probably include specific dorsal and ventral unpaired median (DUM/VUM) neurons such as those activated during flight in locusts (Duch et al. 1999). DUM/VUM neurons are comparatively large and accessible neurons that are typically located on the dorsal, but occasionally ventral, midline of all ventral ganglia of all orthopterans and many other winged insects (Stevenson and Sporhase-Eichmann 1995; Bräunig and Pflüger 2001; Pflüger and Stevenson 2005). Though absent in the brain, a small group of several DUM/VUM neurons in the anterior of the suboesophageal ganglia ascend via the connectives to the brain, where they form widespread projections within all major brain neuropiles (e.g. locusts, Bräunig 1991; honeybees, Schröter et al. 2007; fruit flies, Busch and Tanimoto 2010), including regions where Franz Huber demonstrated, in his now classic experiments, that local electrical stimulation can elicit components of aggressive behaviour (Huber 1955, 1960). Elegant genetic techniques in fruit flies led to the identification of a small subset of octopaminergic cells, possibly VUM type, in the suboesophageal ganglion that are "functionally important for expressing aggression" (Zhou et al. 2008). The precise function of these neurons in aggression remains, however, to be established. Another subset of VUM cells in fruit flies express the sex-determining factor fruitless, and these appear to be involved in mediating the choice between courtship and aggression (Certel et al. 2007, 2010). Very little is known about DUM/VUM neurons in crickets (Gras et al. 1990; Bräunig et al. 1990; Stevenson and Sporhase-Eichmann 1995).

12.3.2 Octopamine and the Winner Effect

In practically all animals investigated, winning an aggressive encounter against a conspecific promotes an individual's aggressiveness, thereby rendering it more likely to win a subsequent encounter (Hsu et al. 2005). Although comparatively little is known of the proximate causes, recent studies implicate the involvement of androgens in rodents (Fuxjager and Marler 2010) and octopamine in crickets (Rillich and Stevenson 2011; Fig. 12.4).

In crickets, winning increases the probability of a cricket subsequently defeating an inexperienced opponent (Khazraie and Campan 1999) and is also associated with increased mating success (Tachon et al. 1999). When winners are matched against each other in knockout tournaments, the fights become progressively more severe and longer with each win scored (Rillich and Stevenson 2011). This winner effect is transient and persists for less than 20 min after winning, which is far shorter than in rodents, where it can last for days (Fuxjager and Marler 2010). Changes in social status in crickets are thus not necessarily associated with (long term) learning and memory, as suggested for fruit flies (Yurkovic et al. 2006). As found for the flight effect, the cricket winner effect is mediated by the amine octopamine. It is prohibited by treatment with the selective octopamine receptor blocker epinastine (cf. Roeder et al. 1998), but not by propranolol, a ß-adrenergic receptor antagonist, by yohimbine, an insect tyramine receptor blocker, nor by fluphenazine, an insect dopamine receptor blocker (Rillich and Stevenson 2011).

Insights into what actually constitutes a win were gained by interrupting fights between two contestants before either won (Rillich and Stevenson 2011). In a subsequent encounter, these crickets exhibit hyperaggressiveness, indicating that a winner effect can result alone from the physical exertion of fighting, without actually scoring a win. This is feasible considering that octopaminergic neurons are

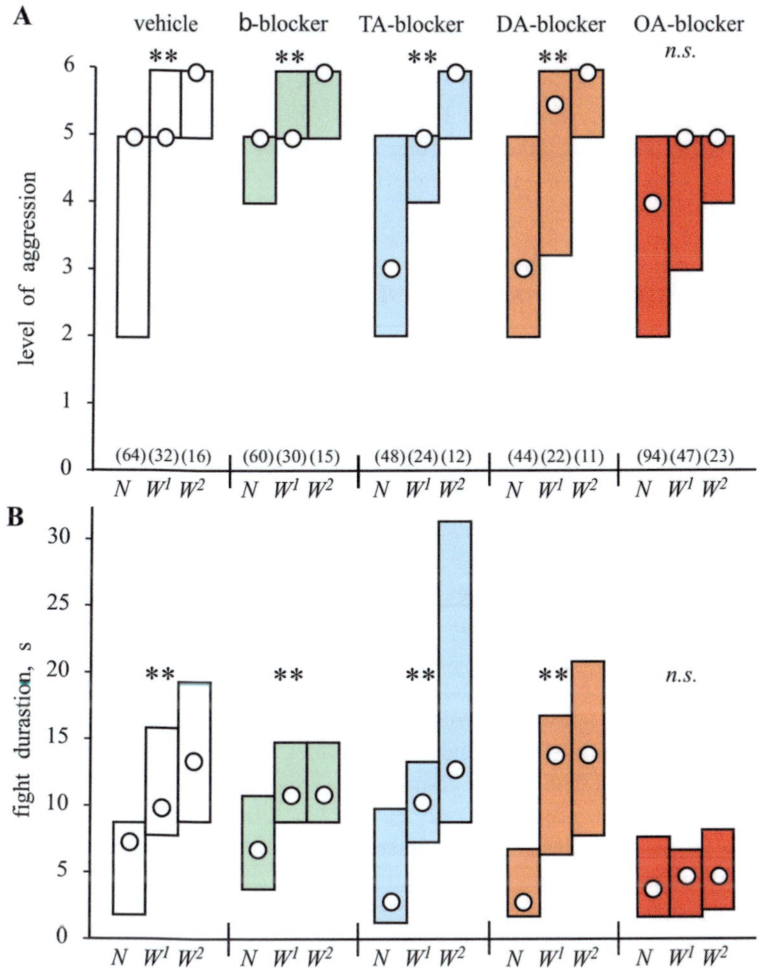

Fig. 12.4 Octopamine and the winner effect. (**a**) Level of aggression and (**b**) fight duration (*circles* medians, *bars* interquartile ranges) for encounters between pairs of weight-matched male crickets that were both either socially inexperienced (naive, N), winners of one previous encounter (W^1) or winners of two previous encounters (W^2) for an inter-fight interval of 5 min. Before the initial fight, the animals were injected with either vehicle (*white bars*), β-adrenergic blocker propranolol (*green bars*), tyramine (TA) blocker yohimbine (*blue bars*), dopamine (DA) blocker fluphenazine (*brown bars*) or octopamine (OA) blocker epinastine (*red bars*). Numbers in parenthesis in (**a**) give the pairs for each round. Significant differences between tournament rounds are indicated (Kruskal-Wallis one-way variance test, ** $p < 0.01$, *n.s.* not significant) (Adopted from Rillich and Stevenson 2011)

activated during walking and by a wide variety of mechanosensory signals (Gras et al. 1990; Morris et al. 1999) and that fighting itself leads to an almost fivefold increase in haemolymph levels of octopamine in winners and losers alike (Adamo et al. 1995). On the other hand, a winner effect also develops in crickets that experience wins against submissive opponents that retreat prior to any physical engagement (Rillich and Stevenson 2011). While it is known in humans that merely watching previous victories can elevate levels of hormones with aggression-promoting properties (Carré and Putnam 2010), this is a surprising finding for crickets. The following gives a further example of an essentially non-physical experience with aggression-promoting effects.

12.3.3 Octopamine and the Residency Effect

Regardless of species, animals that possess a key resource are more likely to win disputes against contenders, but it is hotly debated how this is controlled (reviews: Kemp and Wiklund 2004; Hsu et al. 2005). For male field crickets, burrows are valuable assets offering shelter from predators and an aid in attracting females. Females mate preferentially with burrow owners, and these owners zealously fight off any intruding male (Alexander 1961; Simmons 1986; Rodriguez-Munoz et al. 2011). Cricket species with burrowing males also tend to be more aggressive than non-burrowing species (Bertram et al. 2011).

In the laboratory, crickets that are submissive after having just lost a fight become highly aggressive when given an artificial shelter to occupy, and frequently defeat aggressive intruders (Rillich et al. 2011). This residency effect does not depend on the initial sensory experience of shelter acquisition, since it becomes first evident after 2–15 min of residency and declines within 15 min after taking the shelter away, i.e. on a similar time course as the winner effect. Furthermore, shelters of wire or with a transparent roof are far less effective, and darkness alone ineffective (personal observations). It seems the cricket must reside in a dark, burrow-like structure. Whatever its proximate cause, the residency effect is clearly octopamine-dependent. It is not evident in crickets depleted of octopamine and dopamine, while being unaffected by serotonin depletion, but selectively blocked by treatment with octopamine receptor antagonists (Rillich et al. 2011; Fig. 12.5).

12.3.4 Octopamine and Reward

The paradoxical question posed by our studies is how experiences as diverse as physical exertion (flying), fighting and residency – which span the entire range of energy expenditure – all promote aggression via a seemingly common mechanism involving octopamine? An intriguing idea is that all these experiences are in some way associated as being positive or in some way rewarding (Rillich et al. 2011; Rillich and Stevenson 2011). Physical exercise in mammals, including humans, seems to be equated with reward (Raichlen et al. 2011) and can act as a mood elevator that alleviates symptoms of depression by invoking changes in a variety of neurotransmitter systems including dopamine (Craft and Perna 2004). Aggression

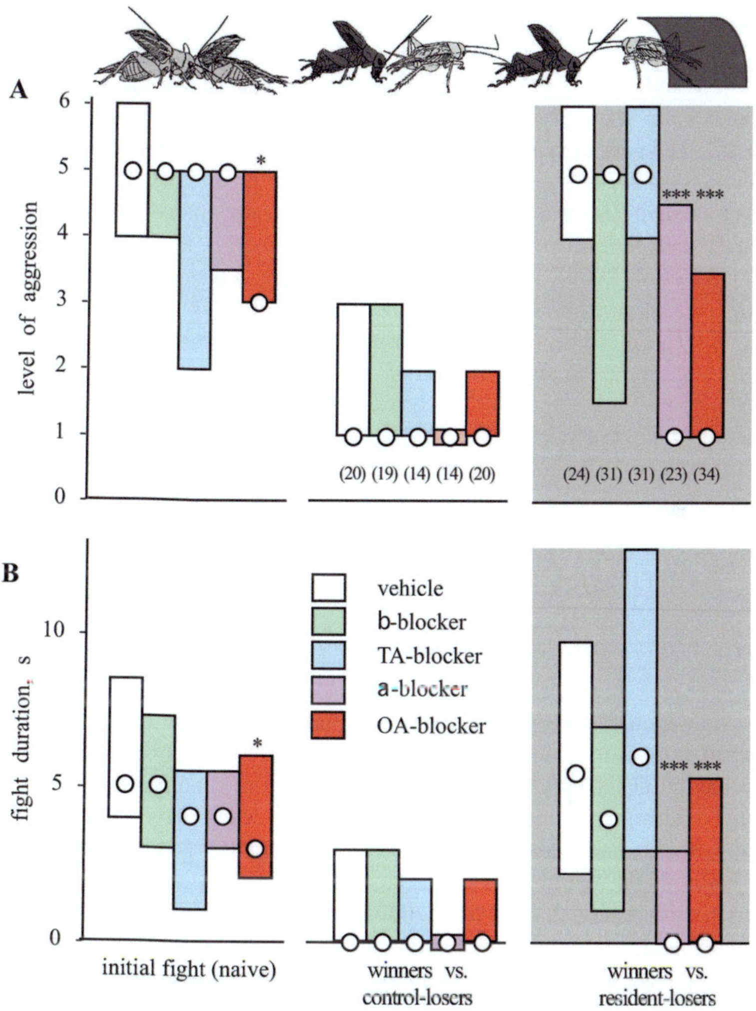

Fig. 12.5 Octopamine and the residency effect. (**a**) Level of aggression and (**b**) fight duration (*circles* medians, *bars* interquartile ranges) for initial fights between weight-matched socially inexperienced (naive) male crickets and for a subsequent interaction between winners and losers which previously remained in the arena without a shelter for 15 min (winners vs. control losers) or occupied a shelter in the arena for 15 min (winners vs. resident losers, *grey* background). Prior to the initial fight, the crickets were treated with either a vehicle (*white bars*), a β-adrenergic blocker (*green bars*), a tyramine (TA) blocker (*blue bars*), an α-adrenergic blocker (*violet bars*) or an octopamine (OA) blocker (*red bars*). The number of contests evaluated (n) is given in parenthesis beneath each column, excepting initial fight, which is pooled. *Asterisks* denote statistically significant differences (Mann–Whitney *U*-test *, **, ***: $p < 0.05$, 0.01, 0.001, respectively) (Adopted from Rillich et al. 2011)

in mammals also leads to increased activity in dopaminergic pathways and increases the expression of androgen receptors in regions of the brain mediating motivation and reward (O'Connell and Hofmann 2011). In insects, rewarding experiences are primarily associated with octopamine (review: Perry and Barron 2013). In honeybees the value of food sources appears to be encoded by octopamine modulating associative reward pathways (Barron et al. 2010). Octopamine is known to convey reward signals in appetitive learning paradigms in honeybees (Hammer and Menzel 1995), fruit flies (Schwärzel et al. 2003) and crickets (Mizunami et al. 2009). It has even been demonstrated that the activity of only one of the group of 15 DUM/VUM neurons ("VUMmx1") in honeybees (cf. Schröter et al. 2007) can substitute for the sucrose reward in an associative, appetitive learning paradigm (Hammer 1993). Considering the possible role of members of this cell group in the expression of aggression and courtship in fruit flies (Zhou et al. 2008; Certel et al. 2007, 2010), we clearly now need to learn more about the related DUM/VUM neurons in the suboesophageal ganglion of crickets.

12.4 Experience-Dependent Promotion of Submission

12.4.1 Opponent Assessment and the Decision to Flee

Fights are concluded the moment one contestant submits or retreats. The events leading to submission and its maintenance are, however, poorly understood. Various behavioural theories agree that information from ritualised agonistic signals exchanged during fighting are assessed to determine when to fight or flee, but differ regarding which individual evaluates these cues (sender, receiver or both: Payne 1998; Elwood and Arnott 2012).

By experimentally manipulating information exchange, it was revealed that crickets evaluate only the opponent's signals and that these signals promote the "decision to flee" (Rillich et al. 2007). In one key experiment, crickets with either blackened eyes ("blinded") or lamed mandibles were found to fight against untreated, equally sized opponents with almost unabated vigour and chance of winning, whereas the "blinded" crickets won practically all fights against crickets with lamed mandibles (Fig. 12.6). This unusual finding is fully conformed to the core prediction of the cumulative assessment model of Payne (1998) that an animal persists in fighting until the accumulated sum of the opponent's actions surpasses some critical threshold to flee. Hence, the "blinded" cricket persists since it receives no visual and only limited physical input from the opponent with lamed mandibles, whereas the latter accumulates the full brunt of his adversary's actions and thus becomes the first to flee (Rillich et al. 2007). The cumulative assessment model also accounts for effects of physical disparities (e.g. size strength and weaponry: Dixon and Cade 1986; Judge and Bonanno 2008; Hall et al. 2010) or energy status (Briffa 2008) on fight outcome. For example, an animal with any physical or energetic advantage will have a greater sensory impact on its opponent, which will hence be

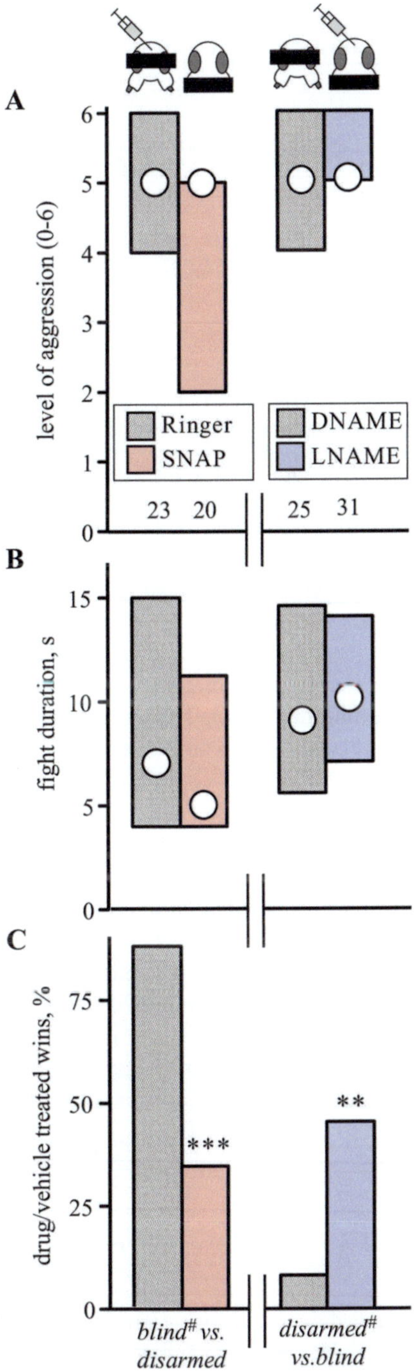

Fig. 12.6 Effects of impaired agonistic signalling and nitridergic drugs on cricket aggression: (**a**) level of aggression, (**b**) fight duration (*circles* medians, *bars* interquartile ranges) and (**c**) win chances (%). In each case male crickets deprived of visual information (*blind*) were matched against weight-matched males with disabled mandibles (*disarmed*), but one opponent (indicated by pictogram and # in the x-axis label) received either control solutions (*grey bars*, ringer or DNAME) or the NO donor S-nitroso-N-acetyl-DL-penicillamine (SNAP, *red bars*) or the NO synthase inhibitor nitro-L-arginine methyl ester hydrochloride (LNAME, *blue bars*). The number of fighting pairs is given above the top axis. *Asterisks* in (**c**) denote statistically significant differences (chi-square test **, ***: $p < 0.01$, 0.001, respectively) (Adopted from Stevenson and Rillich 2015)

more likely to flee first. But how do crickets add up the sensory impact of their opponents, and how does the perception of this lead to retreat?

12.4.2 Adding Up the Odds: The Role of Nitric Oxide

Recent studies have revealed that the gaseous neuromodulator nitric oxide (NO) plays a key role in mediating the effect of an opponent's agonistic actions (Stevenson and Rillich 2015). In both mammals and insect, this unconventional neuromodulator can be synthesised by neurons bearing the enzyme nitric oxide synthase and, once produced, traverses by diffusion to activate the intracellular receptor molecule soluble guanylyl cyclase, which in turn initiates production of the second messenger cyclic guanosine monophosphate (cGMP, review: Müller 1997). In mammals NO can act to suppress aggression, at least partly by influencing serotonergic signalling (Nelson and Trainor 2007), but its specific behavioural function in normal aggressive behaviour is unknown. Earlier work on crickets indicated that NO can either promote or inhibit the expression of aggression depending on circumstances. For example, inhibition of NO synthesis has been reported to prohibit the aggression-promoting effects of flying (Dyakonova and Krushinskyii 2006) or have no effect on socially naive crickets, but increase aggression in submissive losers (Iwasaki et al. 2007). More recently, we found that inhibitors of the NO/cGMP pathway increase aggressiveness in socially naive crickets while activators suppress it (Stevenson and Rillich 2015), i.e. in effect NO suppresses aggression as in mammals. However, rather than simply suppressing the tendency to fight, i.e. aggressive motivation, the application of nitridergic drugs to animals with manipulated signalling abilities revealed that NO mediates the impact of the opponent's aggressive signals during fighting. To take one example, when treated with an NO donor, crickets deprived of visual inputs (blinded) escalate and persist normally, but no longer have a win advantage over opponents rendered unable to inflict force with their mandibles. Conversely, when treated with an NOS inhibitor, crickets with lamed mandibles no longer have a win disadvantage against blinded crickets (Fig. 12.6; for supporting data, see Stevenson and Rillich 2015). Taken together, the data suggest that any aversive stimulus perceived in the context of aggression leads to activation of the NO signalling pathway, which in turn increases the probability of fleeing in response to further aversive stimuli.

12.4.3 The Loser Effect

Social defeat, i.e. losing an agonistic dispute with a conspecific, is followed by a period of suppressed aggressiveness in many animal species (Hsu et al. 2005; Rutte et al. 2006) and is generally regarded as a major stressor, which in humans may play a role in psychiatric disorders (Huhman 2006). Although accompanied by numerous changes in brain chemicals and gene expression (Miczek et al. 2011), the underlying cause of the loser effect is unknown.

Once a cricket has decided to flee, it will subsequently retreat on contact with any conspecific male (Alexander 1961; Adamo and Hoy 1995; Khazraie and Campan 1999; Hofmann and Stevenson 2000) and requires on average some 3 h to fully regain its initial level of aggressiveness (Stevenson and Rillich 2013). Confirming work of earlier authors (Iwasaki et al 2007), treatment with nitridergic drugs revealed that this loser effect in crickets results from activation of the NO/cGMP pathway (Stevenson and Rillich 2015). In males treated with nitridergic agonists, recovery was delayed by up to 24 h, whereas the majority of those receiving antagonists recovered far earlier (within 15 min). It is important to stress that socially subjugated crickets are still potentially aggressive. Losers will often attack other losers when they retreat first, and they will fight vigorously when their eyes are blackened, which eliminates the visual impact of the approaching opponent (Rillich et al. 2007). Losers are also equally responsive to antennal stimulation, the releasing stimulus for aggression, as socially naive crickets (Rillich and Stevenson 2015). Accordingly, losers are more susceptible to aversive stimulation, rather than motivationally depressed.

Nitric oxide is unlikely to act alone in controlling the decision to flee and loser submissiveness – for one it occurs in neurons that can be expected to contain more conventional neuromodulators (see e.g. Bullerjahn et al. 2006), and amines in particular are likely to be involved. Recovery from the loser effect is prohibited in crickets following depletion of octopamine and dopamine after treatment with the synthesis inhibitor AMT, and octopamine or dopamine receptor agonists are sufficient to fully restore aggression (Rillich and Stevenson 2014). However, while loser crickets still regain their aggressiveness after octopamine receptor blockade, they are prevented from doing so by dopamine receptor blockade. Hence, dopaminergic signalling is necessary for the normal recovery of aggression after social defeat in crickets (Rillich and Stevenson 2014). Finally, a mathematical model for recovery from defeat has been developed for crickets on the assumption that serotonin is involved (Yano et al. 2012), but solid experimental evidence for this is lacking (see also Sect. 6 on serotonin below).

12.5 Social Isolation, Biogenic Amines and Aggression

Social isolation results in dramatic behavioural and physiological changes in a wide variety of animal species from insects to man (Cacioppo and Hawkley 2009; Lihoreau et al. 2009; Simpson and Sword 2009; Sokolowski 2010). A wealth of studies have noted that isolation leads to increased aggressive behaviour in vertebrates (Hsu et al. 2005) and in insects such as solitary wasps (Pfennig and Reeve 1989), fruit flies (Zhou et al. 2008; Johnson et al. 2009) and crickets (Alexander 1961; Adamo and Hoy 1995; Iba et al. 1995). Isolation and crowding in insects are also associated with dramatic changes in the levels of biogenic amines (Iba et al.

1995; Rogers et al. 2004; Wada-Katsumata et al. 2011). However, a recent study on crickets revealed that reduced aggression of grouped individual results from social subjugation and resultant submissive behaviour of most group members by one or two dominant males, while heightened aggression in isolates is simply due to recovery from the loser effect and a return to a default aggressive state (Stevenson and Rillich 2013). While the effects of social isolation in different animal groups will no doubt differ depending on social structure, the possibility that recovery from social subjugation may contribute to heightened aggressiveness in social isolates appears to have been neglected in many studies.

12.6 Serotonin and Aggression

The actions of octopamine in arthropods are often functionally antagonised by serotonin. Examples of this antagonism can be seen in antennal scanning in honeybees (Erber et al. 1993), escape in cockroaches Goldstein and Camhi 1991) and crickets (Dyakonova et al. 1999), mating interval in male crickets (Nagao et al. 1991) and aggression in crustaceans (Kravitz and Huber 2003). Serotonin is renowned for its restraining effect on aggression in numerous animals including man (Kravitz and Huber 2003; Nelson and Trainor 2007; Passamonti et al. 2012). In locusts it clearly promotes grouping and swarm formation by subduing mutual avoidance or promoting attraction (Anstey et al. 2009). The role of serotonin in insect aggression is, however, not yet clear.

A promoting effect of serotonin on cricket aggression is suggested by the observation that reduced aggression, after losing and antennal ablation (cf. Hofmann and Schildberger 2001), is correlated with decreased serotonin brain content (Murakami and Itoh 2001, 2003). However, the loss of serotonin from the cricket nervous system following treatment with the synthesis inhibitor alpha-methyltryptophan (AMTP) has been found to induce hyperactivity and enhances startle responses in crickets, but does not have any obvious effects on aggression (Stevenson et al. 2000, 2005; Rillich and Stevenson 2015). Serotonin depletion does, however, appear to reduce the chance of winning (Dyakonova et al. 1999), though this may be a non-selective effect of hyperactivity. Nonetheless, a more recent study (Dyakonova and Krushinskii 2013) revealed clear, but in part functionally conflicting, effects of elevating serotonin levels by treatment with its precursor 5-hydroxytryptophan (5HTP). On the one hand, 5HTP induces a raised "aggressive-like" body posture (see also Kravitz and Huber 2003 on crustaceans), enhanced general activity, more frequent rival song production and longer fights that do not resolve clear losers. On the other hand 5HTP-treated crickets exhibit a delayed latency to spread their mandibles, launch fewer attacks and have an unchanged chance of winning.

Similarly conflicting findings have been reported for fruit flies. While Baier et al. (2002) reported that aggression in fruit flies was unaffected by blockade of serotonin biosynthesis, or 5HTP treatment, Dierick and Greenspan (2007) found aggression was promoted by 5HTP. Similarly, Alekseyenko et al. (2010), using molecular genetic techniques, report that acute activation of serotonergic neurons resulted in flies that escalated faster and fought at higher intensities, while selective disruption of serotonergic neurotransmission yielded flies that fought with reduced ability to escalate fights.

These inconsistencies can be expected to at least partly result from serotonin acting on different receptor subtypes. For example, pharmacological activation of 5HT2-type receptors reduces total aggression in *Drosophila*, while activating 5HT1A-type receptors increased it (Johnson et al. 2009). Also in mammals, different serotonin receptor subtypes influence different aspects of the total aggressive behavioural repertoire (de Boer and Koolhaas 2005). These and other findings challenge the dogmatic view of serotonin acting simply to suppress aggression in mammals, where it is currently thought to limit impulsivity (Nelson and Trainor 2007) or promote the drive to withdraw (Tops et al. 2009). An analogous scenario is conceivable for crickets, where evidence suggests that serotonergic signalling depresses escape responses in aggressive individuals (Dyakonova et al. 1999), while losers show enhanced escape behaviour due to lower brain levels of serotonin after defeat (Murakami and Itoh 2001). In crayfish, the effects of serotonin on escape and aggressive–submissive body posture change with social status due to a shift in the relative expression of different serotonin receptor subtypes to a pattern more appropriate for the new status (Cattaert et al. 2010). In conclusion, some features of dominant behaviour involve activation of the serotonergic system, while a decrease in serotonergic signalling is functionally important for the control of loser behaviour (Dyakonova and Krushinsky 2013). Considering findings in other animal groups, it is conceivable that the different actions of serotonin are mediated via different receptor subtypes, which may change in their relative expression after social defeat.

12.7 Neuropeptides and Aggression

Compared to vertebrates very little is known about the roles of peptides in controlling aggression in invertebrates. In crickets, treatment with the opiate antagonist naloxone elevates aggressiveness in losers, without affecting winners or socially naive animals (Dyakonova et al. 2002). In *Drosophila* aggression is increased following genetic silencing of circuitry employing neuropeptide F, the invertebrate homologue of neuropeptide Y (Dierick and Greenspan 2007).

12.8 Conclusions and Future Directions

Work on crickets have revealed that biogenic amines and nitric oxide signalling play key roles in mediating the effects of a wide variety of experiences on the expression of aggression. In this respect the cricket has advanced to the status of a model system for investigating experience-dependent plasticity of social behaviour.

Octopamine, the invertebrate analogue of noradrenaline, increases aggression by promoting the tendency to fight (Stevenson et al. 2005) and exhibition of agonist behaviours such as lunging (Hoyer et al. 2008; Zhou et al. 2008) and mandible spreading (Rillich and Stevenson 2011). Experiences as diverse as physical exertion (flying, fighting), winning and possession of resources (shelter) may all be evaluated as being in some way positive or rewarding, and these experiences promote aggression via a mechanism dependent on activation of the octopaminergic system (reviews: Stevenson and Rillich 2012; Stevenson and Schildberger 2013; Simpson and Stevenson 2015). In this respect, octopamine can be considered as representing the motivational component of aggression. Candidate neurons for mediating these effects are members of the group of DUM/VUM neurons with somata in the suboesophageal ganglion that project to the brain. Related neurons have already been shown to mediate reward in associative learning of honeybees and are possibly important for the expression of aggression in fruit flies (Zhou et al. 2008). The function of these cells in crickets will not be easy to analyse due to their irregular and variable localization (Sporhase-Eichmann et al. 1992; Stevenson and Spörhase-Eichmann 1995) as well as the current lack of genetic silencing and activation techniques that have been firmly established for fruit flies.

Submissive behaviour and the timing of the decision to flee result from the assessment of agonistic signals exchanged during fighting. Crickets conform to the cumulative assessment hypothesis of Payne (1998) in that they persist in fighting until the sum of the perceived adversary's actions surpasses some threshold to flee (Rillich et al. 2007). Recent studies have revealed that aversive stimuli, such as an opponent's agonist signals, promote the tendency to flee via activation of the NO/cGMP pathway (Stevenson and Rillich 2015). Although defeated crickets have a reduced tendency to fight, they are still potentially aggressive (Rillich et al. 2007). It appears now that losing increases the tendency to flee rather than reduces aggressiveness per se (Stevenson and Rillich 2015). Activation of the NO/cGMP pathway also results in the reduced expression of aggression after losing. While both octopamine and dopamine can each readily restore aggressiveness in losers, alone dopamine is necessary for the normal recovery of aggressiveness (Rillich and Stevenson 2014). The effects of social subjugation have been shown to be responsible for the reduced aggressiveness of grouped individuals, while recovery from the loser effect is the main cause of heightened aggressiveness in isolates (Stevenson and Rillich 2013).

While the role of serotonin in aggression remains unclear, recent findings indicate that this amine may promote some aspects of dominant behaviour while also being functionally important for controlling submissive behaviour after social defeat (Dyakonova and Krushinskii 2013). In view of findings in other animal species, the different actions of serotonin may be mediated via different receptor subtypes, which may have different patterns of expression depending on social status. Genes encoding serotonin receptor subtypes have recently been identified in crickets (Watanabe et al. 2011; Watanabe and Aonuma 2012) and we now need to know more of their distribution in nervous tissue and their specific pharmacology. Candidate serotonergic neurons for influencing aggression have only been identified in *Drosophila* (Alekseyenko and Kravitz 2015).

In addition to biogenic amines, the expression of aggressive behaviour in insects is also modulated by some peptides. Further work is necessary to understand the exact roles of such modulators and how they interact with aminergic pathways. Peptides and nitric oxide often occur as co-transmitters in aminergic neurons (cf. Bullerjahn et al. 2006), so we also need to know more about their distribution in relation to biogenic amines in the cricket brain. We in fact know very little about the neuronal substrates for aggression in crickets, and future effort must be devoted to performing chronic recordings from different regions of the brain during behaviour.

It should also be mentioned that the effects of experiences on aggressive behaviour outlined here are relatively short lived, and yet aggression can have longer-term changes on the operation of the nervous system than those discussed here. Agonistic behaviour can trigger neurogenesis (Ghosal et al. 2009) and FOS-like protein expression in the male cricket brain (Ghosal et al. 2010), but it is not known whether this leads to changes in behaviour. A hint of the possible complexities involved is given by the finding that aggressive behaviour in *Drosophila* is affected by over 50 novel genes with widespread pleiotropic effects (Edwards et al. 2009).

In effect, the biogenic amine octopamine and nitric oxide control the expression of aggression by modulation the respective behavioural thresholds to fight and to flee. These modulators no doubt operate in concert with the amines dopamine and serotonin as well as selected neuropeptides. In essence, octopamine, serotonin and nitric oxide appear to have similar roles to those emerging for their counterparts in the control of aggression in mammals.

Acknowledgements We thank the German Research Council (DFG) for generous funding (DFG Research Group, FOR 1363, grant STE 714/4-1).

References

Adamo SA, Baker JL (2011) Conserved features of chronic stress across phyla: the effects of long-term stress on behavior and the concentration of the neurohormone octopamine in the cricket, *Gryllus texensis*. Horm Behav 60:478–483

Adamo SA, Hoy RR (1995) Agonistic behavior in male and female field crickets, *Gryllus bimaculatus*, and how behavioural context influences its expression. Anim Behav 49:1491–1501

Adamo SA, Linn CE, Hoy RR (1995) The role of neurohormonal octopamine during 'fight or flight' behaviour in the field cricket *Gryllus bimaculatus*. J Exp Biol 198:1691–1700

Albert D, Walsh M, Jonik R (1993) Aggression in humans: what is its biological foundation? Neurosci Biobehav Rev 17:405–425

Alekseyenko OV, Kravitz EA (2015) Serotonin and the search for the anatomical substrate of aggression. Fly 8(4):1–6

Alekseyenko OV, Lee C, Kravitz EA (2010) Targeted manipulation of serotonergic neurotransmission affects the escalation of aggression in adult male *Drosophila melanogaster*. PLoS One 5:e10806

Alexander RD (1961) Aggressiveness, territoriality, and sexual behaviour in field crickets (Orthoptera: Gryllidae). Behavior 17:130–223
Anstey ML, Rogers SM, Ott SR, Burrows M, Simpson SJ (2009) Serotonin mediates behavioral gregarization underlying swarm formation in desert locusts. Science 323:627–630
Aonuma H, Watanabe T (2012) Octopaminergic system in the brain controls aggressive motivation in the ant. Formica Jpn Acta Biol Hung 63(Suppl 2):63–68
Archer J (1988) The behavioural biology of aggression. Cambridge University Press, Cambridge
Baier A, Wittek B, Brembs B (2002) *Drosophila* as a new model organism for the neurobiology of aggression? J Exp Biol 205:1233–1240
Barron AB, Sovik E, Cornish J (2010) The roles of dopamine and related compounds in reward-seeking behavior across animals phyla. Front Behav Neurosci 4:1–9
Bertram SM, Rook VLM, Fitzsimmons JM, Fitzsimmons LP (2011) Fine- and broad-scale approaches to understanding the evolution of aggression in crickets. Ethology 117:1067–1080
Bhatia N, Maiti PP, Choudhary A, Tuli A, Masih D, Khan MMU et al (2011) Animal models in the study of stress: a review. NSHM J Pharm Healthc Manag 2:42–50
Bidaye SS, Machacek C, Wu Y, Dickson BJ (2014) Neuronal control of Drosophila walking direction. Science 344:97–101
Blenau W, Baumann A (2001) Molecular and pharmacological properties of insect biogenic amine receptors: lessons from *Drosophila melanogaster* and *Apis mellifera*. Arch Insect Biochem Physiol 48:13–38
Böhm H, Schildberger K (1992) Brain neurones involved in the control of walking in the cricket *Gryllus bimaculatus*. J Exp Biol 166:113–130
Brace RC, Purvey J (1978) Size-dependant dominance hierarchy in the anemone *Actinia equina*. Nature 273:752–753
Bräunig P (1991) Suboesophageal DUM neurons innervate the principal neuropiles of the locust brain. Philos Trans R Soc Lond B 332:221–240
Braunig P, Pflüger HJ (2001) The unpaired median neurons of insects. Adv Insect Physiol 28:185–266
Bräunig P, Allgäuer C, Honegger HW (1990) Suboesophageal DUM neurones are part of the antennal motor system of locusts and crickets. Experientia 46:259–261
Briffa M (2008) Decisions during fights in the house cricket, Acheta domesticus: mutual or self assessment of energy, weapons and size? Anim Behav 75:1053–1062
Brown WD, Smith AT, Moskalik B, Gabriel J (2006) Aggressive contests in house crickets: size, motivation and the information content of aggressive songs. Anim Behav 72:225–233
Brown WD, Chimenti AJ, Siebert JR (2007) The payoff of fighting in house crickets: motivational asymmetry increases male aggression and mating success. Ethology 113:457–465
Bubak AN, Grace JL, Watt MJ, Renner KJ, Swallow JG (2014) Neurochemistry as a bridge between morphology and behavior: perspectives on aggression in insects. Curr Zool 60:778–790
Buhl E, Schildberger K, Stevenson PA (2008) A muscarinic cholinergic mechanism underlies activation of the central pattern generator for locust flight. J Exp Biol 211:2346–2357
Bullerjahn A, Mentel T, Pflüger HJ, Stevenson PA (2006) Nitric oxide: a co-modulator of efferent peptidergic neurosecretory cells including a unique octopaminergic neurone innervating locust heart. Cell Tissue Res 325:345–360
Busch S, Tanimoto H (2010) Cellular configuration of single octopamine neurons in *Drosophila*. J Comp Neurol 518:2355–2364
Cacioppo JT, Hawkley LC (2009) Perceived social isolation and cognition. Trends Cogn Sci 13:447–454
Cannon WB (1915) Bodily changes in pain, hunger, fear and rage: an account of recent researches into the function of emotional excitement. Appleton, New York
Carre JM, Putnam SK (2010) Watching a previous victory produces an increase in testosterone among elite hockey players. Psychoneuroendocrinology 35:475–479
Cattaert D, Delbecque JP, Edwards DH, Issa FA (2010) Social interactions determine postural network sensitivity to 5-HT. J Neurosci 30:5603–5616

Certel SJ, Savella MG, Schlegel DC, Kravitz EA (2007) Modulation of *Drosophila* male behavioral choice. Proc Natl Acad Sci U S A 104:4706–4711

Certel SJ, Leung A, Lin CY, Perez P, Chiang AS, Kravitz EA (2010) Octopamine neuromodulatory effects on a social behavior decision-making network in *Drosophila* males. PLoS One 5:e13248

Chan YB, Kravitz E (2007) Specific subgroups of FruM neurons control sexually dimorphic patterns of aggression in *Drosophila melanogaster*. Proc Natl Acad Sci U S A 104:9577–9582

Comer C, Baba Y (2011) Active touch in orthopteroid insects: behaviours, multisensory substrates and evolution. Philos Trans R Soc Lond B 366:3006–3015

Craft LL, Perna FM (2004) The benefits of exercise for the clinically depressed. J Clin Psychiatry 6:104–111

Davenport AP, Evans PD (1984) Changes in haemolymph octopamine levels associated with food deprivation in the locust *Schistocerca gregaria*. Physiol Entomol 9:269–274

de Boer SF, Koolhaas JM (2005) 5-HT1A and 5-HT1B receptor agonists and aggression: a pharmacological challenge of the serotonin deficiency hypothesis. Eur J Pharmacol 526:125–139

Dierick HA, Greenspan RJ (2007) Serotonin and neuropeptide F have opposite modulatory effects on fly aggression. Nat Genet 39:678–682

DiRienzo N, Pruitt JN, Hedrick AV (2012) Juvenile exposure to acoustic sexual signals from conspecifics alters growth trajectory and an adult personality trait. Anim Behav 84:861–868

Dixon KA, Cade WH (1986) Some factors influencing male-male aggression in the field Cricket *Gryllus integer* (time of day, age, weight and sexual maturity). Anim Behav 34:340–346

Duch C, Mentel T, Pflüger HJ (1999) Distribution and activation of different types of octopaminergic DUM neurons in the locust. J Comput Neurol 403:119–134

Dyakonova VE, Krushinkskii AL (2006) Effects of an NO synthase inhibitor on aggressive and sexual behavior in male crickets. Neurosci Behav Physiol 36:565–571

Dyakonova VE, Krushinkskii AL (2013) Serotonin precursor (5-hydroxytryptophan) causes substantial changes in the fighting behavior of male crickets, *Gryllus bimaculatus*. J Comp Physiol A 199:601–609

Dyakonova VE, Krushinskii AL (2008) Previous motor experience enhances courtship behaviour in male cricket *Gryllus bimaculatus*. J Insect Behav 21:172–180

Dyakonova VE, Schurmann F, Sakharov DA (1999) Effects of serotonergic and opioidergic drugs on escape behaviors and social status of male crickets. Naturwissenschaften 86:435–437

Dyakonova VE, Schurmann FW, Sakharov DA (2002) Effects of opiate ligands on intraspecific aggression in crickets. Peptides 23:835–841

Edwards AC, Zwarts L, Yamamoto A, Callaerts P, Mackay TF (2009) Mutations in many genes affect aggressive behavior in *Drosophila melanogaster*. BMC Biol 7:29

Elwood RW, Arnott G (2012) Understanding how animals fight with Lloyd Morgan's canon. Anim Behav 84:1095–1102

Erber J, Kloppenburg P, Scheidler A (1993) Neuromodulation by serotonin and octopamine in the honeybee: behaviour, neuroanatomy and electrophysiology. Experientia 49:1073–1083

Evans PD (1985) Octopamine. In: Kerkut GA, Gilbert LI (eds) Comprehensive insect physiology biochemistry and pharmacology. Pergamon, Oxford, pp 499–530

Fernandez MP, Chan YB, Yew JY, Billeter JC, Dreisewerd K, Levine JD et al (2010) Pheromonal and behavioral cues trigger male-to-female aggression in *Drosophila*. PLoS Biol 8:e1000541

Fuxjager MJ, Marler CA (2010) How and why the winner effect forms: influences of contest environment and species differences. Behav Ecol 21:37–45

Getting PA, Dekin MA (1985) Tritonia swimming. A model system for integration within rhythmic motor systems. In: Selverston AI (ed) Model networks and behavior. Plenum Press, New York, pp 3–20

Ghosal K, Gupta M, Killian KA (2009) Agonistic behavior enhances adult neurogenesis in male *Acheta domesticus* crickets. J Exp Biol 212:2045–2056

Ghosal K, Naples SP, Rabe AR, Killian KA (2010) Agonistic behavior and electrical stimulation of the antennae induces Fos-like protein expression in the male cricket brain. Arch Insect Biochem Physiol 74:38–51

Giurfa M (2012) Social learning in insects: a higher-order capacity? Front Behav Neurosci 6:57

Goldstein RS, Camhi JM (1991) Different effects of the biogenic amines dopamine, serotonin and octopamine on the thoracic and abdominal portions of the escape circuit in the cockroach. J Comp Physiol A 168:103–112

Gras H, Hörner M, Runge L, Schürmann FW (1990) Prothoracic DUM neurons of the cricket *Gryllus bimaculatus* respond to natural stimuli activity in walking behaviour. J Comp Physiol A 166:901–914

Haden SC, Scarpa A (2007) The noradrenergic system and its involvement in aggressive behaviors. Aggress Violent Behav 12:1–15

Hall MD, McLaren L, Brooks RC, Lailvaux SP (2010) Interactions among performance capacities predict male combat outcomes in the field cricket. Funct Ecol 24:159–164

Hammer M (1993) An identified neurone mediates the unconditioned stimulus in associative olfactory learning in honeybees. Nature 366:59–63

Hammer M, Menzel R (1995) Learning and memory in the honey bee (Review). J Neurosci 15:1617–1630

Hauser F, Cazzamali G, Williamson M, Blenau W, Grimmelikhuijzen CJP (2006) A review of neurohormone GPCRs present in the fruitfly *Drosophila melanogaster* and the honey bee *Apis mellifera*. Prog Neurobiol 80:1–19

Henry J, Scherman D (1989) Radioligands of the vesicular monoamine transporter and their use as markers of monoamine storage vesicles. Biochem Pharmacol 28:2395–2404

Hofmann HA (1996) The cultural history of Chinese fighting crickets: a contribution not only to the history of biology. [German]. Biologisches Zbl 115:206–213

Hofmann HA, Schildberger K (2001) Assessment of strength and willingness to fight during aggressive encounters in crickets. Anim Behav 62:337–348

Hofmann HA, Stevenson PA (2000) Flight restores fight in crickets. Nature 403:613

Homberg U (2002) Neurotransmitters and neuropeptides in the brain of the locust. Microsc Res Tech 56:189–209

Hoyer SC, Eckart A, Herrel A, Zars T, Fischer SA, Hardie SL et al (2008) Octopamine in male aggression of *Drosophila*. Curr Biol 18:159–167

Hsu Y, Earley RL, Wolf LL (2005) Modulation of aggressive behaviour by fighting experience: mechanisms and contest outcomes. Biol Rev Camb Philos Soc 81:33–74

Huber F (1955) Sitz und Bedeutung nervöser Zentren für Instinkthandlungen beim Männchen von *Gryllus campestris*. Z Tierpsychol 12:12–48

Huber F (1960) Untersuchungen über die Funktion des Zentralnervensystems und insbesondere des Gehirnes bei der Fortbewegung und der Lauterzeugung der Grillen. Z Vergleichende Tierphysiologie 44:60–132

Huber F, Moore TE, Loher W (1989) Cricket behavior and neurobiology. Cornell University, New York

Huhman K (2006) Social conflict models: can they inform us about human psychopathology? Horm Behav 50:640–646

Hurd PL (2006) Resource holding potential, subjective resource value, and game theoretical models of aggressiveness signalling. J Theor Biol 241:639–648

Iba M, Nagao T, Urano A (1995) Effects of population density on growth, behavior and levels of biogenic amines in the cricket, *Gryllus bimaculatus*. Zool Sci 12:695–702

Iwasaki M, Katagiri C (2008) Cuticular lipids and odors induce sex-specific behaviors in the male cricket *Gryllus bimaculatus*. Comput Biochem Physiol A 149:306–313

Iwasaki M, Delago A, Nishino H, Aonuma H (2006) Effects of previous experience on the agonistic behaviour of male crickets, *Gryllus bimaculatus*. Zool Sci 23:863–872

Iwasaki M, Nishino H, Delago A, Aonuma H (2007) Effects of NO/cGMP signaling on behavioral changes in subordinate male crickets, *Gryllus bimaculatus*. Zool Sci 24:860–868

Johnson O, Becnel J, Nichols CD (2009) Serotonin 5-HT(2) and 5-HT(1A)-like receptors differentially modulate aggressive behaviors in *Drosophila melanogaster*. Neuroscience 158:1292–1300

Judge KA, Bonanno VL (2008) Male weaponry in a fighting cricket. PLoS One 3:e3980
Judge KA, Ting JJ, Schneider J, Fitzpatrick MJ (2010) A lover, not a fighter: mating causes male crickets to lose fights. Behav Ecol Sociobiol 64:1971–1979
Jung SN, Borst A, Haag J (2011) Flight activity alters velocity tuning of fly motion-sensitive neurons. J Neurosci 31:9231–9237
Kemp DJ, Wiklund C (2004) Residency effects in animal contests. Proc R Soc Lond 271:1707–1711
Khazraie K, Campan M (1999) The role of prior agonistic experience in dominance relationships in male crickets *Gryllus bimaculatus* (Orthoptera: Gryllidae). Behav Processes 44:341–348
Killian KA, Allen JR (2008) Mating resets male cricket aggression. J Insect Behav 21:535–548
Kravitz E, Huber R (2003) Aggression in invertebrates. Curr Opin Neurobiol 13:736–743
Lihoreau M, Brepson L, Rivault C (2009) The weight of the clan: even in insects, social isolation can induce a behavioural syndrome. Behav Processes 82:81–84
Marder E (2012) Neuromodulation of neuronal circuits: back to the future. Neuron 76:1–11
Maynard Smith J, Price GR (1973) The logic of animal conflict. Nature 246:15–18
McFarland DJ, Sibly RM (1975) The behavioural final common path. Philos Trans R Soc B 270:265–293
Mendl M, Paul ES, Chittka L (2011) Animal behaviour: emotions in invertebrates? Curr Biol 21:R463–R465
Miczek KA, Nikulina EM, Takahashi A, Covington HE III, Yap JJ, Boyson CO et al (2011) Gene expression in aminergic and peptidergic cells during aggression and defeat: relevance to violence, depression and drug abuse. Behav Genet 41:787–802
Mizunami M, Unoki S, Mori Y, Hirashima D, Hatano A, Matsumoto Y (2009) Roles of octopaminergic and dopaminergic neurons in appetitive and aversive memory recall in an insect. BMC Biol 7:46
Morris OT, Duch C, Stevenson PA (1999) Differential activation of octopaminergic (DUM) neurones via proprioceptors responding to flight muscle contractions in the locust. J Exp Biol 202:3555–3564
Müller U (1997) The nitric oxide system in insects. Prog Neurobiol 51:363–381
Murakami S, Itoh MT (2001) Effects of aggression and wing removal on brain serotonin levels in male crickets, *Gryllus bimaculatus*. J Insect Physiol 47:1309–1312
Murakami S, Itoh MT (2003) Removal of both antennae influences the courtship and aggressive behaviors in male crickets. J Neurobiol 57:110–118
Nagao T, Tanimura T (1993) Distribution of biogenic amines in the cricket central nervous system. Anal Biochem 171:33–40
Nagao T, Tanimura T, Shimozawa T (1991) Neurohormonal control of the mating interval in the male cricket, *Gryllus bimaculatus* DeGeer. J Comp Physiol A 168:159–164
Nelson RJ, Trainor BC (2007) Neural mechanisms of aggression. Nature reviews. Neuroscience 8:536–546
Nilsen SP, Chan YB, Huber R, Kravitz EA (2004) Gender-selective patterns of aggressive behavior in *Drosophila melanogaster*. Proc Natl Acad Sci U S A 101:12342–12347
Nosil P (2002) Food fights in house crickets, *Acheta domesticus*, and the effects of body size and hunger level. Can J Zool 80:409–417
O'Connell LA, Hofmann HA (2011) The vertebrate mesolimbic reward system and social behavior network: a comparative synthesis. J Comp Neurol 519:3599–3639
Passamonti L, Crockett MJ, Apergis-Schoute AM, Clark L, Rowe JB, Calder AJ et al (2012) Effects of acute tryptophan depletion on prefrontal-amygdala connectivity while viewing facial signals of aggression. Biol Psychiatry 71:36–43
Payne RJH (1998) Gradually escalating fights and displays: the cumulative assessment model. Anim Behav 56:651–662
Perry CJ, Barron AB (2013) Neural mechanisms of reward in insects. Annu Rev Entomol 58:543–562
Pfennig DW, Reeve HK (1989) Neighbor recognition and context dependent aggression in a solitary wasp, *Sphecius speciosus* (Hymenoptera, sphecidae). Ethology 80:1–18

Pflüger H, Stevenson P (2005) Evolutionary aspects of octopaminergic systems with emphasis on arthropods. Arthropod Struct Dev 34:379–396
Raichlen DA, Foster AD, Gerdeman GL, Seillier A, Giuffrida A (2011) Wired to run: exercise-induced endocannabinoid signaling in humans and cursorial mammals with implications for the 'runner's high'. J Exp Biol 215:1331–1336
Rillich J, Stevenson PA (2011) Winning fights induces hyperaggression via the action of the biogenic amine octopamine in crickets. PLoS One 6:e28891
Rillich J, Stevenson PA (2014) A fighter's comeback: dopamine is necessary for recovery of aggression after social defeat. Horm Behav 66:696–704
Rillich J, Stevenson PA (2015) Releasing stimuli and aggression in crickets: octopamine promotes escalation and maintenance but not initiation. Front Behav Neurosci 9(95):1–11
Rillich J, Schildberger K, Stevenson PA (2007) Assessment strategy of fighting crickets revealed by manipulating information exchange. Anim Behav 74:823–836
Rillich J, Buhl E, Schildberger K, Stevenson PA (2009) Female crickets are driven to fight by the male courting and calling songs. Anim Behav 77:737–742
Rillich J, Schildberger K, Stevenson PA (2011) Octopamine and occupancy – an aminergic mechanism for intruder-resident aggression in crickets. Proc R Soc Lond B 278:1873–1880
Rillich J, Stevenson PA, Pflüger HJ (2013) Flight and walking in locusts – cholinergic co-activation, temporal coupling and its modulation by biogenic amines. PLoS One 8(5):e62899
Rodriguez-Munoz R, Bretman A, Tregenza T (2011) Guarding males protect females from predation in a wild insect. Curr Biol 21:1716–1719
Roeder T (1999) Octopamine in invertebrates. Prog Neurobiol 59:533–561
Roeder T, Degen J, Gewecke M (1998) Epinastine, a highly specific antagonist of insect neuronal octopamine receptors. Eur J Pharmacol 349:171–177
Rogers SM, Matheson T, Sasaki K, Kendrick K, Simpson SJ, Burrows M (2004) Substantial changes in central nervous system neurotransmitters and neuromodulators accompany phase change in the locust. J Exp Biol 207:3603–3617
Rutte C, Taborsky M, Brinkhof MW (2006) What sets the odds of winning and losing? Trends Ecol Evol 21:16–21
Sakura M, Aonuma H (2013) Aggressive behavior in the antennectomized male cricket *Gryllus bimaculatus*. J Exp Biol 216:2221–2228
Schöneich S, Schildberger K, Stevenson PA (2011) Neuronal organization of a fast-mediating cephalo-thoracic pathway for antennal-tactile information in the cricket. J Comp Neurol 519:1677–1690
Schröter U, Malun D, Menzel R (2007) Innervation pattern of suboesophageal ventral unpaired median neurones in the honeybee brain. Cell Tissue Res 327:647–667
Schwärzel M, Monastirioti M, Scholz H, Friggi-Grelin F, Birman S, Heisenberg M (2003) Dopamine and octopamine differentiate between aversive and appetitive olfactory memories in *Drosophila*. J Neurosci 23:10495–10502
Simmons LW (1986) Inter-male competition and mating success in the field cricket, *Gryllus bimaculatus* (de Geer). Anim Behav 34:567–579
Simpson SJ, Stevenson PA (2015) Neuromodulation of social behaviour in insects. In: Canli T (ed) The Oxford handbook of molecular psychology. Oxford University Press, Oxford, pp 27–51
Simpson JS, Sword GA (2009) Phase polyphenism in locusts: mechanisms, population consequences, adaptive significance and evolution. In: Whitman DW, Ananthakrishnan TN (eds) Phenotypic plasticity of insects, mechanisms and consequences. Science Publishers, Enfield, pp 147–189
Sokolowski MB (2010) Social interactions in "simple" model systems. Neuron 65:780–794
Sombati S, Hoyle G (1984) Generation of specific behaviors in a locust by local release into neuropil of the natural neuromodulator octopamine. J Neurobiol 15:481–506
Spörhase-Eichmann U, Vullings HGB, Buijs RM, Hörner M (1992) Octopamine-immunoreactive neurones in the central nervous system of the cricket *Gryllus bimaculatus*. Cell Tissue Res 268:287–304

Staudacher E, Schildberger K (1998) Gating of sensory responses of descending brain neurones during walking in crickets. J Exp Biol 201:559–572
Staudacher EM, Gebhardt M, Dürr V (2005) Antennal movements and mechanoreception: neurobiology of active tactile sensors. Adv Insect Physiol 32:49–205
Stevenson PA, Kutsch W (1987) A reconsideration of the central pattern generator concept for locust flight. J Comp Physiol A 161:115–129
Stevenson PA, Rillich J (2012) The decision to fight or flee – insights into underlying mechanism in crickets. Front Neurosci 6(118):1–12
Stevenson PA, Rillich J (2013) Isolation associated aggression in crickets is a result of recovery from social subjugation. Plos ONE 8(9):e74965
Stevenson PA, Rillich J (2015) Adding up the odds – nitric oxide underlies the decision to flee and post conflict depression. Sci Adv 1:e1500060
Stevenson PA, Schildberger K (2013) Mechanisms of experience dependant control of aggression in crickets. Curr Opin Neurobiol 23:318–323
Stevenson PA, Spörhase-Eichmann U (1995) Localization of octopaminergic neurons in insects. Comp Biochem Physiol B 11:203–215
Stevenson PA, Hofmann HA, Schoch K, Schildberger K (2000) The fight and flight responses of crickets depleted of biogenic amines. J Neurobiol 43:107–120
Stevenson PA, Dyakonova VE, Rillich J, Schildberger K (2005) Octopamine and experience-dependent modulation of aggression in crickets. J Neurosci 25:1431–1441
Tachon G, Murray A-M, Gray DA, Cade WH (1999) Agonistic displays and the benefits of fighting in the field cricket, *Gryllus bimaculatus*. J Insect Behav 12:533–543
Tops M, Russo S, Boksem MA, Tucker DM (2009) Serotonin: modulator of a drive to withdraw. Brain Cogn 71:427–436
Vrontou E, Nilsen SP, Demir E, Kravitz EA, Dickson BJ (2006) Fruitless regulates aggression and dominance in *Drosophila*. Nat Neurosci 9:1469–1471
Wada-Katsumata A, Yamaoka R, Aonuma H (2011) Social interactions influence dopamine and octopamine homeostasis in the brain of the ant *Formica japonica*. J Exp Biol 214:1707–1713
Wang L, Anderson DJ (2010) Identification of an aggression-promoting pheromone and its receptor neurons in *Drosophila*. Nature 463:227–231
Watanabe T, Aonuma H (2012) Identification and expression analyses of a novel serotonin receptor gene, 5-HT2β, in the field cricket, *Gryllus bimaculatus*. Acta Biol Hung 63:58–62
Watanabe T, Sadamoto H, Aonuma H (2011) Identification and expression analysis of the genes involved in serotonin biosynthesis and transduction in the field cricket *Gryllus bimaculatus*. Insect Mol Biol 20:619–635
Wingfield JC, Hegner RE, Dufty AM Jr, Ball GF (1990) The challenge hypothesis—theoretical implications for patterns of testosterone secretion, mating systems, and breeding strategies. Am Nat 136:829–846
Yano S, Ikemoto Y, Aonuma H, Asama H (2012) Forgetting curve of cricket, *Gryllus bimaculatus*, derived by using serotonin hypothesis. Robot Auton Syst 60:722–728
Yoritsune A, Aonuma H (2012) The anatomical pathways for antennal sensory information in the central nervous system of the cricket, *Gryllus bimaculatus*. Invert Neurosci. doi:10.1007/s10158-012-0137-6
Yurkovic A, Wang, Basu C, Kravitz EA (2006) Learning and memory associated with aggression in *Drosophila* melanogaster. Proc Natl Acad Sci U S A 103:17519–17524
Zhou C, Rao Y, Rao Y (2008) A subset of octopaminergic neurons are important for *Drosophila* aggression. Nat Neurosci 11:1059–1067
Zorovic M, Hedwig B (2012) Descending brain neurons in the cricket *Gryllus bimaculatus* (de Geer): auditory responses and impact on walking. J Comp Physiol A 199:25–34

Chapter 13
Fighting Behavior: Understanding the Mechanisms of Group-Size-Dependent Aggression

Hitoshi Aonuma

Abstract Aggressive behavior is a common behavior in animals. In most cases, an animal's behavior toward an opponent is a violent attack. Male crickets (*Gryllus bimaculatus*) exhibit intensively aggressive behavior toward other males, most often culminating in fighting. The detection of conspecific male cuticular substances initiates aggressive behavior in male crickets. After a fight, a loser no longer exhibits aggressiveness in a second bout or in separate encounters with another male; rather the defeated male exhibits avoidance behavior.

Aggressive behavior in crickets provides an excellent model system to understand neuronal mechanisms underlying real-time control of sophisticated behavior and social adaptability of animals. Animals alter their behavior in order to respond to the demands of changing social environments. Society and crowding conditions are dynamic environments. In this chapter, we focus on how crickets determine their behavior depending on their social interactions, focusing on behavioral and physiological aspects. Whether the nitric oxide (NO) system and octopaminergic (OAergic) system in the central nervous system of crickets could mediate aggressive behavior of the crickets is discussed. Based on these results, a neurophysiological model is designed to elucidate the mechanisms of social adaptability. This model demonstrates that a multiple feedback structure, composed of a feedback loop in the nervous systems and individual interactions with other crickets, may be a key to aggression influenced by group size (group-size-dependent aggressive behavior).

Keywords Aggressive behavior • Loser's effect • Internal state • Neuromodulator • Brain • Nitric oxide • Biogenic amine

---

H. Aonuma (✉)
Research Center of Mathematics for Social Creativity, Research Institute for Electronic Science, Hokkaido University, Sapporo 060-0812, Japan

CREST, Japan Science and Technology Agency, Kawaguchi, Japan
e-mail: aon@es.hokudai.ac.jp

H.W. Horch et al. (eds.), *The Cricket as a Model Organism*,
DOI 10.1007/978-4-431-56478-2_13

13.1 Aggressive Behavior of Male Crickets

It is common in animals that a dominant–subordinate hierarchy is established after agonistic aggressive behavior. Agonistic behavior is the result of complex interactions among physiological, motivational, and behavioral systems. Since social and physical environments are two of the most important factors that release aggressive behavior in animals, it is difficult to fully understand aggressive behavior. The cricket provides us one of the greatest model systems to investigate neuronal mechanisms underlying aggressive behavior. In particular, the subordinate cricket (loser of a fight) helps us to understand how aggressive motivation changes depending on social status.

Male crickets exhibit intensive aggressive behavior when they encounter another male, whereas they show courtship behavior to a conspecific female (Alexander 1961). A battle starts out slowly and escalates into a fierce struggle (Fig. 13.1). In crickets, sex discrimination by male individuals is based on cuticular chemical substances on the body surface. In *Gryllus bimaculatus*, studies have shown that the components of the cuticular hydrocarbons are clearly different between males and females (Tregenza and Wedell 1997; Nagamoto et al. 2005). The fact that an isolated female antenna elicits courtship behavior in conspecific males while isolated male antenna elicits aggressive behavior suggests that males detect distinct chemical signals from other crickets. However, both the identity of the signals and

Fig. 13.1 Cricket fights. Fighting between male crickets escalates until an opponent gives up attacking and escapes by moving away. Male crickets start antennal fencing when they perceive cuticular substances through antennal contact. Mandible spreading follows

sensory pathways by which chemical signals alter aggressive and reproductive behaviors have not yet been completely elucidated.

Aggressive behavior of male crickets toward conspecific males has been extensively studied (Alexander 1961; Adamo and Hoy 1995; Hofmann and Stevenson 2000; Stevenson et al. 2005). Males battle with each other over the acquisition of living spaces and mating partners (Simmons 1986). A fight consists of a sequentially escalating series of behaviors beginning with antennal fencing, which leads to the spreading and finally the engaging of the mandibles. The fight continues until one of the crickets gives up attacking his opponent. Then pairs establish a dominant–subordinate hierarchy, with the winner initiating aggressive song and chasing after the loser. The loser, in subsequent encounters with the winner, exhibits avoidance behavior to avoid additional fights (Adamo and Hoy 1995; Hofmann and Stevenson 2000; Funato et al. 2011). This suggests that cuticular substances may be a kind of pheromone that triggers a particular behavior in the cricket. Most pheromone-induced behaviors in insects have been thought to be hardwired: a behavior that could be turned on and off but that lacks plasticity. However, fighting behavior in the cricket is modified by their previous experiences. For example, while cuticular pheromones induce dominant crickets to be aggressive, subordinate crickets respond to cuticular pheromones with avoidance instead of aggression (Sakura and Aonuma 2013). The period in which the subordinate cricket is unwilling to fight continues for more than several hours (Hofmann and Stevenson 2000). Recent experiences such as copulation, flight, the opponent's size and behavior, and population density can also alter aggressive behavior of males (Alexander 1961; Hofmann and Stevenson 2000; Rillich et al. 2007; Funato et al. 2011). Thus, whether males decide to engage in aggressive behaviors with each other may depend on their ability to compare their own motivation to fight with their perception of a potential opponent's fighting performance (Hack 1997; Rillich et al. 2007).

13.2 Antennal Inputs Initiating Aggressive Behavior

13.2.1 Defensive Aggression

Chemical and tactile information from antennae are important for male crickets to express proper aggressive behavior, identifying conspecific males and initiating an attack. However, antennal information is not always necessary to elicit defensive aggression (Sakura and Aonuma 2013). Antennal sensory information that consists of chemical and tactile sensory information is important for crickets when seeking food and mating partners, as well as when detecting threats. Males recognize a conspecific's sex using sensory information from antennae. Both chemical and tactile cues from antennae are integrated in the brain to decide whether to act either in an aggressive or courtship manner. Antennectomized males, for example, express less aggression toward antennectomized opponents, whereas they continue to exhibit typical fighting to an intact opponent. In addition, antennectomized losers showed significantly higher aggressiveness toward the opponent than intact losers do in a second bout. These suggest to us that defensive aggression could be elicited without antennal information. Antennectomized crickets need not use visual or palpal sensory input to elicit defensive aggression. In contrast, intact males showed aspects of aggressive behavior to male cuticular substances before and after winning a fight; however, if these males lost a fight, they showed avoidance behavior. Visual inputs play an important role, since blinded males exhibit more intense fighting than sighted males. Visual information concerning body size, weight, and mandible display behavior of the opponent can suppress aggressiveness (Rillich et al. 2007). Male crickets make decisions whether to fight or flee from a potential opponent based on information from multiple sensory modalities.

13.2.2 Neuronal Pathways Processing Antennal Information

Antennae convey both tactile and chemical sensory inputs to the central nervous system, and components of male aggressive behavior can be elicited by chemical signals released from body parts of another male, such as the forewing and antennae (Nagamoto et al. 2005). On the other hand, male crickets exhibit only weak aggressive behavior toward anesthetized males (Adamo and Hoy 1995), suggesting that chemical signals alone are not sufficient for males to initiate proper aggressive behavior. Thus, multiple sensory cues (i.e., visual, olfactory, and tactile) from a conspecific should be required to elicit intensive aggressive behavior.

Chemical sensory afferents terminate in the antennal lobe, whereas most of exteroceptive mechanosensory afferents terminate in the dorsal lobe of the brain (Homberg et al. 1989). In crickets, seven antennal sensory tracts (assigned as T1-7) were identified (Yoritsune and Aonuma 2012). Tracts T1–T4 project into the antennal lobe, while tracts T5 and T6 course into the dorsal region of the deutocerebrum or the subesophageal ganglion, and finally tract T7 terminates in the ventral area of flagellar afferents. The antennal lobe of the cricket is composed of 49 sexually isomorphic glomeruli (Fig. 13.2). In the protocerebrum ten tracts originate in the antennal lobe, and at least eight tracts arise from the ventral area where flagellar afferents terminate. Projection neurons originating from the antennal lobe terminate in the anterior calyx of the mushroom body and/or the lateral horn, which are secondary centers of chemical processing. The projection neurons originating from the antennal lobe projecting through the inner antenno-cerebral tract terminate in the anterior calyx of the mushroom body. In contrast, the projection neurons originating from the antennal lobe that project through

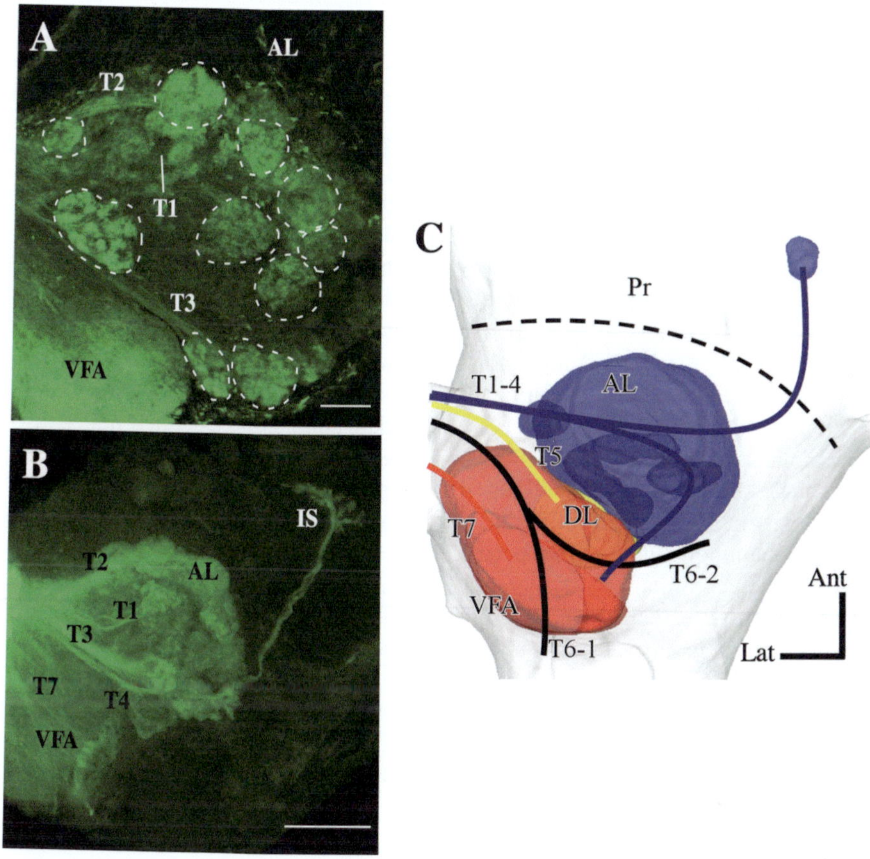

Fig. 13.2 Projection of antennal nerve terminations in the brain. (**a**) Stacked images of the antennal lobe obtained from optical sections made with a confocal scanning laser microscope. Areas surrounded by *dashed lines* indicate glomeruli where antennal sensory afferents terminate. Scale bar = 40 μm. (**b**) A ventral view of the antennal lobe. Scale bar = 100 μm. (**c**) Summary of the projection of antennal sensory afferents. *Dashed lines* indicate the boundary between the deutocerebrum and protocerebrum (Pr). *AL* antennal lobe, *Ant* anterior, *IS* isthmus, *Lat* lateral, *Pr* protocerebrum, *T1–T7* tact 1–7, *VFA* ventral area of flagellar afferents

other antenno-cerebral tracts or accessory antenno-cerebral tracts do not terminate in the anterior calyx of the mushroom body. The neurons projecting through the inner antenno-cerebral tract make direct inputs onto Kenyon cells in the anterior calyx of the mushroom body. Therefore, both in the primary centers of the deutocerebrum and in higher centers of the protocerebrum, chemical and mechanical information should be represented in spatially segregated neuropils. It is not yet clear which center of information processing elicits aggression. Further investigation of the molecular and physiological basis of aggressive behavior is needed.

13.3 Group-Size-Dependent Aggressiveness in Male Crickets

Population density strongly influences cricket behavior including aggressive behavior (Funato et al. 2011). In order to investigate how local population density regulates male aggressiveness in a group, the size of a group can be varied (Fig. 13.3). In a group, pairs of male crickets fight locally to establish a dominant–subordinate relationship. Once this relationship is established, victorious males (dominant) retain their aggressiveness, but defeated males (subordinate) exhibit reduced aggressiveness. If victorious males encounter other victorious males, they fight again to establish dominant–subordinate relationships. However, when defeated males encounter the same victorious males within a short period, they exhibit avoidance behavior and retreat from their opponents. Defeated males also show avoidance if they encounter males victorious in other fights, naive males, or even males defeated in other fights. After repeated fights, a repeatedly defeated male loses his aggressiveness. In due course, one or more males are established as the dominant individual(s) in the group. Depending on the number of individuals in a group, the loss of individual aggressiveness limits the number of fighting crickets at any given time. Thus, overall aggressive behavior in a group is changed as a consequence of group size, but is ultimately determined by individual fights. Higher population density increases the probability that individuals might encounter others and start fighting. Subordinates mainly exhibit avoidance behavior when they encounter other males in a high population group. On the other hand, the highest-ranked males could maintain the motivation to be aggressive more than other males,

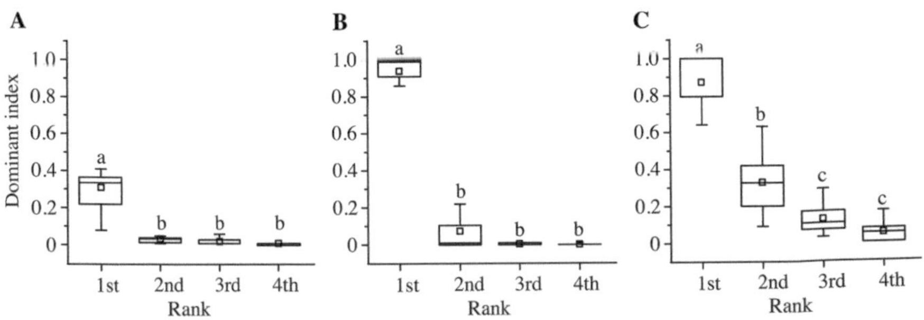

Fig. 13.3 Population density-dependent aggressiveness in cricket. The aggressiveness is evaluated by using a dominance index. The dominance index, Xi, is defined as $Xi = \frac{Di}{Di+Si+Ni}$ $(i = 1, 2, 3, \ldots, \text{num})$, where i is the social rank of the cricket. The total number of crickets is *Num*, and Di indicates the total number of dominance behaviors performed by cricket i. Si indicates the total number of subordinate behaviors performed by cricket i, and Ni indicates the number of nonresponsive behaviors when cricket i encounters another male. (**a**) Four male crickets are placed in a 75 cm^2 arena ($N = 12$). (**b**) Four male crickets are placed in a 300 cm^2 arena ($N = 12$). (**c**) Four male crickets are placed in a 1200 cm^2 arena. (Different *letters* denote significant differences between treatments, $*P < 0.05$, two-tailed Kruskal-Wallis ANOVA with Bonferroni-type multiple nonparametric comparison) (Modified from Funato et al. 2011)

but they suppress aggressiveness in high population density conditions. Therefore, all crickets in high population density group behave with lower aggression.

A decrease in local population density decreases the chance of individuals interacting. The dominant cricket always shows aggressive behavior if it encounters other crickets in a lower population density environment. In contrast, all subordinate crickets show avoidance behavior when they encounter other males, even if that male is also a subordinate. Dominant males establish a despotic hierarchy within a group. This relationship likely depends on local population density. As the local population density decreases, an increasing number of dominant crickets that show aggression will appear in the group.

13.4 Brain Neuromodulators Regulating Cricket Aggression

The male cuticular substances function as releaser pheromones in male crickets. However, the chemicals do not always initiate aggressive behavior. The aggressive motivation is regulated by the previous aggressive interaction. How do subordinate crickets switch to avoidance behavior toward cuticular substances? One of the likeliest possibilities is neuromodulation within the brain.

13.4.1 Role of Nitric Oxide in Aggressive Behavior

NO is thought to function as a neuromodulator in insect nervous systems. Endogenous nitric oxide (NO) is a free radical signaling molecule that diffuses across cell membranes in the nervous system of a wide variety of animals, both vertebrate and invertebrate. It is generated from L-arginine by NO synthase (NOS) (Moncada et al. 1991) and activates the heme-containing enzyme soluble guanylate cyclase (SGC) to generate the second messenger cyclic guanosine monophosphate (cGMP) in target cells (Bredt and Synder 1989). This results in activation of cGMP-activated protein kinase (PKG) that phosphorylates downstream target proteins and evokes a cellular response (Bicker 2001). Cyclic nucleotide-gated channels (Kaupp and Seifert 2002) and cyclic nucleotide phosphodiesterase (Bender and Beavo 2006) are also major targets of cGMP. Furthermore, evidence for cGMP-independent NO signaling has also been found. NO can act directly, without producing cGMP, on ion channels through S-nitrosylation, which regulates the electrical activity of the target cells (Ahern et al. 2002; Wilson et al. 2007). S-Nitrosylation is thought to require a higher concentration of NO than does activation of soluble guanylate cyclase and tends to proceed with slower kinetics than cGMP-mediated actions.

NO is continuously released at a basal level in the cricket nervous system, and cholinergic activation accelerates NO generation (Aonuma et al. 2008). Therefore, relative changes in the concentration of NO likely mediate physiological properties of local circuits in nervous systems. Experimental manipulations of NO signaling and histochemical experiments suggest that NO works as a multifunctional mediator in the nervous systems of invertebrate animals. In mollusks, NO modulates the synaptic efficacy of cholinergic synapses (Mothet et al. 1996) and mediates oscillatory neuronal activities underlying chemical information processing in the central nervous system (Gelperin 1994). In crustaceans, components of the NO/cGMP signaling pathway contribute to neuronal network functions that underlie escape behavior (Aonuma et al. 2000), mechanosensory processing (Aonuma and Newland 2001, 2002; Aonuma et al. 2008), and olfactory processing (Johansson and Mellon 1998). In insects, NO is believed to function as a crucial component in motor control (Qazi and Trimmer 1999), pheromone processing (Seki et al. 2005), olfactory processing (Wilson et al. 2007), and olfaction-related learning behavior (Müller 1997; Matsumoto et al. 2006).

The presence of NO/cGMP signaling has been demonstrated in the cricket brain (Aonuma and Niwa 2004). Pharmacological and behavioral experiments demonstrate that NO signaling mediates the motivation of aggressive behavior in the male cricket (Iwasaki et al. 2007). Inhibition of NO/cGMP signaling pathways using the NOS inhibitor L-NAME and SGC inhibitor ODQ restores the aggressiveness of subordinates. The aggressiveness of a cricket that is head injected with L-NAME or with ODQ shows normal aggression at the first engagement. However, the loser that is head injected with L-NAME or with ODQ shows significantly increased aggressiveness.

Interestingly, the behavior of subordinates whose antennae are removed is similar to the behavior of subordinates whose NO/cGMP signaling is inhibited (Sakura and Aonuma 2013). This suggests that NO/cGMP signaling in the antennal sensory processing pathway could participate in the neuronal mechanism underlying aggressive behavior. Indeed, there are putative NO donor and target neurons in the antennal lobe of the cricket (Aonuma et al. 2004).

13.4.2 Role of Brain Octopamine in Aggressive Behavior

Octopamine (OA) is another chemical that affects aggressive behavior in crickets. It has been demonstrated that the biogenic amine level in hemolymph mediates cricket aggression (Adamo et al. 1995). Brain OA titer of the defeated cricket decreases after fighting (Aonuma et al. 2009). In order to elucidate if OA regulates aggressiveness of male crickets, pharmacological experiments are necessary. Mianserin and epinastine are widely used as antagonists of OA in insects (Roeder 1990; Roeder et al. 1998). The octopaminergic (OAergic) system is involved in the motivation of aggressive behavior in cricket (Rillich et al. 2011). Brain OA has been demonstrated to increase aggressive motivation in ants as well (Aonuma and Watanabe 2012). In crickets, head injection of the OA antagonists mianserin decreases the aggressiveness of male crickets (Aonuma et al. 2009). In a sequential encounter, the mianserin-injected crickets show significantly lower aggressiveness in the first encounter as compared to controls. In a second encounter, after a few hours, the mianserin-injected cricket still shows significantly lower aggressiveness. Therefore, OA antagonist slows recovery of aggressiveness in male crickets.

The depression of the subordinates' aggressiveness is alleviated by an NOS inhibitor, whereas an OA inhibitor prolongs the reduction of the subordinates' aggressiveness. The results of these pharmacological experiments suggest that the NOergic and OAergic systems have mutually opposite effects on initiating aggressive behavior in the cricket. In order to confirm if the effect of NO is countered by OA and vice versa, we co-injected a mixture of L-NAME and mianserin and found that L-NAME rescues the effects of mianserin (Aonuma et al. 2009). The NOergic and OAergic systems regulate experience-dependent aggressive behavior of male crickets. We thus hypothesize that NO/cGMP system mediates the OAergic system in the cricket brain.

13.4.3 Modeling of NOergic and OAergic Modulation During Aggressive Behavior

In order to understand the dynamic activities of the cricket brain, a neuromodulatory model based on pharmacological experiments has been proposed (Kawabata et al. 2007, 2012). The results of neuroanatomical, pharmacological, and behavioral experiments suggest to us the importance of NOergic and OAergic neuromodulation in the brain during aggressive behavior. However, there are still gaps between physiology and behavior. Modeling is a powerful approach that can bridge the gaps between them (see Chap. 20).

According to pharmacological experiments, changing the levels of NO and OA in the brain is suggested to modulate cricket aggressive behavior. Cuticular pheromones from male crickets trigger aggressive behavior in conspecific males. The NOergic system in the chemical information processing center of the brain (i.e., antennal lobe) must be involved in the initiation of aggressive behavior. Because NO is a gaseous molecule, it diffuses about 100–200 μm/s, and its lifetime is several seconds in duration (Philippides et al. 2000). A physiological model based on the dynamics of NO/cGMP signaling and OA can be considered. This model consists of a diffusion equation for NO, differential equations for cGMP and OA levels, and a threshold model for behavioral choice, which is based on OA levels (Fig. 13.4a). These components are connected in series. Given the hypothesis that brain OA levels can shift a cricket's behavior from attacking an opponent to avoiding an opponent, it is assumed that fighting behavior is selected when OA is above a given threshold. Likewise, avoidance behavior is selected when OA levels are below this threshold. Computer simulations of the model demonstrate that increased levels of cGMP, caused by the increase in brain NO level, lead to a decrease in OA levels (Kawabata et al. 2007). Since experiencing a defeat during fighting depresses the initiation of aggressive behavior for some time, it assumes that the OA levels in the winner's brain increases to a certain value. Concurrently, the OA level in the loser is

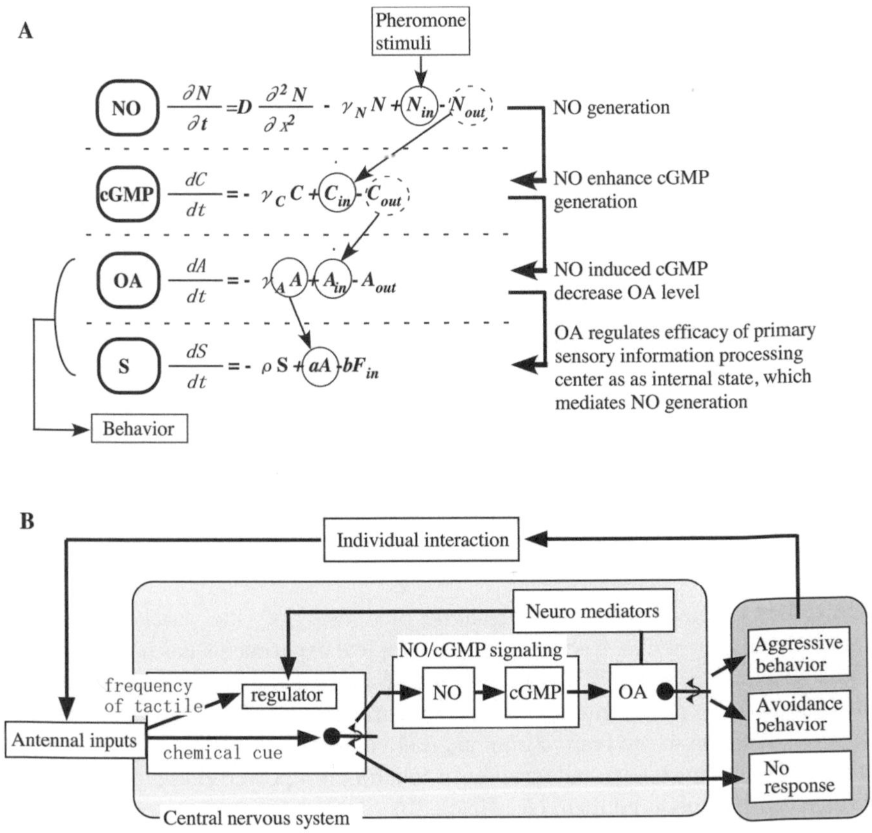

Fig. 13.4 Neuromodulatory model for cricket aggressive behavior. (**a**) Neuromodulation model consists of a reaction–diffusion equation of the dynamics of the gaseous molecule NO and differential equations for cGMP and OA levels. (**b**) Multi-feedback structure that regulates behavior changes in aggressive behavior in the cricket

reduced to a lower value and is in proportion to the length of the fighting period. On the other hand, cricket aggression plays out locally in a group, and the aggressiveness of each cricket changes depending on the group size (Funato et al. 2011).

Previous fight experience can influence a cricket's decision to fight or flee. Furthermore, the brain OA level is modified by the density of the group (Iba et al. 1995). Thus the effect of individual interactions between crickets on the physiological changes in the brain should be considered along with the efficacy of sensory inputs from the antennae. Our hypothesis is that the OAergic system affects the mechanisms for processing sensory information. OA activates the processing of this information and individual interactions suppress this processing. In light of this, the physiological model describes behavior that is determined by two input sources, sensory inputs due to individual interactions between crickets and feedback due to the internal state of the brain (Kawabata et al. 2012). Sensory inputs received from another individual might be modulated at the primary center for sensory information processing. Thus the model is designed to account for this effect by incorporating a value “S” for the efficacy of antennal inputs. The value “S” expresses the efficiency of processing of antennal sensory information and influences behavior selection. The influence of the internal state is likely due to factors related to the production and output quantity of OA in the cascade model. The efficacy (“S”), which is modulated by the internal state in the brain, as well as individual interactions between crickets could regulate the generation of NO. This would, in turn, influence fight-related decision-making. When the efficacy of sensory inputs is above a particular threshold, aggressive behavior occurs, in keeping with the NO/cGMP-OA model. When efficacy is low, the cricket continues wandering and gives no response to the tactile and chemical cues. The duration is designed to represent the temporary reduction in the efficacy of sensory input, which is caused by individual interactions. Here, the influences are assumed to be proportional to OA levels and stimuli from individual interactions between crickets. The results of computer simulations in multi-individual environments show that this model could be considered appropriate for swarm activities of the cricket. Thus we hypothesize that the main mechanism underlying behavioral adaptability is a multiple feedback structure that is composed of feedback loops in the nervous systems combined with input from the social environment (Fig. 13.4b).

References

Adamo S, Hoy R (1995) Agonistic behaviour in male and female field crickets, *Gryllus bimaculatus*, and how behavioural context influences its expression. Anim Behav 49:1491–1501

Adamo SA, Linn CE, Hoy RR (1995) The role of neurohormonal octopamine during ‘fight or flight’ behaviour in the field cricket *Gryllus bimaculatus*. J Exp Biol 198:1691–1700

Ahern GP, Klyachko VA, Jackson MB (2002) cGMP and S-nitrosylation: two routes for modulation of neuronal excitability by NO. Trends Neurosci 25:510–517

Alexander DR (1961) Aggressiveness, territoriality, and sexual behavior in field crickets (*Orthoptera*: *Gryllidae*). Behavior 17:130–223

Aonuma H, Newland P (2001) Opposing actions of nitric oxide on synaptic inputs of identified interneurones in the central nervous system of the crayfish. J Exp Biol 204:1319–1332

Aonuma H, Newland P (2002) Synaptic inputs onto spiking local interneurons in crayfish are depressed by nitric oxide. J Neurobiol 52:144–155

Aonuma H, Niwa K (2004) Nitric oxide regulates the levels of cGMP accumulation in the cricket brain. Acta Biol Hung 55:65–70

Aonuma H, Watanabe T (2012) Octopaminergic system in the brain controls aggressive motivation in the ant, *Formica japonica*. Acta Biol Hung 63(Suppl 2):63–68

Aonuma H, Nagayama T, Takahata M (2000) Modulatory effects of nitric oxide on synaptic depression in the crayfish neuromuscular system. J Exp Biol 203:3595–3602

Aonuma H, Iwasaki M, Niwa K (2004) Role of NO signaling in switching mechanisms in the nervous system of insect. In: Proceedings of the SICE annual conference on CD-ROM, pp 2477–2482 (ISBN 2474-907764-907722-907767)

Aonuma H, Kitamura Y, Niwa K, Ogawa H, Oka K (2008) Nitric oxide-cyclic guanosine monophosphate signaling in the local circuit of the cricket abdominal nervous system. Neuroscience 157:749

Aonuma H, Sakura M, Ota J, Asama H (2009) Social adaptive functions in animals -learning from insect social behaviors-". In: Proceedings of the 2009 IEEE/RSJ international conference on intelligent robots and systems (Workshops/Tutorials, CD), pp 10–15

Bender AT, Beavo JA (2006) Cyclic nucleotide phosphodiesterases: molecular regulation to clinical use. Pharmacol Rev 58:488–520

Bicker G (2001) Sources and targets of nitric oxide signalling in insect nervous systems. Cell Tissue Res 303:137–146

Bredt DS, Synder SH (1989) Nitric oxide mediates glutamate-linked enhancement of cGMP levels in the cerebellum. Proc Natl Acad Sci U S A 86:9030–9033

Funato T, Nara M, Kurabayashi D, Ashikaga M, Aonuma H (2011) A model for group-size-dependent behaviour decisions in insects using an oscillator network. J Exp Biol 214:2426–2434

Gelperin A (1994) Nitric oxide mediates network oscillations of olfactory interneurons in a terrestrial mollusc. Nature 369:61–63

Hack MA (1997) Assessment strategies in the contests of male crickets, *Acheta domesticus* (L.). Anim Behav 53:733–747

Hofmann HA, Stevenson PA (2000) Flight restores fight in crickets. Nature 403:613

Homberg U, Christensen TA, Hildebrand JG (1989) Structure and function of the deutocerebrum in insects. Annu Rev Entomol 34:477–501

Iba M, Nagao T, Akihisa U (1995) Effects of population density on growth, behavior and levels of biogenic amines in the cricket, *Gryllus bimaculatus*. Zool Sci 12:695–702

Iwasaki M, Nishino H, Delago A, Aonuma H (2007) Effects of NO/cGMP signaling on behavioral changes in subordinate male crickets, *Gryllus bimaculatus*. Zool Sci 24:860–868

Johansson KU, Mellon D Jr (1998) Nitric oxide as a putative messenger molecule in the crayfish olfactory midbrain. Brain Res 807:237–242

Kaupp UB, Seifert R (2002) Cyclic nucleotide-gated ion channels. Physiol Rev 82:769–824

Kawabata K, Fujiki T, Ikemoto Y, Aonuma H, Asama H (2007) A neuromodulation model for adaptive behavior selection of the cricket. J Robot Mech 19:388–394

Kawabata K, Fujii T, Aonuma H, Suzuki T, Ashikaga M, Ota J, Asama H (2012) A neuromodulation model of behavior selection in the fighting behavior of male crickets. Robot Auton Syst 60:707–713

Matsumoto Y, Unoki S, Aonuma H, Mizunami M (2006) Critical role of nitric oxide-cGMP cascade in the formation of cAMP-dependent long-term memory. Learn Mem 13:35–44

Moncada S, Palmer RMJ, Higgs EA (1991) Nitric oxide: physiology, phathophysiology, and pharmacology. Pharmacol Rev 43:109–142

Mothet JP, Fossier P, Tauc L, Baux G (1996) NO decreases evoked quantal ACh release at a synapse of *Aplysia* by a mechanism independent of Ca2+ influx and protein kinase G. J Physiol (Lond) 493:769–784

Müller U (1997) The nitric oxide system in insects. Prog Neurobiol 51:363–381

Nagamoto J, Aonuma H, Hisada M (2005) Discrimination of conspecific individuals via cuticular pheromones by males of the cricket *Gryllus bimaculatus*. Zool Sci 22:1079–1088

Philippides A, Husbands P, O'Shea M (2000) Four dimensional neuronal signaling by nitric oxide: a computational analysis. J Neurosci 20:1199–1207

Qazi S, Trimmer BA (1999) The role of nitric oxide in motoneuron spike activity and muscarinic-evoked changes in cGMP in the CNS of larval *Manduca sexta*. J Comp Physiol A 185:539–550

Rillich J, Schildberger K, Stevenson PA (2007) Assessment strategy of fighting crickets revealed by manipulating information exchange. Anim Behav 74:823–836

Rillich J, Schildberger K, Stevenson PA (2011) Octopamine and occupancy: an aminergic mechanism for intruder-resident aggression in crickets. Proc R Soc B 278:1873–1880

Roeder T (1990) High-affinity antagonists of the locust neuronal octopamine receptor. Eur J Pharmacol 191:221–224

Roeder T, Degen J, Gewecke M (1998) Epinastine, a highly specific antagonist of insect neuronal octopamine receptors. Eur J Pharmacol 349:171–177

Sakura M, Aonuma H (2013) Aggressive behavior in the antennectomized male cricket *Gryllus bimaculatus*. J Exp Biol 216:2221–2228

Seki Y, Aonuma H, Kanzaki R (2005) Pheromone processing center in the protocerebrum of Bombyx mori revealed by nitric oxide-induced anti-cGMP immunocytochemistry. J Comp Neurol 481:340–351

Simmons LW (1986) Inter-male competition and mating success in the field cricket, *Gryllus bimaculatus* (De Geer). Anim Behav 34:567–579

Stevenson PA, Dyakonova V, Rillich J, Schildberger K (2005) Octopamine and experience-dependent modulation of aggression in crickets. J Neurosci 25:1431–1441

Tregenza T, Wedell N (1997) Definitive evidence for cuticular pheromones in a cricket. Anim Behav 54:979–984

Wilson CH, Christensen TA, Nighorn AJ (2007) Inhibition of nitric oxide and soluble guanylyl cyclase signaling affects olfactory neuron activity in the moth, *Manduca sexta*. J Comp Physiol A Neuroethol Sens Neural Behav Physiol 193:715–728

Yoritsune A, Aonuma H (2012) The anatomical pathways for antennal sensory information in the central nervous system of the cricket, *Gryllus bimaculatus*. Invert Neurosci 12:103–117

Chapter 14
Cercal System-Mediated Antipredator Behaviors

Yoshichika Baba and Hiroto Ogawa

Abstract In crickets, the cercal system is a mechanosensory system in which the receptor organ is a pair of antenna-like appendages called cerci located at the rear of the abdomen and covered with mechanosensory hairs, similar to the bristles on a bottlebrush. This system mediates the detection, localization, and identification of air currents surrounding the animal. Owing to the easy accessibility of the nervous system and advantages in physiological techniques, the cercal system of the cricket has been used as a model system for investigations of development, mechanoreceptor biomechanics, neural maps, and neural coding over the past few decades. In contrast, the behavioral significance of this system is still poorly understood. In early studies, the cercal system was regarded to be a simple system for triggering an escape reflex against an approaching predator. However, recent studies suggest that it is a more complex "generalist" system involved in various behaviors. In this chapter, we review studies on the cricket cercal system from a neuroethological point of view and discuss the future direction of neuroethological research on this system.

Keywords Cercal system • Wind-sensitive organ • Identified interneurons • Antipredator behavior • Escape response • Mechanosensory receptors

14.1 Introduction

Most researchers of neuroethology (the study of animal behavior and its underlying control by the nervous system) have studied behaviors that show clear stimulus–response causalities, because they are likely to be driven by reliable neural circuits. An escape behavior in arthropods is one such behavior because it is reliably elicited by a key stimulus, and the input–output relationships in this behavior are clear. For

Y. Baba
Department of Dermatology, Columbia University, New York, NY 10032, USA

H. Ogawa (✉)
Department of Biological Sciences, Faculty of Science, Hokkaido University, Sapporo 060-0810, Japan
e-mail: hogawa@sci.hokudai.ac.jp

H.W. Horch et al. (eds.), *The Cricket as a Model Organism*,
DOI 10.1007/978-4-431-56478-2_14

example, escapes by crayfish elicited by water flow are driven by specific neurons (e.g., Reichert and Wine 1983). The running escape behavior of the cockroach is driven by a few identified giant interneurons (GIs), which are activated by air currents (e.g., Camhi 1980). Both behaviors are simple, repeatable responses, and relatively few identifiable neurons are involved in the initiation of the behavior. These key-triggering neurons are often called command neurons or are said to be part of a command system.

Studies on wind-evoked escape behavior in insects began in the 1970s. The cercal system was shown in early studies to be important for the oriented escape response (Hoyle 1958; Camhi 1980; Boyan et al. 1986). Briefly, the information pathway mediated by the cercal system is as follows: the cerci are mechanoreceptive organs comprising a pair of antenna-like appendages at the rear of the abdomen. Large numbers of mechanoreceptors distributed on the cerci detect the air current surrounding the animal. The receptor (hair) neurons extend their afferent axons to the terminal abdominal ganglion (TAG) and have synaptic contacts with interneurons. The projection interneurons receive signals from the sensory afferents and send the processed information to higher interneurons in the rostral central nerve system (CNS), including thoracic and cephalic ganglia. The leg movement for escape running is orchestrated by motor outputs produced by the interneuron network within the CNS. Many of the findings relating to the pathway come from studies of the escape behavior of cockroaches, pioneered by Roeder (1948) and more recently by Camhi and colleagues (e.g., Camhi et al. 1978; Camhi and Tom 1978), Ritzmann and colleagues (e.g., Ritzmann and Pollack 1986, 1990), and Comer and colleagues (e.g., Comer and Dowd 1987). Other researchers have carried out comparative studies using other insects, such as crickets (e.g., Edwards and Palka 1974; Matsumoto and Murphey 1977; reviewed in Jacobs et al. 2008), locusts (reviewed in Boyan and Ball 1990), and mantis (Boyan and Ball 1986).

Primary neural circuits for the cricket cercal system are located within the TAG, which is relatively easy to access. Because of this technical advantage, in terms of the electrophysiology and neuronal histology, this system has been used as one of the most suitable model systems for various fields of neuroscience. Therefore, many studies have been carried out on cercal system development and neural rearrangement (e.g., Murphey 1985; Kämper and Murphey 1987; Chiba et al. 1992), receptor biomechanics (e.g., Shimozawa and Kanou 1984a, b; Kumagai et al. 1998; Shimozawa et al. 1998; Magal et al. 2006; Cummins et al. 2007; Dangles et al. 2008), and interneuron networks (e.g., Bodnar 1993; Baba et al. 2001). Studies by Miller and colleagues on the cricket cercal system provided significant insights into neural coding (e.g., Theunissen and Miller 1991; Crook et al. 2002; Dimitrova et al. 2002; Aldworth et al. 2005, 2011) and neural maps (e.g., Troyer et al. 1994; Jacobs and Theunissen 1996; Paydar et al. 1999). Jacobs et al. (2008) provide a review of their recent studies focused on neural coding at the interface between the mechanosensory afferents and primary interneurons. Recent imaging studies mentioned in Chap. 18 have also demonstrated dendritic integration of the identified interneuron (Ogawa et al. 2004, 2008) and spatiotemporal representation of the stimulus by afferents (Ogawa et al. 2006).

In early studies, it was assumed that the cercal system in crickets would be involved in a similar escape response to that seen in cockroaches. However, recent studies have suggested that it is a considerable oversimplification to class this system as an "escape system." Rather, the cercal system should be thought of as functioning as a low-frequency, near-field extension of the auditory system of the animal (Jacobs et al. 2008). From a neuroethological point of view, it might be that the cercal-mediated escape behavior in crickets is more complicated than that in the cockroach. For example, there are four types of mechanoreceptors in the cercus of crickets: filiform hairs, bristle hairs, clavate hairs, and campaniform sensilla. Briefly, filiform hairs detect air currents, bristle hairs detect tactile sensation, clavate hairs detect gravity, and campaniform sensilla detect distortions of the cuticle on the cercus. These receptors make contributions to important behaviors; for example, bristle hairs are involved in mating behavior (Sakai and Ootsubo 1988; Shell and Killian 2000); filiform hairs drive defense behavior (Gnatzy and Heußlein 1986; Gnatzy and Kämper 1990) as well as escape behavior (Kanou et al. 1999); and clavate hairs contribute to postural maintenance (Horn and Bischof 1983; Horn and Föller 1985). Other than postural maintenance, these behaviors are all competitive behaviors. The neural pathways for the processing of information sent by the four types of cercal signal are likely to interact in the CNS. In addition, even when focusing on air-current information, crickets have at least four options against an approaching predator: to (1) jump escape, (2) run or walk escape, (3) defend, or (4) freeze (see below). Such complexity makes it difficult to elucidate the neural system of cercal-related antipredator behaviors in crickets.

In this chapter, we review the cercal-mediated antipredator behaviors of crickets and the underlying cercal sensors and neural systems in the CNS. To elucidate the complete neural circuits driving antipredator behavior would result in significant insights into information coding and decision-making in real neural networks. Most of the data reviewed here come from two cricket species, the house cricket, *Acheta domesticus*, and a field cricket, *Gryllus bimaculatus*, with a few from the wood cricket, *Nemobius sylvestris*. However, the differences among the species are not considered in depth here.

14.2 Cercal-Related Antipredator Behaviors

Crickets use their cerci to detect the air current generated by predators and take various actions to survive. According to previous reports, such actions include a jump escape, run escape, walk escape, turn, headstand, stillstand, abdominal lift, kick, antennal swing, withdrawal, cercal cleaning, or no response (Dumpert and Gnatzy 1977; Stabel et al. 1985; Gnatzy and Heißlein 1986; Gras and Hörner 1992; Tauber and Camhi 1995; Baba and Shimozawa 1997; Kanou et al. 1999; Dangles et al. 2006a, 2007; Oe and Ogawa 2013; Table 14.1). In a recent review, Casas and Dangles (2010) described how the air current detected by the cerci can elicit at least 14 distinct responses, including evasion, flight, offensive reactions, scanning,

Table 14.1 Cercal-related antipredator behaviors in the cricket

Type of behavior	Condition, sex, age, and species	Air-current stimulus	References
Kick	Thorax fixed, female, 4–5 day adult, *G. bimaculatus*	$d = 9$ mm $f = 20$ mm $v = 1.9–3.75$ m/s	Dumpert and Gnatzy (1977)
Run	Tethered walking, both sexes, adult, *A. domesticus*	$d = 8$ mm Both sides $v = 0.5–1.5$ m/s	Stabel et al. (1985)
No reaction, jumping, head stand, stillstand, kick	Free moving, both sexes, nymphs and adult, *A. domesticus*	Air current generated by hunting wasp	Gnatzy and Heußlein (1986)
Run	Tethered on treadmill, both sexes, adult, *G. bimaculatus*	$d = 2$ mm $f = 6$ mm Both sides $v = 0.2–2$ m/s	Gras and Hörner (1992)
Jump, turn, turn + jump	Free moving, both sexes, adult, *G. bimaculatus*	$w = 2 \times 25$ cm^2 $f = 20$ mm All directions $v = 2.3$ m/s	Tauber and Camhi (1995)
Jump, run, walk, turn, withdrawal, abdominal lift, kick, cercal cleaning, antennal swing, freeze	Free moving in arena, male, adult (3–30 days), *G. bimaculatus*	$d = 1$ mm $f = 20$ mm Above 2 cm from cerci $v = 38–770$ cm/s	Baba and Shimozawa (1997)
Any movements, but no observe kick, head stand, cercal cleaning	Free moving in arena, female, adult (>1 week), *G. bimaculatus*	$d = 10$ mm $f = 20$ mm All directions $v = 10–390$ cm/s	Kanou et al. (1999)
Run, jump	Free moving in arena, *N. sylvestris*	Air current generated by spider or piston movement	Dangles et al. (2006a, 2007)
Run + turn	Free moving in area, tethered on treadmill, male, adult, *G. bimaculatus*	$d = 15$ mm $f = 75$ mm All directions $v = {\sim}1$ m/s (measured at cerci)	Oe and Ogawa (2013)

d/*w* diameter or width of nozzle, *f* distance from nozzle to animal, *v* velocity of air current

freezing, and various reactions during male stridulation. In addition, the response can depend on the behavioral state of the animal as well as on the context of the environment. We classify these responses into four strategies: the escape strategy (jump run and walk escape), the defense strategy (headstand, stillstand, abdominal lift, and kick), the alert strategy (turn, antennal swing, withdrawal, and cercal cleaning), and the freeze strategy (no response). All strategies except for the alert

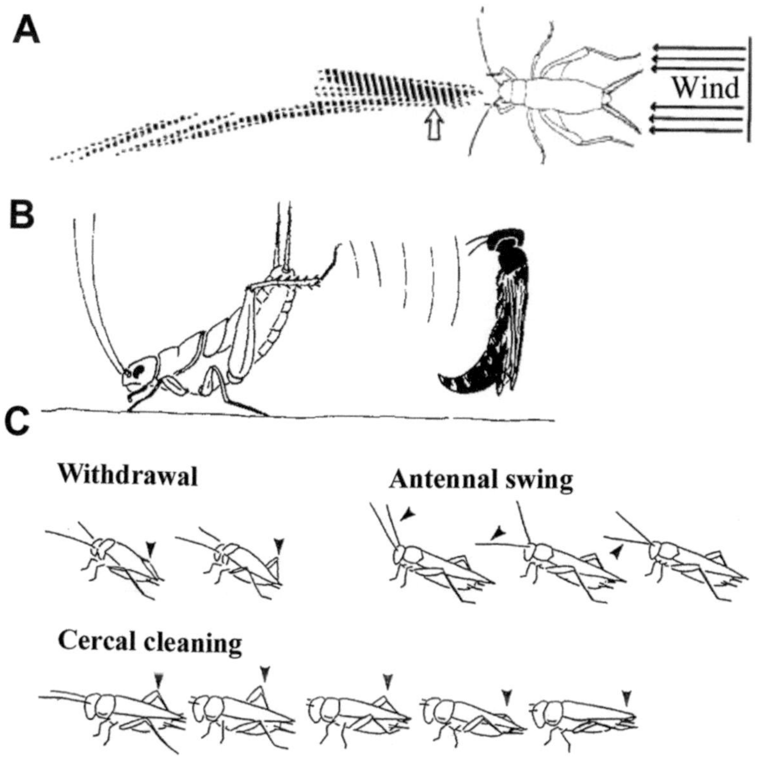

Fig. 14.1 Cercal-related behaviors. (**a**) Jump escape. *Black arrows* show the air current and the *white arrow* shows the jump point. *Dashed lines* represent the long axis of the cricket at 4-ms intervals. (**b**) Defense behavior (stillstand). A cricket lifts its abdomen and hind legs up against a wasp. (**c**) Alert behaviors. *Arrowheads* show the movement of antenna or legs (Reproduced, with permission, as follows: (**a**) from Fig. 3b in Tauber and Camhi (1995); (**b**) from Fig. 1 in Gnatzy and Heußlein (1986); and (**c**) from Fig. 2 in Baba and Shimozawa (1997))

strategy compete against each other. Here, we focus on the three strategies elicited in response to natural and potential predators, as well as artificial air currents.

The first natural predator investigated was a sphecid wasp, *Liris niger*. When *A. domesticus* individuals were attacked by the wasps, they adopted three survival strategies, escape (jump), defense (headstand, stillstand, and kicking; Fig. 14.1b shows kicking), and freeze (Gnatzy and Heißlein 1986). Headstand referred to a behavior in which the crickets suddenly raised their abdomen from a resting position (we also assumed that "stillstand" involved more abdominal lift compared with "headstand"; the "abdominal lift" behavior involved both "headstand" and "stillstand"; thus, we use "abdominal lift" hereafter). These actions were triggered before tactile contact occurred and after the wasp had approached within a radius of 1–3 cm, as measured from the rear end of the abdomen of the cricket. The running speed of the wasp was approximately 20–50 cm/s (Gnatzy and Heißlein 1986; Ganzy and Kämper 1990). The second natural predator investigated was the wolf spider, *Pardosa spp*. The spider attacked *N. sylvestris* only when the cricket came within 2–7 cm, whereupon the spider launched its attack at a median distance of 4.5 cm. In response, the crickets adopted an escape strategy, using a jump or walk escape (Dangles et al. 2007). Typical escape behavior comprised an initial 90° pivot followed by a jump escape, which was observed in more than half of the crickets tested. Escape performances were higher in juvenile than older instars (Dangles et al. 2007). On average, crickets escaped after the spider came within 1.4 cm. Attack velocities by spiders ranged from 2 to 41 cm/s (Dangles et al. 2006a).

Some researchers have observed the behavior of small mammals when hunting crickets (Ivanco et al. 1996; Munz et al. 2010). Baba et al. (2009) reported that *G. bimaculatus* would respond to an approaching mouse, with most crickets choosing the escape strategy (walk and run escape) and some choosing the alert (turn, antennal pointing), defense (abdominal lift, kick), or freeze strategies. However, crickets did not show a jump escape before a mouse had come into contact with the body of the cricket. Velocities of approaching mice to crickets ranged from 2.8 to 617 mm/s. Crickets standing in an arena (40 cm in diameter) began to move when mice entered a circle that was 20 cm in diameter, with the center being the tip of abdomen of the cricket. The mean distance was 83.8 mm for escape, 93.6 mm for alert, and 67.4 mm for defense behaviors. The average velocities of the approaching mice driving the escape and defense posture were 150 mm/s and 177.2 mm/s, respectively. When the freeze strategy was employed and the crickets were touched by a mouse, they started a run and jump escape (Baba, unpublished data). The study about "tongue hunting" of frogs provides additional insights into how the cercal system detects the approaching predator. The mean velocity of air currents generated by the tongues of toads hunting cockroaches was 2 cm/s measured at the location of the prey (Camhi et al. 1978).

In contrast to natural predator studies, three studies addressed the behavior of crickets in response to an artificial air-current stimulus. In the first report, the wind stimulus was produced by a motor that depressed a plastic sheet, which closed off one end of a wind tunnel. This end of the tunnel, which was rectangular in cross section, measured $25 \times 30\ \mathrm{cm}^2$. An air puff was expelled from the opposite, open end of the tunnel, which was a slit 2 cm high and 25 cm wide (Tauber and Camhi 1995). In this experiment, *G. bimaculatus* adopted an escape strategy against the air puff (duration approximately 200 ms, peak velocity 2.3 m/s). The observed behaviors were classified into three categories: turn response, jump response (Fig. 14.1a), and a turn + jump response. In the second report, air current was delivered with a nozzle (10 mm in inner diameter) placed 18 cm away from the cricket (Kanou et al. 1999). The air-current velocities were 0.1, 0.9, 1.5, 3.0, and 3.9 m/s and stimulus duration was 70 ms. In this study, neither defensive behavior nor cercal cleaning were observed. More than half (56 %) of crickets responded to the stimulus of 3.9 m/s velocity. At the stimulus velocities of 1.5 and 3.0 m/s, most escapes were triggered in the opposite direction to the stimulus (Kanou et al. 1999). In the third report, a fine nozzle (1 mm in diameter) was used for stimulus delivery, and the air-current velocity ranged from 38 to 770 cm/s (Baba and Shimozawa 1997). Stimulation was applied 2 cm above the cerci and stimulus duration was 2 s. In this study, the crickets exhibited various responses, which were classified into 11 types: walk, run, jump, turn, antennal swing, withdrawal, cercal cleaning, kick, abdominal lift, leg lift, and freeze. Antennal swing occurred simultaneously with 60 % of the kicks, 63 % of the walks, 70 % of the runs, 75 % of the leg lifts, 80 % of abdominal lifts, 87 % of cercal cleanings, and 90 % of withdrawal behaviors (Baba and Shimozawa 1997; Fig. 14.1c).

When taken together, these results show that crickets have at least three strategies for reacting to a stimulus: escape, defense, and freeze (we assume that the alert strategy is not a final action but is performed to collect more information or to prepare a final action). The escape strategy has two sub-strategies: jump escape or walk and run escape. Tauber and Camhi (1995) highlighted the advantages of the jump escape, explosive or rapidly accelerating quality, and the use of movement in three dimensions, thus incorporating an additional degree of freedom for the running escape. By contrast, the freeze strategy has not received much attention, although it might be an important survival strategy. Major predators of crickets, which include amphibians, reptiles, spiders, and insects, are known to prey mainly upon moving objects mainly. Mice, which we used in a preliminary study of crickets' escape behavior, and mammals tended to aim more at a moving cricket more than one that was in a frozen state (Baba unpublished data). These facts suggest that a freeze strategy must be one of the rational actions of some crickets that have less strong motor abilities. For example, the period of running and its velocity in the escape behavior of the crickets were not as long or as fast as recorded for cockroaches (Gras et al. 1994). Furthermore, Nishino and Sakai (1996) reported that *G. bimaculatus* showed thanatosis and discussed that it could have a role for predator avoidance. Thus, more focus is required on the freeze strategy as an important antipredator behavior in the crickets.

It is likely that there is a robust relationship between input (stimulus) and output (response) in cercal-mediated behavior in response to the small range of stimuli. For example, wide strong airflow applied by the wind tunnel drove a jump escape, whereas a small puff applied from an air nozzle drove a run and walk escape. A smaller (1 mm diameter) or vertical stimulation elicited the defense strategy in addition to other responses. When we studied the cercal system using artificial stimulation, only the stimulus intensity could be controlled. However, it is likely that vertical angle or flow diameter of the air-current stimulus would also affect behavior. Thus, when discussing the behavioral function of the cercal system, we must include specific stimulus parameters.

14.3 Cercal Mechanoreceptors

Crickets use filiform hairs to detect air currents. The filiform hairs are comprised of a rod-type hair, which is movable, and a socket (Murphey 1985). The base of the hair is elliptical and is inserted into a collar of flexible cuticle in the socket (Keil and Steinbrecht 1984). The directional movement axis of each hair is determined by the orientation of a hinge-like structure in the socket. A single receptor neuron innervating each hair at the socket generates action potentials as the hair moves (Edwards and Palka 1974). The amplitude of the response of each sensory receptor cell to any air-current stimulus depends upon the velocity and direction of the air current, and the directional tuning curves of the receptor afferents are well described by cosine functions (Landolfa and Jacobs 1995; Landolfa and Miller 1995). The hair length affects the response properties and thresholds. Longer hairs, which are more flexible and respond to low-frequency air currents with low thresholds, could code wind velocity. In contrast, shorter hairs are stiffer and respond only to high frequencies. Given that their thresholds depend on stimulus frequencies, these hairs would code wind acceleration (Shimozawa and Kanou 1984a, b). The threshold range is 0.03–300 mm/s, with an absolute sensitivity of approximately 0.03 mm/s (Shimozawa and Kanou 1984a, b). Most receptor neurons were saturated by an air current of 3–60 cm/s, and their maximum firing rate was approximately 400 spikes/s (Landolfa and Miller 1995). Conductance velocities of sensory afferents were approximately 1–2 m/s in *G. bimaculatus* (Baba, unpublished data) and 1.87 m/s in *A. domesticus* (Mulder-Rosi et al. 2010). The cercal length is approximately 10 mm. Given that the filiform hairs are distributed over a cercus about 10 mm long, the afferent spikes of the sensory neurons would vary in arrival time from 5 to 10 ms, assuming the conduction velocities of the afferent spikes are constant. Recently, it was reported that small variations in spike-propagation velocities between afferent axons enables the cercal system to function as a “delay line” (Mulder-Roshi et al. 2010).

The number of filiform hairs on a cercus varies depending on species, instar stages, and probably the growth environment. Each cercus is covered with 300–750 filiform hairs in adult *A. domesticus* (Edward and Palka 1974; Palka and Olberg 1977; Jacobs et al. 2008; Miller et al. 2011; Chiba et al. 1992), 400–500 in adult *G. bimaculatus* (Shimozawa and Kanou 1984b), and 322 in adult *N. sylvestris* (Dangles et al. 2006b). Individual filiform hairs vary in length and diameter even on a single cercus. The approximate length of 120 hairs was 800–1,500 μm, with most of the remainder (300–400 hairs) being approximately 400 μm long in adult *A. domesticus* (Edwards and Palka 1974). In adult *G. bimaculatus*, the largest hairs were 1,500 μm long and 9 μm in diameter, and the smallest were 30 μm long and 1.5 μm in diameter (Shimozawa and Kanou 1984b). Filiform hairs can be classified approximately into two or three subtypes depending on their movable direction: transverse (T-hair), longitudinal (L-hair), or oblique (O-hair) relative to the longitudinal axis of the cercus. T-hairs occur on the dorsal and ventral surface, L-hairs on the medial and lateral surface, and O-hairs on the dorsal–lateral, ventral–lateral, and ventral–medial surfaces of the cercus (Landolfa and Jacobs 1995). Additionally, Landolfa and Jacobs (1995) reported a more detailed classification (eight types). Recently, Miller and colleagues (2011) reported the complete distribution of filiform hairs on a cercus and provided a summary of the distribution of preferred direction of each hair (Fig. 14.2c, d). The directional data showed that crickets have hairs arranged in four diagonal directional bands (Miller et al. 2011).

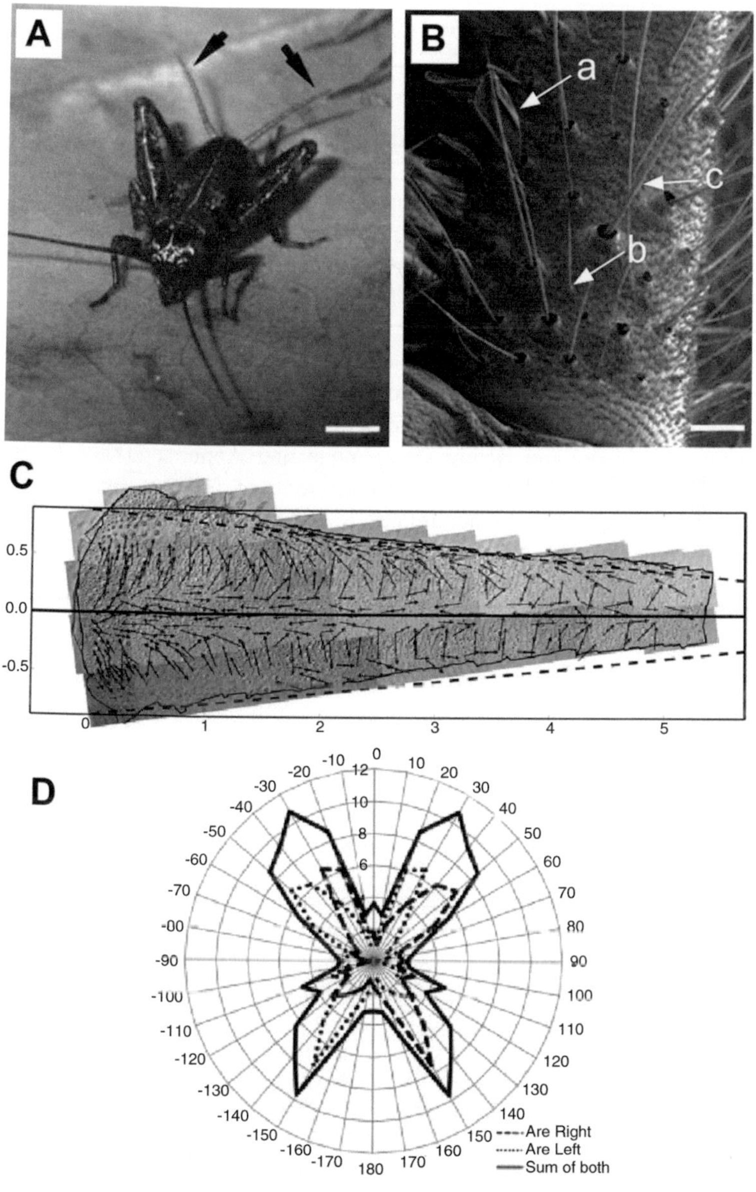

Fig. 14.2 Cercal mechanoreceptors. (**a**) A wood cricket and its cerci (*arrows*). Scale bar = 2 mm. (**b**) Scanning electron micrograph of the proximal region of the cerci. *Arrows* **a**–**c** show a clavate hair, bristle hair, and filiform hair, respectively. Scale = 100 μm. (**c**) Distribution of filiform hairs on filet preparation. *Arrows* show the excitatory direction of each hair. Vertical and horizontal units are in mm. (**d**) Number of filiform hairs in individual directions. Zero refers to the anterior direction (Reproduced, with permission, as follows: (**a, b**) from modified Fig. 1 in Dangles et al. (2006b); and (**c, d**) from modified Figs. 8a and 11d, respectively, in Miller et al. (2011))

14.4 Synaptic Connections from Receptor Afferents to Interneurons

Filiform sensory neurons project to the posterior TAG, which comprises five neuromeres (abdominal neuromere 7–11; Jacobs and Murphey 1987). The seventh to eighth neuromeres have a structure similar to unfused abdominal ganglion, which is the anterior TAG. Neuromeres 9–11 (posterior TAG) are fused and their structure modified (Baba et al. 2010). The TAG in adult *G. bimaculatus* shows sexual dimorphism, but dimorphism of the filiform sensory projection has not been observed.

In *A. domesticus*, the 1500 sensory afferents synapse with a group of approximately 30 local interneurons and approximately 10 pairs of identified ascending interneurons. The putative neurotransmitter of the cercal-to-interneuron synapses is acetylcholine (Meyler and Reddy 1985; Yono and Aonuma 2008). Jacobs and colleagues demonstrated that the sensory afferents of filiform sensilla terminate in an orderly array, which is not related to their location on the cercus but to directional sensitivity in the horizontal plane (Jacobs and Theunissen 1996; Paydar et al. 1999). The functional afferent map was constructed from anatomical data of central projections and physiological response properties of mechanosensory afferents (Jacobs and Teunissen 1996) and visualized by calcium-imaging experiments (Ogawa et al. 2006). The topography of mechanosensory afferents based on their response property is very interesting to note because we have assumed that the projection of mechanosensory afferents is generally organized to reconstruct their topography within the CNS like a mammalian somatosensory system, but the afferent map in the cercal system is based on physiological relationships. Furthermore, the observation that postsynaptic neurons targeted by cercal afferents were changed by partial ablation of the afferents suggests synaptic rearrangement adjusting to the development of filiform hairs in each molt (Chiba et al. 1992).

14.5 Interneurons

Dendrites or neurites of ascending and local interneurons are arborized within the projection area of the filiform sensory neurons. At least eight (in *G. bimaculatus*) to ten (in *A. domesticus*) pairs of ascending interneurons have been identified as giant interneurons (GIs) owing to their relatively large-diameter axons, which enable rapid conduction of action potentials. The axons of the GIs run through the ventral nerve cord contralateral to the somata (hereafter, we used ipsilateral or contralateral to the somata position). In *G. bimaculatus*, it was confirmed that eight pairs of GIs project their axons to motor centers in the thorax and integrative centers in the brain. The GIs are named on the basis of the location of somata in the TAG, such as GI 7-1, 8-1, 9-1, 9-1b, 9-2, 9-3, 10-2, and 10-3 (Fig. 14.3b). The first number represents the soma location of the abdominal neuromere, and the second number

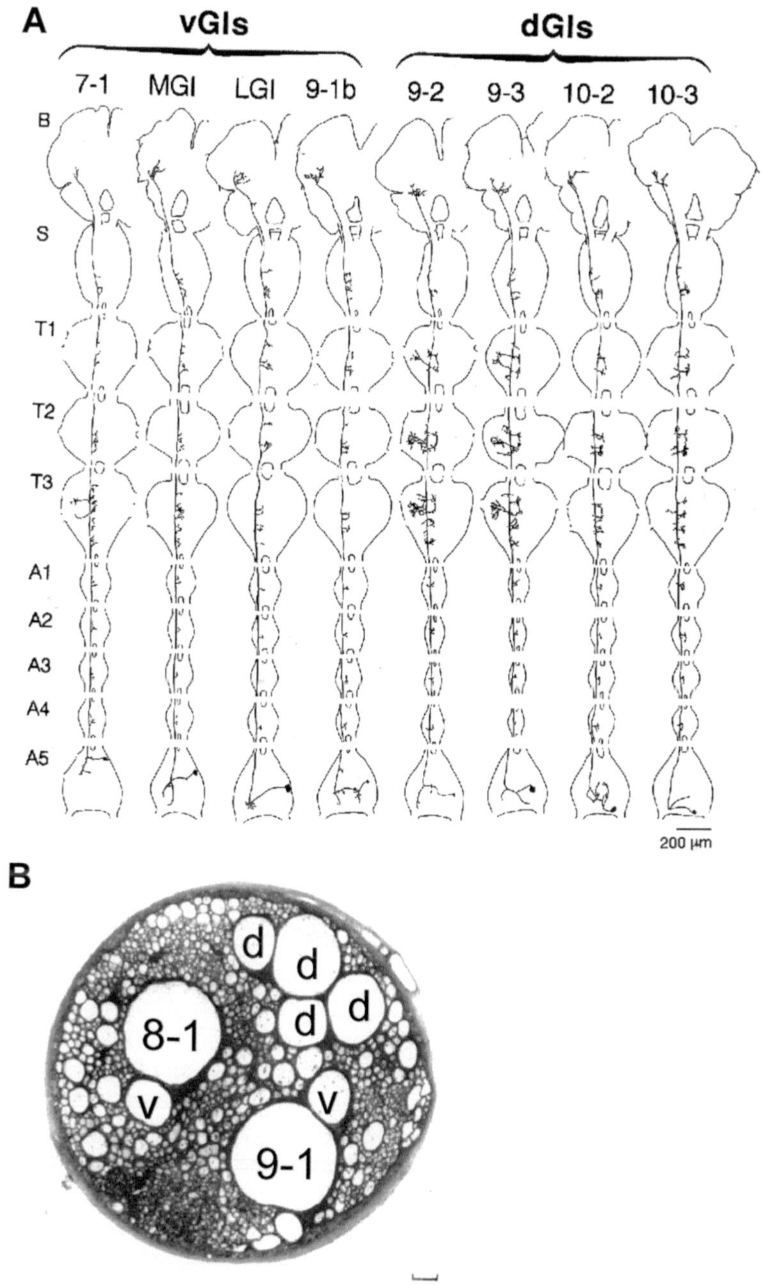

Fig. 14.3 (**a**) Morphology of eight types of giant interneuron (GIs). GIs 7-1, 8-1, 9-1, and 9-1b are ventral GIs (vGIs) and the others are dorsal GIs (dGIs). (**b**) Cross section of ventral nerve cord. The axons labeled 8-1, 9-1, and v are delivered by vGIs. The axons labeled with d are delivered by dGIs. Scale bar = 10 μm (Reproduced, with permission, as follows: (**a**) from Fig. 1 in Hirota et al. (1993) and (**b**) from modified Fig. 6 in Edwards and Palka (1974))

represents the identified number. These are classified into two groups depending on the position of the axons within the ventral nerve cord. The GIs 7-1, 8-1, 9-1, and 9-1b are classified as ventral GIs (vGIs) owing to the middle or ventral position of their axons. The other GIs with axons located at the dorsal side of the ventral nerve cord are classified as dorsal GIs (dGIs). In the wood cricket, eight GIs have also been identified, although these were named differently to the system described above (Insausti et al. 2011). Here, we use the conventional names for GIs in *G. bimaculatus* and *A. domesticus*. In *G. bimaculatus*, eight pairs of GIs project their axons to all segmental ganglia, including thoracic and cephalic ganglia (Fig. 14.3a; Hirota et al. 1993). The axonal projection within thoracic ganglia differed in its branching pattern between dGIs and vGIs. The dGIs arborize medial and lateral branches, whereas the vGIs, except for GI 7-1, have only medial branches. Depolarizing current injection into dGIs elicited neural activities in the thoracic nerve branch and occasionally induced leg movements, such as walking, whereas the activation of vGIs elicited no activity (Gras and Kohstall 1998; Hirota et al. 1993). Directional sensitivity of ascending interneurons, including GIs, would mediate some directional control of the cercal-mediated response.

The response properties, including directional and frequency sensitivity to sinusoidal sounds or air puffs, threshold, and input source of the filiform hairs, have been well studied in the GIs. Integrative studies have described that the directional sensitivity of GIs is primarily based on the location of GI's dendritic arbor within the functional afferent map mentioned above (Jacobs and Theunissen 2000; Ogawa et al. 2008). Most electrophysiological data were recorded from decapitated animals. Recording sites of the intracellular electrode varied between studies, such as axons within the ventral nerve cord, cell body, or dendrite within the TAG. An in vivo calcium-imaging technique has been used to examine dendritic excitability (Ogawa et al. 2000, 2002a, b) and synaptic integration of directional information in GIs (Ogawa et al. 2004, 2008). Furthermore, whole-cell clamp recording of cultured GIs has demonstrated voltage-activated Na^+, K^+, and Ca^{2+} currents (Kloppenburg and Hörner 1998). The Ca^{2+} accumulation in dendrites of GIs modulated wind sensitivity via synaptic depression (Ogawa et al. 2001). Below, we describe the distinct response properties of individual GIs and other interneurons.

GI 8-1, also called the medial giant interneuron (MGI), and GI 9-1, also called the lateral giant interneuron (LGI), have been well studied because they have the largest and second largest axons, respectively, and the responses are easy to record intracellularly. The T-hairs contralateral to the somata of the GIs provide the strongest excitatory input to GI 8-1 and 9-1. The ipsilateral T-hairs inhibit GI 8-1 (Matsumoto and Murphey 1977), and the ipsilateral T-hairs and contralateral L-hairs inhibit GI 9-1. The inhibition reduced the number of outgoing spikes from a level elicited by excitation alone, and it did so in proportion to the level of wind responsiveness displayed in GI 8-1 (Baba et al. 2001). The Gl 8-1 displayed the highest sensitivity to the air currents from a contralateral-rear direction (Kanou 1996). The GIs 8-1 and 9-1 received their main excitatory inputs from rather short cercal hairs but not from the conspicuously long filiform sensory hairs (Kanou and Shimozawa 1984). A simple model of the directional sensitivity of the interneurons based on their monosynaptic inputs from the cercal sensory neurons can account for the major features of the directional sensitivity of the interneurons (Bacon and Murphey 1984; Shephard and Murphey 1986). Laser ablation of GI 8-1 diminished control of the turn angle and decreased locomotion distance in the initial walking response to air-puff stimuli, suggesting that GI 8-1 is involved in turn-angle control and maintenance of wind-elicited walking behavior (Oe and Ogawa 2013).

The GIs 10-2 and GI 10-3 show spontaneous activity and are highly sensitive (the lowest threshold was 0.03 mm/s or less). The long filiform hairs connect to them (Shimozawa and Kanou 1984b). GIs 10-2 and 10-3 are similar in terms of their synaptic inputs, which include excitatory synaptic input from ipsilateral L-hairs and contralateral T-hairs and inhibitory input from contralateral L-hairs and ipsilateral T-hairs (Levine and Murphey 1980). These interneurons are tuned broadly to the direction and velocity of bulk airflow movements and project a coarse-coded image of these parameters to higher centers, which is accurate to approximately 8° (Theunissen and Miller 1991; Theunissen et al. 1996). The four GIs comprising left and right 10-2 and left and right 10-3 serve as a functional unit encoding the direction of the air current in the horizontal plane (Miller et al. 1991). Recent research has demonstrated that the GIs 10-2 and 10-3 also respond with high sensitivity to complex dynamic multidirectional features of air currents on a smaller spatial scale than the physical dimensions of the cerci (Mulder-Rosi et al. 2010), using a coding scheme that is more complex than simple linear encoding (Aldworth et al. 2011).

Relatively few studies have demonstrated the response characteristics of other GIs. Gl 9-2 is the most sensitive to air currents from contralateral-rear directions and is less sensitive to air currents from ipsilateral-front directions (Miller et al. 1991; Kanou 1996). Gl 9-3 is more sensitive to air currents from contralateral-front directions and is less sensitive to those from an ipsilateral-front direction (Miller et al. 1991; Kanou 1996). It has been suggested that GIs 9-2 and 9-3 form a second directional encoding unit sensitive to higher velocities (Miller et al. 1991). GI 7-1 responds to wind stimulation, although its dendrites do not overlap the projection area of sensory afferents (Hirota et al. 1993). GI 9-1b has spontaneous activity, with a threshold of <0.2 cm/s (Kämper 1984). The recent report that ablation of GI 9-1b delays reaction time of the wind-elicited walking implies an important role for G I9-1b in rapid initiation of that response (Oe and Ogawa 2013).

In total, 7–15 pairs of ascending interneurons, other than GIs, have been identified and also determined to respond to air particle displacement of filiform hairs (Kämper 1984; Jacobs and Murphey 1987; Baba et al. 1991). These non-giant ascending interneurons are likely to contribute to antipredator behaviors. In addition, at least 21 pairs of non-spiking local interneurons and 15 pairs of spiking local interneurons have been found in the TAG (Bodnar et al. 1991; Baba et al. 1995). If some of these local interneurons influenced the activities of GIs, it would contribute to signal processing (Bodnar 1993; Baba 2002), but further work is required to determine the mechanisms behind this effect.

14.6 Conclusion and Further Studies

Although there have been several studies on sensory neurons and interneurons in the cercal system of crickets, there is still a gap in knowledge between the neural mechanisms and their behavioral functions. Although basic neural study does not require behavioral relationships, behavioral data would enhance the relevancy of the neural data. For example, the recovery level of neural networks by rearrangement could be tested using a behavioral assay. To understand the contributions of GIs and other ascending interneurons in cercal-mediated behaviors, one should consider the following points. First, decapitated crickets have been used in most physiological studies focused on elucidating directional sensitivity, neural connection, and input source. Lower diversity in physiological responses in this preparation was useful to figure out the basic mechanism of neural coding and information processing in the receptor neurons and interneurons. However, these data are not enough to explain the variety of cercal-mediated behaviors. Second, there are very few studies using interneuron ablation experiments. Although it is known that there are eight pairs of GIs and some pairs of non-GIs and that giant ascending interneurons are assumed to contribute to antipredator behaviors, their roles are not known from a behavioral perspective. In the cockroach, ablation of GIs affected directional sensitivity in escape response (Comer 1985; Comer and Dowd 1987). Recently, it has been reported that selective ablation of GIs modified motor control depending on the stimulus direction and that the directional control in the wind-evoked walking behavior required descending signals from the brain ganglion (Oe and Ogawa 2013). The directional information conveyed by GIs is essential for giving the direction of the escape behavior. If genetic manipulation techniques are established in the crickets as mentioned in other chapters, cell-specific activation or silencing will provide direct evidence of the behavioral function of individual interneurons.

Another point to consider is stimulus intensity. According to behavioral reports, the range of velocity of approaching predators that trigger antipredator behavior is 20–500 mm/s. It is assumed that the air current that crickets detect would be slower than the approaching velocity because the air current reaching the cerci must be smaller than the movement of the predator. As mentioned above, the threshold of the sensory neurons and GIs is more than 0.03 mm/s, and the dynamic range is approximately 0.03–300 mm/s. By contrast, in behavioral analyses, artificial (air puff) stimuli of relatively higher velocities (380–7700 mm/s) than observed in natural conditions or in neural studies have been used. This gap also suggests that the cercal system should not be classified as a simple escape system but as a near field and low-frequency auditory system (Jacobs et al. 2008). Additionally, crickets use three strategies in response to cercal stimulation, but strict directional orientation is only required for the defense strategy. Ascending interneurons including GIs would convey the information for crickets to choose the appropriate strategy in addition to the directional information for performing a particular behavior. Although the above data have shown some input–output relationships, they have not shown that these are robust. The cercal-mediated antipredator behavior is more complicated than was previously thought. Given that the nervous system of the cricket is relatively easy to access and manipulate under free moving conditions, this system is a suitable model to use to analyze the neural basis underlying flexible and situation-dependent behaviors.

References

Aldworth ZN, Miller JP, Gedeon T, Cummins GI, Dimitrov AG (2005) Dejittered spike-conditioned stimulus waveforms yield improved estimates of neuronal feature selectivity and spike-timing precision of sensory interneurons. J Neurosci 25:5323–5332

Aldworth Z, Dimitrov AG, Cummins GI, Gedeon T, Miller JP (2011) Temporal coding in a nervous system. PLoS Comput Biol 7(5):e1002041

Baba Y (2002) Neural connections between giant interneurons and local interneurons of the terminal abdominal ganglion in the cricket. Soc Neurosci Abstr. program No. 60.8

Baba Y, Shimozawa T (1997) Diversity of motor responses initiated by a wind stimulus in the freely moving cricket, *Gryllus bimaculatus*. Zool Sci 14:587–594

Baba Y, Hirota K, Yamaguchi T (1991) Morphology and response properties of wind-sensitive non-giant interneurons in the terminal abdominal ganglion of crickets. Zool Sci 8:437–445

Baba Y, Hirota K, Shimozawa T, Yamaguchi T (1995) Differing afferent connections of spiking and nonspiking wind-sensitive local interneurons in the terminal abdominal ganglion of the cricket *Gryllus bimaculatus*. J Comp Physiol A 176:17–30

Baba Y, Masuda H, Shimozawa T (2001) Proportional inhibition in the cricket medial giant interneuron. J Comp Physiol A 187:19–25

Baba Y, Patel SR, Patel FN, Shah N, Patel D, Tukada A, Comer CM (2009) Integration between mechano- and visual-information for orienting a predator in offence behavior. Neuroscience meeting planner program No. 847.23

Baba Y, Tsukada A, Ogawa H (2010) Sexual dimorphism of the cricket's terminal abdominal ganglion in shape and distribution of GABA-like immuno-reactive neurons. Zool Sci 27:506–513

Bacon JP, Murphey RK (1984) Receptive fields of giant interneurons are related to their dendritic structure. J Physiol 352:601–623

Bodnar DA (1993) Excitatory influence of wind-sensitive local interneurons on an ascending interneuron in the cricket cercal sensory system. J Comp Physiol A 172:641–651

Bodnar DA, Miller JP, Jacobs GA (1991) Anatomy and physiology of identified wind sensitive local interneurons in the cricket cercal sensory system. J Comp Physiol A 168:553–564

Boyan GS, Ball EE (1986) Wind-sensitive interneurones in the terminal ganglion of praying mantids. J Comp Physiol 159:773–789

Boyan GS, Ball EE (1990) Neural organization and information processing in the wind-sensitive cercal receptor/giant interneuron system of the locust and other orthopteroid insects. Progress Neurobiol 35:217–243

Boyan GS, Ashman S, Ball EE (1986) Initiation and modulation of flight by a single giant interneuron in the cercal system of the locust. Naturwissenschaften 73:272–274

Camhi JM (1980) The escape system of the cockroach. Sci Am 243:144–157

Camhi JM, Tom W (1978) The escape behavior of the cockroach *Periplaneta americana*. I. Turning response to wind puffs. J Comp Physiol 128:193–201

Camhi JM, Tom W, Volman S (1978) The escape behavior of the cockroach *Periplaneta americana*. II. Detection of natural predators by air displacement. J Comp Physiol 128:203–212

Casas J, Dangles O (2010) Physical ecology of fluid flow sensing in arthropods. Ann Rev Entomol 55:505–520

Chiba A, KamperKämper G, Murphey RK (1992) Response properties of interneurons of the cricket cercal sensory system are conserved in spite of changes in peripheral receptors during maturation. J Exp Biol 164:205–226
Comer MC (1985) Analyzing cockroach escape behavior with lesions of individual giant interneurons. Brain Res 335:342–346
Comer CM, Dowd JP (1987) Escape turning behavior of the cockroach. Changes in directionality induced by unilateral lesions of the abdominal nervous system. J Comp Physiol A 160:571–583
Crook S, Miller J, Jacobs G (2002) Modeling frequency encoding in the cricket cercal sensory system. Neurocomputing 44–46:769–773
Cummins B, Gedeon T, Klappera I, Cortez R (2007) Interaction between arthropod filiform hairs in a fluid environment. J Theor Biol 247:266–280
Dangles O, Ory N, Steinmann T, Christides J-P, Casas J (2006a) Spider's attack versus cricket's escape: velocity modes determine success. Anim Behav 72:603–610
Dangles O, Pierre D, Magal C, Vannier F, Casas J (2006b) Ontogeny of air-motion sensing in cricket. J Exp Biol 209:4363–4370
Dangles O, Pierre D, Christides JP, Casas J (2007) Escape performance decreases during ontogeny in wild crickets. J Exp Biol 210:3165–3170
Dangles O, Steinmann T, Pierre D, Vannier F, Casas J (2008) Relative contributions of organ shape and receptor arrangement to the design of cricket's cercal system. J Comp Physiol A 194:645–663
Dimitrov AG, Miller JP, Aldworth ZN, Parker AE (2002) Spike pattern-based coding schemes in the cricket cercal sensory system. Neurocomputing 44–46:373–379
Dumpert K, Gnatzy W (1977) Cricket combined mechanoreceptors and kicking response. J Comp Physiol 122:9–25
Edwards JS, Palka J (1974) The cerci and abdominal giant fibers of the house cricket, *Acheta domesticus*. I. Anatomy and physiology of normal adults. Proc R Soc Lond B 185:83–103
Gnatzy W, Heußlein R (1986) Digger wasp against cricket. I. Receptors involved in the antipredator strategies of the prey. Naturewissenschaften 73:212–215
Gnatzy W, Kämper G (1990) Digger wasp against crickets. II. An airborne signal produced by a running predator. J Comp Physiol A 167:551–556
Gras H, Kohstall D (1998) Current injection into interneurones of the terminal ganglion modifies turning behavior of walking crickets. J Comp Physiol A 182:351–361
Gras H, Hörner M (1992) Wind-evoked escape running of the cricket Gryllus bimaculatus. I Behavioural analysis. J Exp Biol 171:189–214
Gras H, Hörner M, Schürman F-H (1994) A comparison of spontaneous and wind-evoked running modes in crickets and cockroaches. J Insect Physiol 40:373–384
Hirota K, Sonoda Y, Baba Y, Yamaguchi T (1993) Distinction in morphology and behavioral role between dorsal and ventral groups of cricket giant interneurons. Zool Sci 10:705–709
Horn E, Bischof H-J (1983) Gravity reception in cricket: the influence of cercal and antennal afferences on the head position. J Comp Physiol 150:93–98
Horn E, Föller W (1985) Tonic and modulatory subsystems of the complex gravity receptor system in crickets, *Gryllus bimaculatus*. J Insect Physiol 31:937–946
Hoyle G (1958) The leap of the grasshopper. Sci Am 198:30–35
Insausti TC, Lazzari CR, Casas J (2011) The morphology and fine structure of the giant interneurons of the wood cricket *Nemobius sylvestris*. Tissue Cell 43:52–65
Ivanco TL, Pellis SM, Whishaw IQ (1996) Skilled forelimb movements in pray catching and in reaching by rats (*Rattus norvegicus*) and opossums (*Monodelphis domestica*): relations to anatomical differences in motor system. Behav Brain Res 79:163–181
Jacobs GA, Murphey RK (1987) Segmental origins of the cricket giant interneuron system. J Comp Neurol 265:145–157
Jacobs GA, Theunissen FE (1996) Functional organization of a neural map in the cricket cercal sensory system. J Neurosci 16:769–784
Jacobs GA, Theunissen FE (2000) Extraction of sensory parameters from a neural map by primary sensory interneurons. J Neurosci 20:2934–2943

Jacobs GA, Miller JP, Aldworth Z (2008) Computational mechanisms of mechanosensory processing in the cricket. J Exp Biol 211:1819–1828

Kämper G (1984) Abdominal ascending interneurons in crickets: responses to sound at the 30-Hz calling-song frequency. J Comp Physiol A 155:507–520

Kämper G, Murphey RK (1987) Sensory neurons after cross-species transplantation in crickets: the role of positional information. Dev Biol 122:492–502

Kanou M (1996) Directionality of cricket giant interneurons to escape eliciting unidirectional air-current. Zool Sci 13:35–46

Kanou M, Shimozawa T (1984) A threshold analysis of cricket cercal interneurons by an alternating air-current stimulus. J Comp Physiol A 154:357–365

Kanou M, Ohshima M, Inoue J (1999) The air-puff evoked escape behavior of the cricket *Gryllus bimaculatus* and its compensational recovery after cercal ablations. Zool Sci 16:71–79

Keil TA, Steinbrecht RA (1984) Mechanosensitive and olfactory sensilla of insects. In: King RC, Akai H (eds) Insect ultrastructure, vol 2. Plenum Press, New York, pp 477–516

Kloppenburg P, Hörner M (1998) Voltage-activated currents in identified giant interneurons isolated from adult crickets *Gryllus bimaculatus*. J Exp Biol 201:2529–2541

Kumagai T, Shimozawa T, Baba Y (1998) Mobilities of the cercal wind-receptor hairs of the cricket, *Gryllus bimaculatus*. J Comp Physiol A 183:7–21

Landolfa MA, Jacobs GA (1995) Direction sensitivity of the filiform hair population of the cricket cercal system. J Comp Physiol A 177:759–766

Landolfa MA, Miller JP (1995) Stimulus/response properties of cricket cercal filiform receptors. J Comp Physiol A 177:749–757

Levine RB, Murphey RK (1980) Pre- and postsynaptic inhibition of identified giant interneurons in the cricket (*Acheta domesticus*). J Comp Physiol 135:269–282

Magal C, Dangles O, Caparroy P, Casas J (2006) Hair canopy of cricket sensory system tuned to predator signals. J Theor Biol 241:459–466

Matsumoto SG, Murphey RK (1977) The cercus-to-giant interneuron system of crickets. IV. Patterns of connectivity between receptors and the Medial giant interneuron. J Comp Physiol 119:319–330

Meyer MR, Reddy GR (1985) Muscarinic and nicotinic cholinergic binding sites in the TAG of the cricket (*Acheta domesticus*). J Neurochem 45:1101–1112

Miller JP, Jacobs GA, Theunissen FE (1991) Representation of sensory information in the cricket cercal sensory system I. Response properties of primary interneurons. J Neurophysiol 66:1680–1689

Miller JP, Krueger S, Heys JJ, Gedeon T (2011) Quantitative characterization of the filiform mechanosensory hair array on the cricket cercus. PLoS ONE 6:e27873

Mulder-Rosi J, Cummins GI, Miller JP (2010) The cricket cercal system implements delay-line processing. J Neurophysiol 103:1823–1832

Munz M, Brecht M, Wolfe J (2010) Active touch during shrew prey capture. Front Behav Neurosci 4:1–11

Murphey RK (1985) Competition and chemoaffinity in insect sensory systems. TINS 8:120–125

Nishino H, Sakai M (1996) Behaviorally significant immobile state of so-called thanatosis in the cricket Gryllus bimaculatus Degree: its characterization, sensory mechanism and function. J Comp Physiol A 179:613–624

Oe M, Ogawa H (2013) Neural basis of stimulus-angle-dependent motor control of wind-elicited walking behavior in the cricket *Gryllus bimaculatus*. PLoS ONE 8:e80184

Ogawa H, Baba Y, Oka K (2000) Spike-dependent calcium influx in dendrites of the cricket giant interneuron. J Neurobiol 44:45–56

Ogawa H, Baba Y, Oka K (2001) Dendritic calcium accumulation regulates wind sensitivity via short-term depression at cercal sensory-to-giant interneuron synapses in the cricket. J Neurobiol 46:301–313

Ogawa H, Baba Y, Oka K (2002a) Spike-triggered dendritic calcium transients depend on synaptic activity in the cricket giant interneurons in directional control. J Neurobiol 50:234–244

Ogawa H, Baba Y, Oka K (2002b) Direction of action potential propagation regulates calcium increases in distal dendrites of wind-evoked walking the cricket giant interneurons. J Neurobiol 53:44–56

Ogawa H, Baba Y, Oka K (2004) Directional sensitivity of dendritic calcium responses to wind stimuli in cricket giant interneurons. Neurosci Lett 358:185–188

Ogawa H, Cummins GI, Jacobs GA, Miller JP (2006) Visualization of ensemble activity patterns of mechanosensory afferents in the cricket cercal sensory system with calcium imaging. J Neurobiol 66:293–307

Ogawa H, Cummins GI, Jacobs GA, Oka K (2008) Dendritic design implements algorism for extraction of sensory information. J Neurosci 28:4592–4603

Palka J, Olberg R (1977) The cereus-to-giant interneuron system of crickets III. Receptive field organization. J Comp Physiol 119(301):317

Paydar S, Doan CA, Jacobs GA (1999) Neural mapping of direction and frequency in the cricket cercal sensory system. J Neurosci 19:1771–1781

Reichert H, Wine JJ (1983) Coordination of lateral giant and non-giant systems in crayfish escape behavior. J Comp Physiol 153:3–15

Ritzmann RE, Pollack AJ (1986) Identification of thoracic interneurons that mediate giant interneuron-to-motor pathways in the cockroach. J Comp Physiol A 159:639–654

Ritzmann RE, Pollack AJ (1990) Parallel motor pathways from thoracic interneurons of the ventral giant interneuron system of the cockroach, Periplaneta americana. J Neurobiol 21:1219–1235

Roeder K (1948) Organization of the ascending giant fiber system in the cockroach (*Periplaneta americana*). J Exp Zool 108:243–261

Sakai M, Ootsubo T (1988) Mechanism of execution of sequential motor acts during copulation behavior in the male cricket Gryllus bimaculatus DeGeer. J Comp Physiol A 162:589–600

Shell LC, Killian KA (2000) The role of cercal sensory feedback during spermatophore transfer in the cricket, *Acheta domesticus*. J Insect Physiol 46:1017–1032

Shepherd D, Murphey RK (1986) Competition regulates the efficacy of an identified synapse in crickets. J Neurosci 6:3152–3160

Shimozawa T, Kanou M (1984a) Varieties of filiform hairs: range fractionation by sensory afferents and cereal interneurons of a cricket. J Comp Physiol A 155:485–493

Shimozawa T, Kanou M (1984b) The aerodynamics and sensory physiology of range fractionation in the cereal filiform sensilla of the cricket *Gryllus bimaculatus*. J Comp Physiol A 155:495–505

Shimozawa T, Kumagai T, Baba Y (1998) Structural scaling and functional design of the cercal wind-receptor hairs of cricket. J Comp Physiol A 183:171–186

Stabel J, Wendler G, Scharstein H (1985) The escape reaction of *Acheta domesticus* under open-loop conditions. In: Gewecke M, Wendler G (eds) Insect locomotion. P. Parey, Berlin, pp 79–85

Tauber E, Camhi JM (1995) The wind-evoked escape behavior of the cricket *Gryllus bimaculatus*: integration of behavioral elements. J Exp Biol 198:1895–1907

Theunissen FE, Miller JP (1991) Representation of sensory information in the cricket cercal sensory system II. Information theoretic calculation of system accuracy and optimal tuning-curve widths of four primary interneurons cercal sensory system. J Neurophysiol 66:1690–1703

Theunissen F, Roddey JC, Stufflebeam S, Clague H, Miller JP (1996) Information theoretic analysis of dynamical encoding by four identified primary sensory interneurons in the cricket cercal system. J Neurophysiol 75:1345–1364

Troyer TW, Levin JE, Jacobs GA (1994) Construction and analysis of a data base representing a neural map. Microsc Res Tech 29:329–343

Yono O, Aonuma H (2008) Cholinergic neurotransmission from mechanosensory afferents to giant interneurons in the terminal abdominal ganglion of the cricket *Gryllus bimaculatus*. Zool Sci 25:517–525

Chapter 15
The Biochemical Basis of Life History Adaptation: *Gryllus* Studies Lead the Way

Anthony J. Zera

Abstract During the past two decades, studies of lipid metabolism in the wing-polymorphic cricket, *Gryllus firmus*, have contributed significantly to our understanding of the biochemical mechanisms underlying life history adaptation. Radiotracer studies of *G. firmus* have documented morph differences in the flow of metabolites through pathways of lipid and amino acid metabolism. These differences result in the preferential diversion of nutrients to lipid flight fuel biosynthesis and away from ovarian growth in the flight-capable morph and vice versa in the flightless/high fecundity morph. These studies verify the widely held but previously undocumented hypothesis that trade-offs between life history traits (e.g., dispersal vs. reproduction) seen at the level of the whole organism result from trade-offs in the flow of metabolites through pathways of intermediary metabolism. These differences in pathway flux are produced by global change in the activities of numerous enzymes of lipid metabolism. Genetic, endocrine, and gene expression studies suggest that these changes are due to evolutionary change in regulators that control the activities of whole blocks of enzymes. Morph differences in enzyme activity are caused primarily by altered gene expression leading to altered enzyme concentration without change in kinetic properties. These results bear on long-standing, but as yet unresolved, questions regarding the mechanisms of enzyme microevolution. Finally, ongoing transcriptome profiling has identified additional morph differences in the expression of genes encoding enzymes of glyceride biosynthesis.

Keywords Life history • Trade-off • Wing polymorphism • Evolutionary physiology • Lipid metabolism • Triglyceride biosynthesis

A.J. Zera (✉)
School of Biological Sciences, University of Nebraska, Lincoln, NE 68588, USA
e-mail: azera1@unl.edu

H.W. Horch et al. (eds.), *The Cricket as a Model Organism*,
DOI 10.1007/978-4-431-56478-2_15

15.1 Introduction

Life history evolution has been a major focus of research in evolutionary biology since at least the 1960s (Stearns 1992; Roff 2002). Life history traits are key organismal features that determine lifetime patterns of growth, reproduction, longevity, etc. These traits are major components of Darwinian fitness, and variation in suites of life history traits often constitute some of the most important adaptations to different habitats, for different functions, etc. A common feature of life history adaptation is trade-offs or negative associations between various life history traits. For example, in many organisms, reproductive output is negatively associated with dispersal, resulting in phenotypes specialized for dispersal at the expense of reproduction or vice versa.

A long-standing issue in life history evolution has been the physiological basis of life history adaptation (Townsend and Calow 1981; Zera and Harshman 2001; Flatt and Heyland 2011). Physiology is used here in a broad sense to denote any aspect of organismal function (neural, hormonal, metabolic, etc.), studied at any biological level (systemic, biochemical, molecular). Life history traits such as the timing and magnitude of egg production, dispersal, growth, etc. all require various products of metabolic pathways, such as fatty acid fuel for flight or protein for egg yolk production. Thus, evolutionary modifications of flux (i.e., metabolite flow) through pathways of metabolism are expected to be a key aspect of life history adaptation. Moreover, interactions among metabolic pathways, a key feature of intermediary metabolism, provide the functional basis for life history trade-offs.

Until the mid-late 1980s, physiological studies of life history adaptation focused almost exclusively on whole-organism aspects of energetics (respiration rate, concentration of energy reserves) at the level of individual species. Very little was known about specific aspects of intermediary metabolism that were altered to produce life history adaptations. Nor was much known about the biochemical-genetic basis of life history adaptation within species. During the past 15 years, this situation has changed dramatically, and studies involving wing-polymorphic *Gryllus* crickets have led the way in investigations of the biochemical basis of life history adaptation (Zera and Harshman 2001, 2009, 2011).

Lipids play a key role in many life history traits (Townsend and Calow 1981; Zera and Harshman 2001). For example, lipid is an important energy reserve, and many life history adaptations (dispersal and migration; diapause; resistance to starvation) involve enhanced production and storage of lipid (Downer 1985). In addition, individuals exhibiting life history traits, such as increased reproductive output, that trade-off with life history traits mentioned above, often exhibit reduced somatic lipid reserves. Thus, evolutionary modification of lipid metabolism appears to play a pivotal role not only in the evolution of individual life history traits but also with respect to trade-offs between traits (Zera and Harshman 2001, 2011). However, the details of the mechanisms involved, at the level of pathways of metabolism and enzymes involved in these pathways, had not been studied in detail prior to the *Gryllus* work described in this chapter. The present chapter will provide a brief overview of recent and ongoing *Gryllus* studies with a focus on integrative, multilevel studies of adaptive modification of lipid metabolism. The main goal of these studies has been to understand the chain of causality that runs from adaptive modification of gene sequence/expression leading to modified activities of enzymes and flux (metabolite flow) through whole pathways of intermediary metabolism that ultimately results in changes in whole-organism lipid reserves.

15.2 Background on and Experimental Advantages of Wing Polymorphism in *Gryllus*

Before discussing biochemical studies of lipid metabolism in wing-polymorphic *Gryllus*, a brief description of this polymorphism will be given. Wing polymorphism is common in natural populations of many insect species, having evolved independently in most major insect orders. The polymorphism involves the existence of morphs (discontinuous phenotypes) in the same population that differ in numerous traits, and which are adapted for flight at the expense of egg production and vice versa (see Fig. 15.1). Wing polymorphism is a classic example of a life history trade-off in which the evolution of enhanced allocation to reproductive output is coupled with the evolution of decreased allocation to somatic function (e.g., flight ability). The polymorphism plays a central role in the life cycle of many species (Harrison 1980; Zera and Denno 1997; Guerra 2011; Zera and Brisson 2011).

Like many other wing-polymorphic insects, wing-polymorphic *Gryllus* species contain one morph (LW(f)) that has fully developed wings and flight muscles and is capable of flight but has delayed and reduced egg production relative to the obligately flightless (SW, short-winged) morph that has vestigial wings and underdeveloped flight muscles. A second flightless morph [LW(h)], that is produced from the LW(f) morph via flight muscle histolysis, has many attributes of the SW morph, such as enhanced ovarian growth (Zera et al. 1997). The LW(h) morph is not a major focus of the present chapter (see Zera et al. 1997, and Chap. 7 of this volume). LW(f) and SW morphs in *Gryllus* can be produced due to differences in genetic as well as numerous environmental factors (e.g., temperature, photoperiod, density), as is the case for morphs in most other wing-polymorphic species. However, the primary focus of study in *Gryllus* has been adaptive genetic differences between morphs in physiology, using artificially selected lines that had been derived from field-collected individuals and which were raised in a single environment.

Wing polymorphism in *Gryllus* has many advantages in evolutionary-physiological studies of life history adaptation. Extensive field studies in this and other groups have clearly identified the adaptive basis of the polymorphism in natural populations (Harrison 1980; Roff 1986a; Zera and Denno 1997; Zera et al. 2007a; Guerra 2011). This is not the case for some other life history phenotypes studied in model genetic organisms, particularly laboratory-generated mutants with

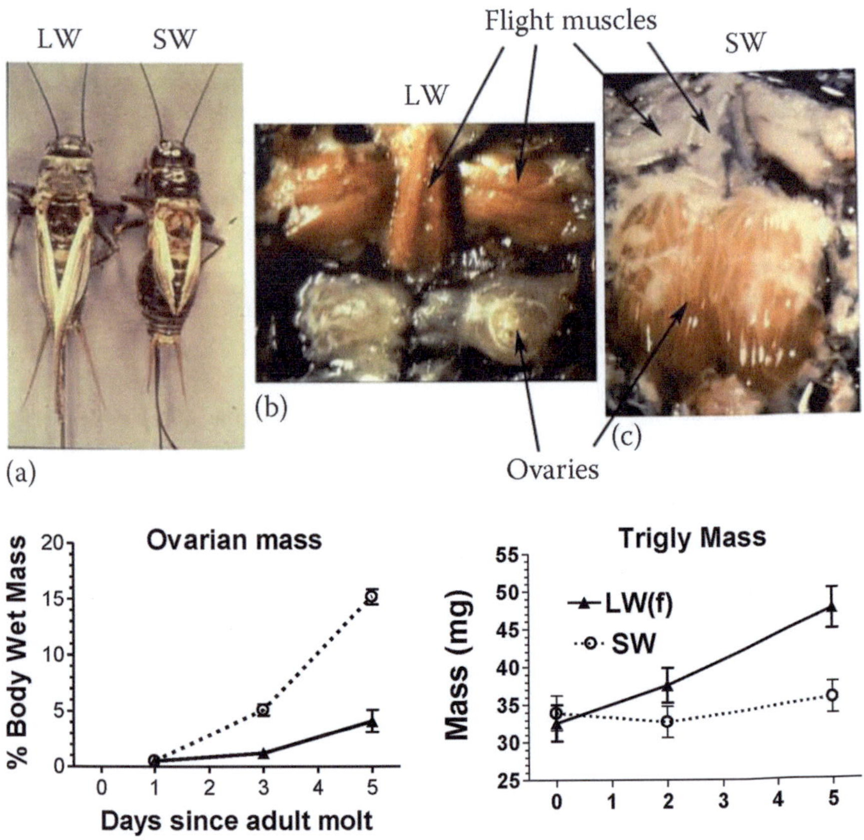

Fig. 15.1 *Top* panel. Day-5 adult female long-winged (LW) and short-winged (SW) morphs of *Gryllus firmus* illustrating (**a**) much larger hind wings and (**b**, **c**) much larger flight muscles, but much smaller ovaries in the LW compared to the SW morph. *Lower* panels. Reduced growth of the ovaries but greater accumulation of whole-body triglyceride in the LW(f) vs. SW morph during the first 5 days of adulthood. In these figures, LW is equivalent to LW(f) (Figure is from Zera 2005)

altered life history traits (e.g., insulin signaling mutants in *Drosophila*). While useful for dissecting the underlying physiological mechanisms of life history trait variation, extrapolating from these mutant studies to evolution in outbred populations is often questionable (Zera and Harshman 2009). This is due to the mutants often exhibiting highly reduced fitness, which is expected to result in their rapid elimination from outbred populations (for additional discussion, see Zera and Harshman 2009, 2011).

Wing-polymorphic crickets are easily reared in the laboratory and are amenable to basic evolutionary-genetic manipulations, such as artificial selection. Individuals of these species are large (ca. 1 g adults), and phenotypic differences between morphs (fecundity, flight muscle mass, enzyme activities) also are very large (see Fig. 15.1). Thus, detailed physiological studies, such as measurement of organ-specific enzyme activities or in vivo incorporation of radiotracers into reproductive vs. somatic organs, can be routinely undertaken in *Gryllus*. These techniques are much more difficult or are not realistic to employ in much smaller organisms such as fruit flies, aphids, or nematodes (e.g., *C. elegans*). Moreover, the current development of various molecular-genetic techniques (see other chapters in this volume) and "omics" techniques (e.g., transcriptome profiling, proteomics) is reducing some of the technical disadvantages of *Gryllus* compared to model genetic species such as *Drosophila melanogaster*.

In addition to biochemical studies of life history adaptation, *Gryllus* has been used to investigate a variety of other important issues in evolutionary physiology, which cannot be discussed in the present chapter because of space constraints. *Gryllus* work on the endocrine basis of life history adaptation is reviewed in Zera and Harshman (2001, 2009) and Zera (2013) and Chap. 7 of the present volume. Experimental evolution of endocrine regulation per se is reviewed in Zera (2006, 2013) and Zera et al. (2007b). Evolution of hormonal circadian rhythms is covered in Zera and Zhao (2009) and in Chap. 7 of the present volume.

15.3 Studies of Morph-Specific Metabolic Adaptation

15.3.1 Whole-Organism Studies of Nutrient Acquisition and Allocation to Lipid Production

Initial studies of life history adaptation in *Gryllus firmus* and *Gryllus rubens* involved artificial selection on the LW(f) vs. the SW morph in laboratory populations, founded from field-collected individuals, to produce multiple genetic stocks that are nearly pure breeding for the LW(f) or SW morphs (Roff 1986a, b; Roff and Fairbairn 2001, 2007). These early genetic studies quantified the genetic basis of individual traits and negative genetic correlations (i.e., genetic trade-offs) between key traits that comprise the polymorphism: large wings, large flight muscles, and small ovaries vs. small wings, underdeveloped flight muscles, and large ovaries. Documenting the genetic basis of trait variation/covariation is of central importance in evolutionary studies, since trait differences must have a genetic basis in order to evolve. Of particular importance to studies described below, the availability of selected lines provided the material to undertake physiological studies to identify the underlying functional causes of differences in components of flight and reproduction.

Early feeding and whole-organism physiological studies in *Gryllus* provided the background physiological context to investigate the biochemical basis of life history adaptation. These studies were the first to demonstrate experimentally that a life history trade-off within a species was indeed caused by the differential allocation of internal resources to competing life history functions (Mole and Zera 1993; Crnokrak and Roff 2002; Zera and Larsen 2001; reviewed in Zera and

Harshman 2001). LW(f) and SW morphs roughly consumed the same amount of food, but the LW(f) morph allocated a greater proportion of ingested resources to aspects of flight capability (e.g., respiratory maintenance of functional flight muscles and production of lipid flight fuel) at the expense of ovarian growth, while the opposite occurred in the SW morph. Thus the evolution of a flightless SW morph freed up nutrients previously required to construct and maintain various aspects of flight ability (i.e., in the absence of flight), allowing these nutrients to be rechanneled into egg production, resulting in the more fecund SW morph.

In artificially selected lines of *G. firmus*, LW(f) and SW female morphs emerge as adults with low and equivalent triglyceride stores (Fig. 15.1). However, during the first week of adulthood, LW(f) adult females accumulate more somatic triglyceride at the expense of ovarian growth, while the opposite situation occurred in SW adult females. Thus, during the same stage of development, there is a clear trade-off between allocation of nutrients to somatic lipid reserves vs. ovarian growth (Zera and Larsen 2001). Analogous trade-offs between lipid accumulation and egg production have been reported in a number of other species and thus appear to be a widespread physiological characteristic of life history trade-offs (reviewed in Townsend and Calow 1981 and in Zera and Harshman 2001, 2009).

15.3.2 Morph-Specific Modifications of Flux Through Pathways of Intermediary Metabolism

The core of the *Gryllus* research on the metabolic basis of life history adaptation focused on morph-specific alterations in flux through whole pathways of lipid metabolism, activities of enzymes of the pathways, and the causes of enzyme activity differences (e.g., kinetics vs. gene expression) that result in elevated lipid accumulation in the LW(f) morph. Radiotracer studies in the LW(f) and SW genetic stocks demonstrated that flux through the de novo pathway of fatty acid and triglyceride biosynthesis was greater in LW(f) adult females (Zera and Zhao 2003a), while fatty acid oxidation was reduced in LW(f) compared to SW lines (Fig. 15.2). Both of these factors accounted for the greater accumulation of triglyceride flight fuel in the LW(f) morph during early adulthood. This was the first documented example of a trade-off at the level of flux through pathways of intermediary metabolism that underlies a life history trade-off, a widely held, but previously unsubstantiated, assumption in life history research. In addition, in the LW(f) morph, newly biosynthesized triglyceride was preferentially allocated to the fat body, while in the SW morph, newly synthesized lipid was preferentially allocated to the ovaries. Lipid is an important component of eggs as well as an important flight fuel and somatic energy reserve. Thus, in studies of life history trade-offs between reproduction and somatic maintenance, it is important to distinguish between somatic and ovarian lipid. However, this has typically not been done

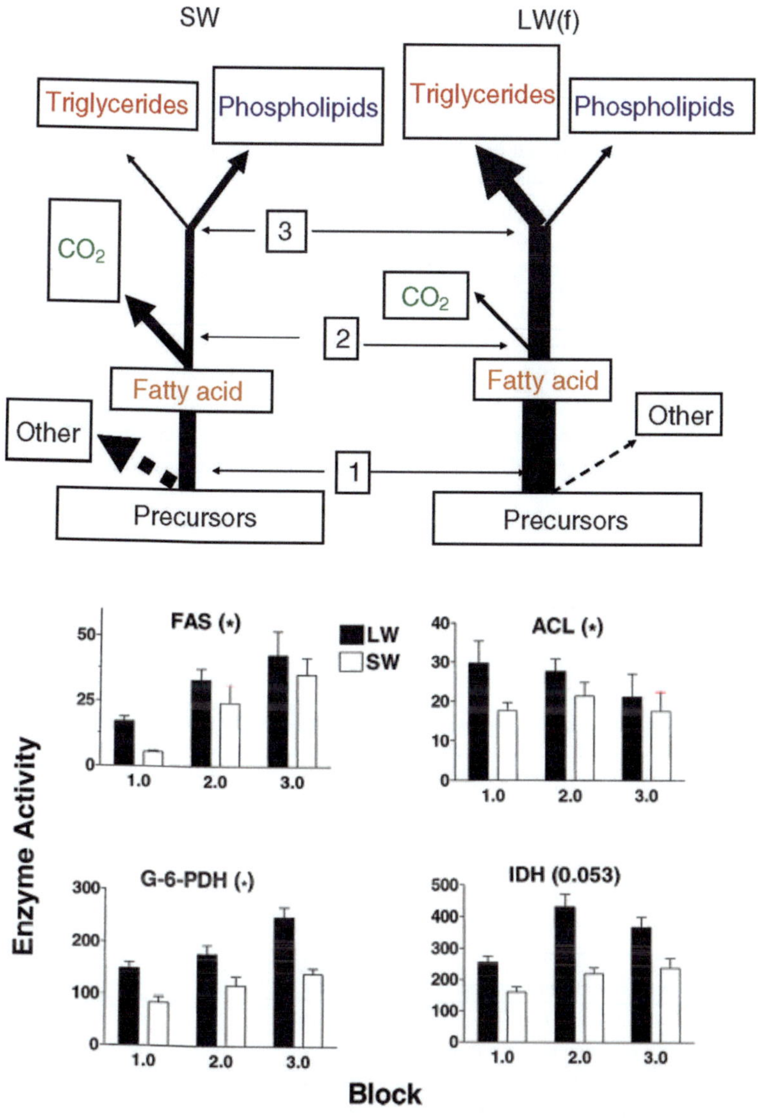

Fig. 15.2 *Top* panel. Results of radiotracer studies illustrating three trade-offs in the biosynthesis or oxidation of lipid classes in the flight-capable (LW(f)) and flightless, reproductive (SW) morphs of *Gryllus firmus*: (*1*) greater incorporation of radiolabel into total lipid vs. other pathways in LW (f), (*2*) greater conversion of fatty acid into triglyceride vs. oxidation to CO_2 in LW(f), and (*3*) greater production of triglyceride vs. phospholipid in LW(f). *Bottom* four panels: higher fat body-specific activities of lipogenic enzymes in three pairs of LW (same as LW(f)) vs. SW artificially selected lines. Block refers to independent selection trial (See Zera 2005). *FAS* fatty acid synthase, *ACL* ATP-citrate lyase, *G-6-PDH* glucose-6-phosphate dehydrogenase, *IDH* NADP$^+$-isocitrate dehydrogenase. *Top* panel is from Zera and Harshman (2011); *bottom* panels are from Zera (2005). Values in *parentheses* refer to the results of paired t-tests comparing LW and SW line means ($* = P < 0.05$)

for small insects such as *Drosophila* and illustrates one of the advantages of the much larger *Gryllus* (Zhao and Zera 2002; Zera 2005).

Another important finding of the *Gryllus* radiotracer studies was the trade-off in production of specific classes of lipids used in particular life history functions. Different classes of lipid can have very different roles. For example, triglyceride is a major energy reserve, while phospholipid is the major structural component of membranes. However, previous life history studies usually did not distinguish between differences in individual classes of lipids between life history phenotypes. Radiotracer studies of lipid biosynthesis mentioned above also identified a trade-off between the production of triglyceride vs. phospholipid: LW(f) females produced more triglyceride at the expense of phospholipid, while the opposite situation occurred in the SW morph (although total lipid biosynthesis is lower in the SW morph) (Fig. 15.2; Table 15.1). The shift to increased production of phospholipid in the SW morph is probably another metabolic adaptation to accommodate increased egg production which requires a greater amount of phospholipid for membranes in developing eggs.

Table 15.1 Examples of morph differences in various biochemical and other traits in adult *Gryllus firmus*

Trait	Morph difference (LW (f) relative to SW)	References
Respiration rate	Higher	A, B, C
Lipid reserves accumulated during adulthood	Higher	D, E
Ovarian growth and egg production	Lower	C, E, F
Rate of total lipid and triglyceride biosynthesis	Higher	E, G, H
Relative rate of triglyceride vs. phospholipid biosynthesis	Higher	E, G
Rate of fatty acid oxidation	Lower	E, I
Rate of amino acid oxidation	Higher	J, K
Conversion of amino acids into ovarian protein	Lower	J, K
Conversion of amino acids into lipid	Higher	J, K
Specific activities of lipogenic enzymes	Higher	E, L
Transcript abundance of lipogenic enzymes	Higher	K, M
Concentration of lipogenic enzymes	Higher	K, M, N
Kinetic properties of purified lipogenic enzymes	Equivalent	K, M, N
Transcript abundance of enzymes of glyceride biosynthesis	Higher	O

LW(f) dispersing morph with long wings and functional flight muscles, *SW* flightless morph with vestigial wings and flight muscles; see Fig. 15.1 and text

References: A = Mole and Zera (1993), B = Crnokrak and Roff (2002), C = Zera and Harshman (2001), D = Zera and Larsen (2001), E = Zera (2005), F = Zera et al. (2007a, b), G = Zhao and Zera (2002), H = Zera and Harshman (2009), I = Zera and Zhao (2003a), J = Zera and Zhao (2006), K = Zera and Harshman (2011), L = Zera and Zhao (2003b), M = Schilder et al. (2011), N = Zera et al. (2014), O = Vellichirammel et al. (2014)

Other large-magnitude differences in flux through pathways of intermediary metabolism were identified between the morphs, most notably, through pathways of amino acid metabolism. For example, compared with the LW(f) females, SW females converted a greater amount of amino acids into egg protein, oxidized a smaller proportion of amino acids, and converted a smaller amount of amino acids into somatic lipid reserves (Zera and Zhao 2006; reviewed in Zera and Harshman 2009; Table 15.1). There results collectively show that specialization for flight vs. reproduction has evolved by extensive remodeling of metabolite flow through various pathways of intermediary metabolism.

15.3.3 Enzyme Activities

Subsequent studies documented large-magnitude differences between the morphs in activities of numerous lipogenic enzymes that give rise to the elevated flux through the lipogenic pathway in the LW(f) morph (Zera and Zhao 2003b; Fig. 15.2). A similar positive association was observed between enzyme activity and life history traits and/or triglyceride concentration in *Drosophila melanogaster* (Luckinbill et al. 1990; Harshman and Schmidt 1998; Merritt et al. 2006; reviewed in Zera and Harshman 2009, 2011). The emerging picture from these biochemical studies is the global evolutionary modification of the activities of many, possibly most, enzymes involved in intermediate metabolism between the morphs of *G. firmus*. This likely has occurred via evolutionary changes in as yet unidentified metabolic regulators that coordinately control the activities of blocks of enzymes in whole pathways of metabolism (Zera and Zhao 2003b; Zera 2011), as has occurred in laboratory evolution of glucose metabolism in yeast (Ferea et al. 1999).

This hypothesis is supported by the essentially perfect co-segregation of physiological, morphological, and reproductive traits in adult F_2 female morphs produced by interstock crosses between LW(f) and SW lines and backcrosses (Zera and Zhao 2003a, b). Thus, F_2 LW(f) females had higher lipogenic enzyme activities, higher flux through the lipogenic pathway, reduced fatty acid oxidation, larger wings and flight muscles and smaller ovaries compared with F_2 SW females. Differences in these traits between F_2 LW(f) and SW morphs were of the same magnitude as differences between LW(f) and SW female parents used to produce the F_1 generation. This very strong co-segregation is inconsistent with the various morphological, reproductive, and biochemical features being inherited independently of each other. Rather, their coordinate expression appears to be most reasonably explained by a master polymorphic regulator or set of regulators (hormonal or other regulators) with pleiotropic effects on these various traits (Zera and Zhao 2003b). Hormone manipulation experiments also support this hypothesis: application of a juvenile hormone analogue to LW(f) female adults changed each trait mentioned above (except for wing length which cannot be modified in adults) to values seen in SW, unmanipulated females (Zera and Zhao 2004). A major question for future research is the nature of the regulator(s) that control the suite of variable traits that differ between the morphs. This intensive study of modification of intermediary metabolism in the context of life history adaptation remains one of the most important contributions of *Gryllus* to our knowledge of the physiological mechanisms underlying life history adaptation (Zera 2005, 2009; Zera and Harshman 2009, 2011). There are currently no comparable studies of other life history adaptations that occur in natural populations.

15.3.4 Enzyme Kinetics and Expression

Ongoing research is identifying the specific biochemical and molecular causes of increased activity of lipogenic enzymes that underlie the increased triglyceride production of the LW(f) morph (Schilder et al. 2011; Zera et al. 2014; Table 15.1). These studies are focusing on three enzymes, each of which plays an important role in lipogenesis: $NADP^+$-isocitrate dehydrogenase ($NADP^+$-IDH), an important producer of NADPH required for de novo fatty acid biosynthesis; 6-phosphogluconate dehydrogenase (6-PGDH), a pentose-shunt enzyme which is another important producer of NADPH; and ATP-citrate lyase (ACL), an enzyme in the de novo pathway of fatty acid biosynthesis.

Investigations of $NADP^+$-IDH, the most extensively studied of the three enzymes to date, documented no significant differences in kinetic constants (e.g., substrate or cofactor Michaelis constants (K_M) or turnover number (k_{cat})) between LW(f) and SW enzymes (Schilder et al. 2011). Nor were any non-synonymous DNA sequence differences (with one exception) observed between the coding region of the $NADP^+$-IDH gene in multiple sequences from each of three LW (f) vs. three SW lines. Thus, with one exception in one of 24 sequences, amino acid sequences were identical in $NADP^+$-IDHs in all LW(f) and SW lines. Most of the difference in fat body $NADP^+$-IDH activity between the morphs was due to the difference in transcript abundance which leads to a corresponding difference between the morphs in enzyme concentration (Fig. 15.3, Table 15.1). A similar situation was observed for the pentose-shunt enzyme 6-phosphogluconate dehydrogenase, another important producer of NADPH (Zera et al. 2014). Like $NADP^+$-IDH, higher 6-PGDH enzyme activity in LW(f) female fat body was due to elevated enzyme concentration, with no significant differences in kinetic properties of the enzyme from LW(f) vs. SW lines. In this case, however, elevated enzyme concentration in LW(f) fat body was primarily due to greater enzyme stability and, possibly, elevated gene expression (Zera unpublished). The third enzyme, ACL, also exhibits higher transcript abundance in the fat body of LW(f)-selected lines and no kinetic differences between the enzyme from LW(f) and SW lines. Enzyme concentration has yet to be measured for this enzyme.

These studies not only represent the most intensive investigations of the enzymological differences underlying life history adaptation in an outbred population, they also represent one of the handful of detailed investigations of intraspecific enzyme adaptation. Specifically, studies of $NADP^+$-IDH and 6-PGDH in *Gryllus*

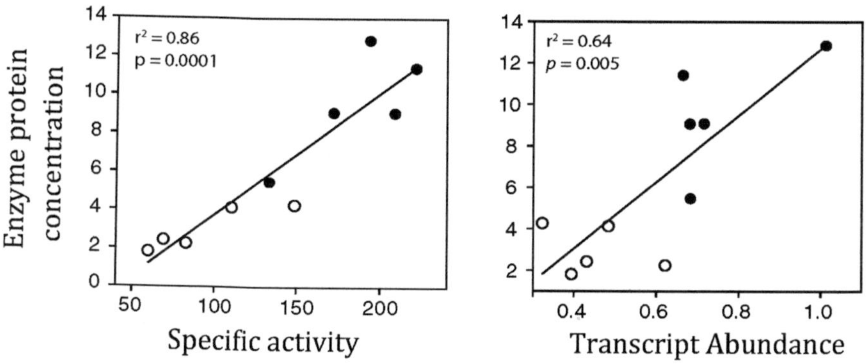

Fig. 15.3 Associations between enzyme protein concentration and specific activity or relative transcript abundance for NADP$^+$-isocitrate dehydrogenase in individuals from one pair of LW (f) (*filled circles*) or SW (*open circles*) selected lines of *G. firmus*. Note the strong covariation between enzyme activity and enzyme protein concentration and enzyme concentration and transcript abundance (Data from Schilder et al. 2011)

have identified the relative importance of evolutionary change in enzyme concentration vs. kinetic properties of the enzyme to adaptive difference in enzyme activity, a central topic of study in population genetics since the 1970s (Storz and Zera 2011; Schilder et al. 2011). NADP$^+$-IDH investigations in *Gryllus* also are among the few that provide information on the functional significance of variation in transcript abundance and the causal connection between variation in protein function and changes in flux through pathways of metabolism.

15.3.5 Morph-Specific Transcriptome Profiling

Previous work on the biochemical basis of life history adaptation and trade-offs in *G. firmus*, discussed above, focused on candidate enzymes and pathways of lipid and amino acid metabolism. Ongoing transcriptome profiling using RNA-Seq is focusing on morph-specific global changes in gene regulation, measured in the fat body (Vellichirammel et al. 2014, unpublished data). Only a few results of that study will be mentioned here. First, the LW(f) morph exhibited elevated transcript levels compared to the SW morph for a number of enzymes of glyceride biosynthesis, such as 1-acylglycerol-3-phosphate acyltransferase, phosphatidate phosphatase, and glycerol-3-phosphate dehydrogenase. These data are consistent with the rate of triglyceride biosynthesis being elevated in the LW(f) compared to the SW morph, mentioned above. Indeed, differences in transcript abundance for these enzymes are much greater in magnitude (e.g., greater than tenfold for 1-acylglycerol-3-phosphate) than morph differences in either transcript abundance (less than twofold) or specific activity (twofold, Fig. 15.2) for enzymes of the de novo fatty acid biosynthetic pathway. This suggests that the elevated rate of triglyceride biosynthesis in the LW(f) morph might be primarily due to upregulation of the glyceride portion of lipogenesis (in which fatty acids are linked to glycerol phosphate) and less to the biosynthesis of fatty acids, which has been the main focus of previous enzyme studies of lipid biosynthesis. These transcriptome data provide additional tools to investigate molecular aspects of morph-specific triglyceride biosynthesis in *G. firmus*.

Second, the fat body of the reproductive SW morph exhibited a substantially (tenfold) elevated level of an insulin-related peptide transcript compared to the LW (f) morph. Insulin-related peptides are thought to be important regulators of reproduction, metabolism, and trade-offs between these traits in a variety of insects (Zera and Harshman 2009; Zera et al. 2007b; Flatt and Heyland 2011). The homologue of the transcript elevated in SW *Gryllus* also is substantially elevated in the gregarious phase of the desert locust that exhibits earlier sexual maturation and egg production compared with the solitary phase (Badisco et al. 2008). Identification of this important endocrine regulator opens up the possibility of investigating insulin-like peptide regulation of morph-specific reproduction and metabolism in *Gryllus*.

15.4 Summary, Conclusions, and Future Directions

Over several decades, *Gryllus* crickets have made important contributions to the nascent field of evolutionary physiology and continue to be at the forefront of several areas of research in this field (Zera and Harshman 2001, 2009, 2011). Detailed studies of morph-specific differences in lipid metabolism in *G. firmus* currently constitute the most detailed analysis of evolutionary modification of intermediary metabolism in the context of life history adaptation in outbred populations. In particular, identification of trade-offs in flux through several pathways of intermediary metabolism (Zhao and Zera 2002; Zera and Zhao 2006; Zera and Harshman 2011) provided that first direct confirmation of the widespread assumption in life history physiology that life history trade-offs at the level of whole organisms result from trade-offs in the flow of metabolites through pathways of metabolism. These studies set the stage for future investigations of the regulators that control the differential flux through interacting pathways in morphs. Moreover, enzymological-genetic studies of lipogenic enzymes from LW(f) and SW morphs (Schilder et al. 2011; Zera et al. 2014) constitute one of the most detailed studies of intraspecific enzyme adaptation. This was a prominent area of research in population genetics, but has languished since the 1980s as the field moved to statistical analyses of DNA sequence variation of enzymes to identify the role of natural selection in enzyme microevolution. The recent *Gryllus* studies may help to spur a renewal in functional investigations of enzyme microevolution. Finally, transcriptome studies are identifying global changes in gene expression that contribute to adaptive differences between wing morphs (Vellichirammel et al. 2014, unpublished data). Some data, such as the greater expression of genes encoding enzymes of the triglyceride pathway in the LW(f) morph, verify previous radiotracer studies of pathway flux. Important new findings are also emerging from these studies, such as morph-specific differences in components of the insulin signaling. The transcriptome study may also be important for identifying morph-specific differences in regulators responsible for the differences in metabolism between morphs. Finally, the ongoing development of various molecular tools, discussed in various chapters of this volume, will also be invaluable with respect to conducting more sophisticated experiments that test various ideas regarding endocrine and biochemical aspects of morph adaptation discussed in this chapter.

Acknowledgments The author gratefully acknowledges the support of the National Science Foundation which has continuously supported research in the author's laboratory for over 20 years (most recently awards IBN-0212486, IOS-0516973, IOS-1122075).

References

Badisco L, Claeys I, Van Hiel M, Clynen E et al (2008) Purification and characterization of an insulin-related peptide in the desert locust, *Schistocerca gregaria*: immunolocalization, cDNA cloning, transcript profiling and interaction with neuroparsin. J Mol Entomol 40:137–150

Crnokrak P, Roff D (2002) The trade-off to flight capability in *Gryllus firmus*: the influence of whole-organism respiration rate on fitness. J Evol Biol 15:388–398

Downer RGH (1985) Lipid metabolism. In: Kerkut GA, Gilbert LI (eds) Comprehensive insect physiology, biochemistry and pharmacology, vol 10. Pergamon, Oxford, pp 77–114

Ferea TL, Botstein D, Brown PO, Rosensweig RF (1999) Systemic changes in gene expression patterns in yeast. Proc Natl Acad Sci U S A 96:9721–9726

Flatt T, Heyland A (eds) (2011) Mechanisms of life history evolution. Oxford University Press, Oxford

Guerra PA (2011) Evaluating the life history trade-off between dispersal capability and reproduction in wing dimorphic insects: a meta-analysis. Biol Rev 86:813–835

Harrison RG (1980) Dispersal polymorphisms in insects. Annu Rev Ecol Syst 11:95–118

Harshman LG, Schmidt JL (1998) Evolution of starvation resistance in *Drosophila melanogaster*: aspects of metabolism and counter-impact selection. Evolution 52:1679–1685

Luckinbill LS, Riha V, Rhine S, Grudzein TA (1990) The role of glucose-6-phosphate dehydrogenase in the evolution of longevity in *Drosophila melanogaster*. Heredity 65:29–38

Merritt TJS, Sezgin E, Zhu C-T, Eanes WF (2006) Triglyceride pools, flight and activity variation at the Gpdh locus in *Drosophila melanogaster*. Genetics 172:293–304

Mole S, Zera AJ (1993) Differential allocation of resources underlies the dispersal-reproduction trade-off in the wing-dimorphic cricket, *Gryllus rubens*. Oecologia 93:121–127

Roff DA (1986a) The evolution of wing dimorphism in insects. Evolution 40:1009–1020

Roff DA (1986b) The genetic basis of wing dimorphism in the sand cricket, *Gryllus firmus*, and its relevance to the evolution of wing dimorphism in insects. Heredity 57:221–231

Roff DA (2002) Life history evolution. Sinauer Associates, Sunderland

Roff DA, Fairbairn DJ (2001) The genetic basis of dispersal and migration, and its consequences for the evolution of correlated traits. In: Clobert E, Danchin E, Dhondt AA, Nichols JD (eds) Dispersal. Oxford University Press, Oxford, pp 191–202

Roff DA, Fairbairn DJ (2007) The evolution and genetics of migration in insects. Bioscience 57:155–164

Schilder RJ, Zera AJ, Black C, Hoidel M, Wehrkamp C (2011) The biochemical basis of life history adaptation: molecular/enzymological causes of NDAP$^+$-isocitrate dehydrogenase activity differences between morphs of *Gryllus firmus* that differ in lipid biosynthesis and life history. Mol Biol Evol 28:3381–3393

Stearns SC (1992) The evolution of life histories. Oxford University Press, Oxford

Storz J, Zera AJ (2011) Experimental approaches for evaluating the contributions of candidate protein-encoding mutations to phenotypic evolution. In: Orgogozo V, Rockman MV (eds) Methods in evolutionary genetics. Springer, New York, pp 377–396

Townsend CR, Calow P (eds) (1981) Physiological ecology. An evolutionary approach to resource use. Blackwell Scientific Publications, Oxford

Vellichirammel NN, Zera AJ, Schilder RJ, Wehrkamp C, Riethoven J-JM, Brisson JA (2014) De novo transcriptome assembly and morph-specific gene expression profiling of the wing-polymorphic cricket, *Gryllus firmus*. PLoS ONE. doi:10.1371/journal.pone.0082129

Zera AJ (2005) Intermediary metabolism and life history trade-offs: lipid metabolism in lines of the wing-polymorphic cricket, *Gryllus firmus*, selected for flight capability vs. early-age reproduction. Integr Comp Biol 45:511–524

Zera AJ (2006) Evolutionary genetics of juvenile hormone and ecdysteroid regulation in *Gryllus*: a case study in the microevolution of endocrine regulation. Comp Biochem Physiol Part A 144A:365–379

Zera AJ (2009) Wing polymorphism in *Gryllus* (Orthoptera: Gryllidae): proximate endocrine, energetic and biochemical mechanisms underlying morph specialization for flight vs. reproduction. In: Whitman DW, Ananthakrishnan TN (eds) Phenotypic plasticity of insects. Mechanisms and consequences. Science Publishers, Enfield, pp 609–653

Zera AJ (2011) Microevolution of intermediary metabolism: evolutionary genetics meets metabolic biochemistry. J Exp Biol 214:179–190

Zera AJ, Brisson JA (2011) Quantitative, physiological, and molecular genetics of dispersal/migration. In: Baguette M, Benton TG, Bullock JM, Clobert J (eds) Dispersal ecology and evolution. Oxford University Press, Oxford, pp 63–94

Zera AJ (2013) Morph-specific JH titer regulation in wing-polymorphic *Gryllus* crickets. Proximate mechanisms underlying adaptive genetic modification of JH regulation. In: Devillers J (ed) Juvenile hormone and juvenoids: modeling biological effects and environmental fate. CRC Press, Boca Raton, pp 31–64

Zera AJ, Denno RF (1997) Physiology and ecology of dispersal polymorphism in insects. Ann Rev Entomol 42:207–231

Zera AJ, Harshman LG (2001) Physiology of life history trade-offs in animals. Annu Rev Ecol Syst 32:95–126

Zera AJ, Harshman LG (2009) Laboratory selection studies of life-history physiology in insects. In: Garland T Jr, Rose MR (eds) Experimental evolution: methods and applications. University of Califronia Press, Berkeley, pp 217–262

Zera AJ, Harshman LG (2011) Intermediary metabolism and the biochemical-molecular basis of life history variation and trade-offs in two insect models. In: Flatt T, Heyland A (eds) Mechanisms of life history evolution. Oxford University Press, Oxford, pp 311–328

Zera AJ, Larsen A (2001) The metabolic basis of life history variation: genetic and phenotypic differences in lipid reserves among life history morphs of the wing-polymorphic cricket, *Gryllus firmus*. J Insect Physiol 47:1147–1160

Zera AJ, Zhao Z (2003a) Life history evolution and the microevolution of intermediary metabolism: activities of lipid-metabolizing enzymes in life-history morphs of a wing-dimorphic cricket. Evolution 57:568–596

Zera AJ, Zhao Z (2003b) Morph-dependent fatty-acid oxidation in a wing-polymorphic cricket: implications for morph specialization for dispersal vs. reproduction. J Insect Physiol 49:933–943

Zera AJ, Zhao Z (2004) Effect of a juvenile hormone analogue on lipid metabolism in a wing-polymorphic cricket: implications for the biochemical basis of the trade-off between reproduction and dispersal. Biochem Physiol Zool 77:255–266

Zera AJ, Zhao Z (2006) Intermediary metabolism and life-history trade-offs: differential metabolism of amino acids underlies the dispersal-reproduction trade-off in a wing-polymorphic cricket. Am Nat 167:889–900

Zera AJ, Zhao Z (2009) Morph-associated JH titer diel rhythm in *Gryllus firmus*: experimental verification of its circadian basis and cycle characterization in artificially-selected lines raised in the field. J Insect Physiol 55:450–458

Zera AJ, Sall J, Grudzinski K (1997) Flight-muscle polymorphism in the cricket *Gryllus firmus*: muscle characteristics and their influence on the evolution of flightlessness. Physiol Zool 70:519–529

Zera AJ, Harshman LG, Williams T (2007a) Evolutionary endocrinology: the developing synthesis between endocrinology and evolutionary genetics. Annu Rev Ecol Evol Syst 38:793–817

Zera AJ, Zhao Z, Kaliseck K (2007b) Hormones in the field: evolutionary endocrinology of juvenile hormone and ecdysteroids in field populations of the wing-dimorphic cricket *Gryllus firmus*. Physiol Biochem Zool 80:592–606

Zera AJ, Wehrkamp C, Schilder RJ, Black C, Berkheim D, Gribben P (2014) Purification and characterization of 6-phosphogluconate dehydrogenase from the wing-polymorphic cricket *Gryllus firmus*: implications for the biochemical basis of life-history adaptation. Comp Biochem Physiol Part B 172B:29–38

Zhao Z, Zera AJ (2002) Differential lipid biosynthesis underlies a tradeoff between reproduction and flight capability in a wing-polymorphic cricket. Proc Natl Acad Sci U S A 99:16829–16834

Chapter 16
Reproductive Behavior and Physiology in the Cricket *Gryllus bimaculatus*

Masaki Sakai, Mikihiko Kumashiro, Yukihisa Matsumoto, Masakatsu Ureshi, and Takahiro Otsubo

Abstract *Gryllus bimaculatus* males have a reproductive cycle consisting of a mating stage and a sexually refractory stage. During the mating stage, the male exhibits distinct behavior that encompasses three main stages: calling, courtship, and copulation. The last stage, copulation, is carried out in a fixed manner by the stimulus-response chain. The final copulatory act, spermatophore extrusion, is caused by stimulation of mechano-sensilla in the epiphallus during genitalia coupling, which terminates the mating stage. The sexually refractory stage starts with spermatophore extrusion, during which the male is rather aggressive and does not exhibit any mating behavior. A male first shows spermatophore preparation when stimulated by the female, then forms the new spermatophore, and finally recommences the calling song, i.e., the start of the mating stage. Physiological investigations reveal that the male mating behavior is mainly controlled by the brain and the terminal abdominal ganglion (TAG), which exerts three types of inhibition on the pattern generators for mating behavior. The brain also facilitates sexual excitation via octopamine. One of the conspicuous features of the reproductive behavior in *Gryllus bimaculatus* is that the sexually refractory stage between spermatophore preparation and the start of calling song is time fixed at around 1 h. Experiments utilizing the targeted cooling of the central nervous system indicate a presence of a time-measuring mechanism ("timer") that is located within the TAG. Long-term spike recordings of neurons also support the presence of such a timer within the TAG. Finally, the occurrence of mating-like actions in larval nymphs is discussed. All of these findings have now generated a large body of work that will

---

M. Sakai (✉) • M. Kumashiro • T. Otsubo
Division of Bioscience, Graduate School of Natural Science and Technology, Okayama University, Tsushima-Naka-3-1-1, Kita-ku, Okayama 700-8530, Japan
e-mail: masack@cc.okayama-u.ac.jp

Y. Matsumoto
College of Liberal Arts and Science, Tokyo Medical and Dental University, 2-8-30 Kounodai, Ichikawa 272-0827, Japan

M. Ureshi
Faculty of Culture and Education, Saga University, Honjyo-cho 1, Saga 840-8502, Japan

H.W. Horch et al. (eds.), *The Cricket as a Model Organism*,
DOI 10.1007/978-4-431-56478-2_16

help establish *Gryllus* as a new experimental system for studying reproductive behavior and physiology in crickets and other insect species.

Keywords Male cricket • *Gryllus bimaculatus* • Reproductive behavior • Nervous system • Postembryonic development

16.1 Introduction

There are about 2500 species of crickets known in the world. Some of them live near human habitats and have aroused human interest since the ancient days (Hsu 1929). The male crickets have complex behavioral repertoires such as calling, courtship, copulation, and fighting, and they have been extensively studied since the 1950s (Choperd 1951; Huber 1955; Hörmann-Heck 1957; Alexander 1961; Alexander and Otte 1967; Beck 1974; Loher and Rence 1978; Evans 1988). More recently, crickets have been used in neuroethological research (Huber et al. 1989) because of their comparatively large-sized nervous system and their suitability for surgical experimentation. The range of these studies is very broad, encompassing song production (Bentley 1969), song recognition and sound localization (Weber et al. 1981; Loher et al. 1993), startle response to wind (Palka et al. 1977; Pollack and Hoy 1989; Shimozawa and Kanou 1984), reproduction (Sakai and Kumashiro 2004a), death feigning (Nishino and Sakai 1996), and fighting and status decision (Simmons 1986; Adamo and Hoy 1995; Stevenson et al. 2000; Iwasaki et al. 2006). Here, we focus on cricket mating behavior and its physiological mechanisms, and we describe our work using *Gryllus bimaculatus* as a model system.

16.2 Reproductive Behavior

16.2.1 Male Cycle

In order to manifest reproductive behavior, crickets should possess sexually mature ovaries and the testes. The adult male performs mating behavior consisting of calling, courtship, and copulation when it is paired with the female. However, these actions depend upon the internal state of the male, which changes cyclically via copulation. That is, the male has a reproductive cycle consisting of two stages (Fig. 16.1). The period between the start of the calling and the end of copulation is defined as the mating stage. Immediately after copulation, the male becomes sexually inactive and aggressive even toward the female. This period between the end of copulation and the restart of the calling song is named the sexually refractory stage. Only after this refractory period is over can the calling behavior start again. As in many animals, the male has the dominant, active role in mating behavior, so

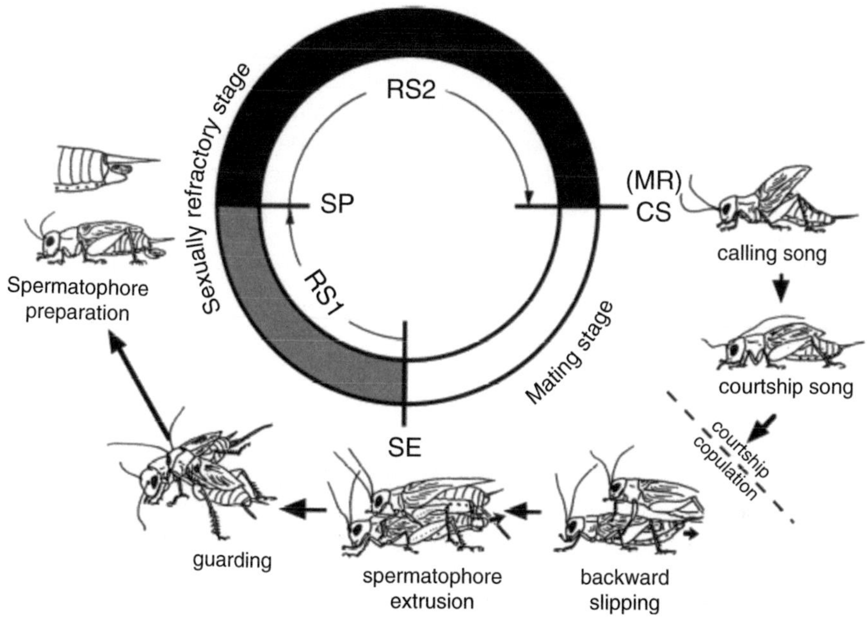

Fig. 16.1 Behavior of the male cricket in the reproductive cycle. The cycle consists of the mating stage, beginning at the onset of the calling song (CS) and ending with spermatophore extrusion (SE), and the sexually refractory stage beginning at SE and ending at CS. The sexually refractory stage is further divided into two stages: the first refractory stage (RS1) between SE and spermatophore preparation (SP), and the second refractory stage (RS2) between SP and CS. The latter interval (SPCS) is time fixed and thus called the time-fixed sexually refractory stage. In the mating stage, the male exhibits courtship and copulation with the female, while in the sexually refractory stage, the male exhibits guarding and aggressive behavior toward the female. The mating response (MR) consists of copulatory actions to a model mimicking the abdomen of the female. The MR can be elicited nearly at the same time as the CS is re-emitted. Thus, the first occurrence of CS or MR after SP indicates the end of the sexually refractory stage or the start of the mating stage

most studies focus almost exclusively on males. Hence, any reference to female crickets is brief and descriptions are included only when necessary.

16.2.2 Mating Stage

Under experimental conditions, when a male and a female are placed in a small space, the male will soon start to call for and court the female. When the female responds to the male by mounting the male's back, the male engages in copulation and transfers the spermatophore to the female. Then, the mating stage ends and the sexually refractory stage starts (Fig. 16.1). However, there are some cases in which the mating stage is terminated without spermatophore transfer (see Sect. 16.2.2.5).

16.2.2.1 Sexual Recognition

The male recognizes the sex of other individuals exclusively by the chemical contact with its antennae; olfactory and visual stimuli are not important (Rence and Loher 1977; Ogawa and Sakai 2009). When the male encounters a female, male courtship of her begins. In contrast, if a male encounters another male, both will engage in aggressive behavior. The ensuing fight is finished rather quickly and its outcome will determine the hierarchical relationship between the males (Adamo and Hoy 1995; Stevenson et al. 2000). Interestingly, if two males are confined to a small space, the dominant male will begin courting the subordinate male (Ogawa and Sakai 2009).

16.2.2.2 Courtship

The details of courtship are described in Fig. 16.2, and they encompass a total of six stages. First, searching males encounter a female (Fig. 16.2(1)). Second, if the male makes antennal contact with the female, he will recognize the other cricket as a *bona fide* female (Fig. 16.2(2)). Third, the male will initiate a large turn in the opposite direction from the female and will spread his antennae wide (Fig. 16.2(3)). Following this, the male then makes several small successive turns in the same direction (lasting 4.5 s). This action allows the male to orient its abdominal end

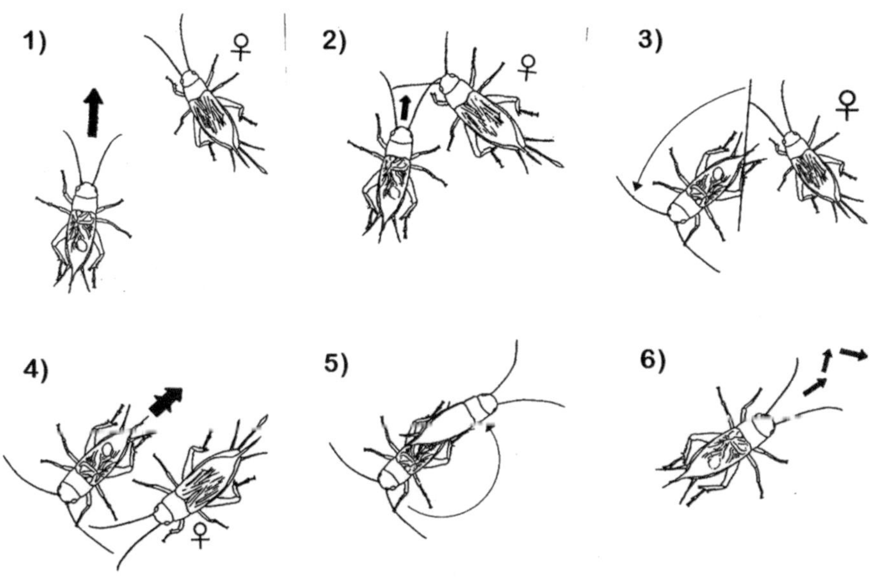

Fig. 16.2 Courtship sequence. (*1*) Search. (*2*) Encounter with the female. (*3*) Turn away from the female. (*4*) Backward walking. (*5*) Quick turn toward posterior direction. (*6*) Search with an intense calling song

toward the female. This, in turn, will provide the female with an opportunity to mount onto the male's back. Fourth, the male walks backward slowly while singing the courtship song (Fig. 16.2(4)). The copulation starts when the female mounts the male (which typically occurs about 15–20 s from the initial contact). Fifth, if the female does not proceed with mounting, the male makes a quick 180° turn to face the female and try to re-engage her (Fig. 16.2(5)). In situations when no female is present, the male will walk in a zigzag manner, singing an aggressive calling for 5–10 s (Fig. 16.2(6)). If the female still cannot be found, the male will cease its calling and walk for 70 s before starting a new search.

All the actions performed by the male are well organized and aimed toward the successful completion of copulation. Note that while engaged in searching (Fig. 16.2 (6)), a male may engage in additional activities, such as spiral walking, an effective search strategy that has been found in desert ants (Muller and Wehner 1994).

16.2.2.3 Copulation

The sequence of copulation behavior has been extensively studied in *Gryllus bimaculatus* and is summarized in Fig. 16.3 (Sakai et al. 1991). First, the male slips under the female in response to female's mounting and walks backward (lasting 3.2 s). Then, the male hangs the epiphallus (hook) onto the female subgenital plate (5.6 s). Following this, the genitalia of the male and female are united (4 s), which leads to spermatophore extrusion. Finally, the spermatophore is pushed up into the female genital chamber where it becomes fixed (9.1 s). On average, the entire copulation is completed in 20 s.

16.2.2.4 Chain Reaction

Every copulatory action of the male cricket proceeds in the manner of a chain reaction as explained for the mechanism of mating behavior in the sticleback by Tinbergen (1951). That is, a stimulation of sensilla in a certain region causes a fixed action response, and as a result of such response, another region will be stimulated. This in turn causes yet another response and so on. The input and output relationships were analyzed by using a model of the female (key stimulus) and surgical ablation of mechano-sensilla on the dorsal surface, i.e., the cerci and the epiphallus (Sakai and Ootsubo 1988). The chain reaction (Fig. 16.4) proceeds as follows: (1) When the female steps on the male dorsal region (S1), the male shows the intense posture with the abdomen slightly raised (R1). (2) The male's posterior cerci make contact with the female ventral region (S2). In response to cercal stimulation, the male walks backward with the cerci vibrating (R2). This cercal vibration seems to be useful not only for the male to help orient his body axis parallel to that of the female but also to make the female quiet. After the start of copulation, the female falls into a thanatotic state due to the vibratory stimulation caused by the male's body thrust. (3) The male stops backward walking when his abdominal end reaches the end of the female's abdomen. The male stops because

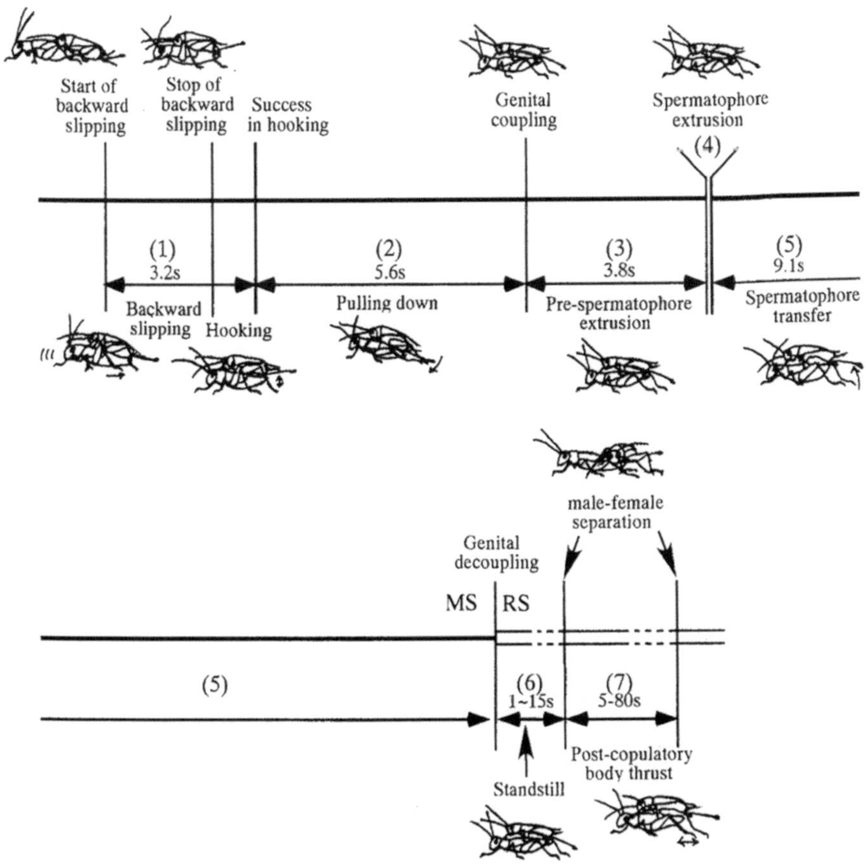

Fig. 16.3 Copulatory sequence and average periods for sequential behavior patterns. Onset at *upper left*. (*1–7*) indicates each step. Time is given in seconds

the posterior region of his cerci loses contact with the female. At that moment, the anterior part of the cerci and the last (tenth) abdominal tergite (see Fig. 16.5) come in contact with the female (S3). This triggers the hooking movement, which is to hang the epiphallus onto the subgenital plate of the female (R3). (4) When the male succeeds in hooking, the sensilla on the hook are stimulated (S4), and it pulls down the subgenital plate of the female (R4). (5) In response to the pulling down of the subgenital plate (S5), the female's copulatory papilla of the genitalia protrudes backward, which naturally enters the epiphallus (R5). Here, the unification of genitalia is attained. (6) If the unification continues at least for 4 s (S6), the dorsal pouch of the male's phallus is suddenly distorted so that the attachment plate of the spermatophore is extruded (R6). (7) Then, the median pouch and ventral lobes inflate with fluid and push up the spermatophore (R7). Finally, the attachment plate is pushed into the genital chamber of the female and fixed onto the convolutions of the upper surface of the genital chamber (R8) (Sakai and Kumashiro 2004a).

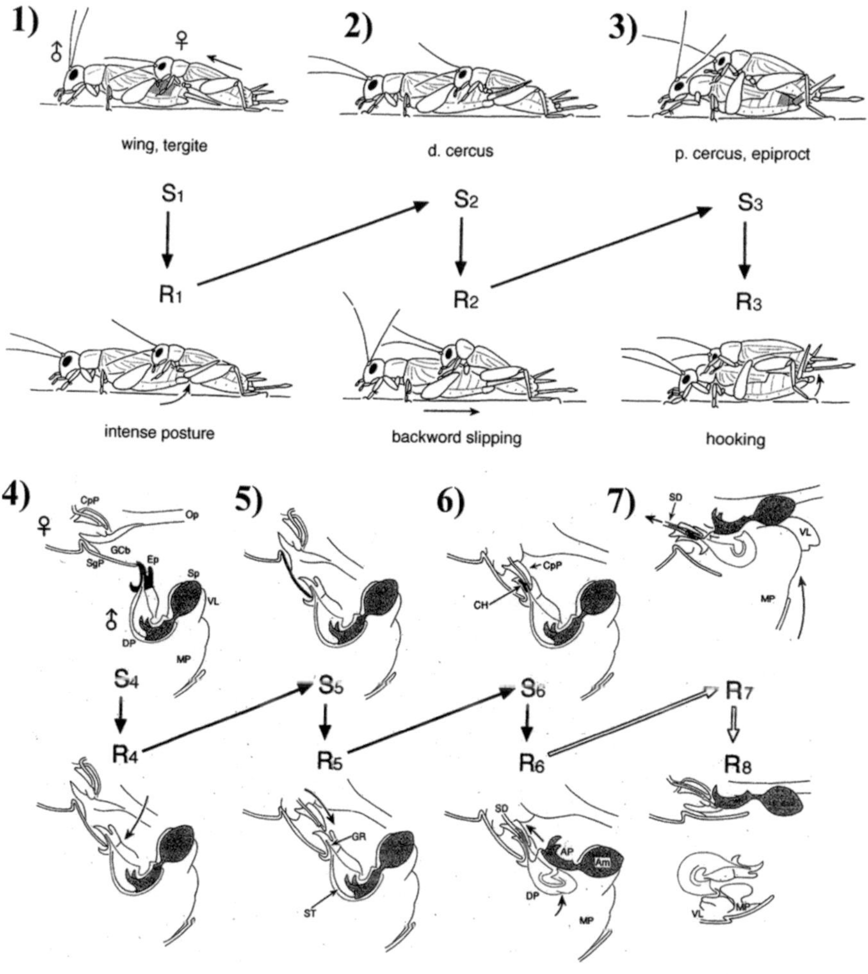

Fig. 16.4 Chain reaction in cricket copulation. Three steps (1–3) leading to success in hooking. Upper three figures (S1, S2, and S3) show locations (*gray*) of key stimulus detection by male, and lower three figures (R1, R2, and R3) show male responses to respective stimulation. In S1, tergite and wings are stimulated. In S2, distal region of cercus (d. cercus) is stimulated. In S3, proximal region of cercus (p. cercus) and epiproct (last abdominal tergite) are stimulated. *Large black arrows* show stimulus-response chain. Three steps (4–6) occur after success in hooking. Upper figures (S4, S5, and S6) show locations (*black*) of key stimulus detection. Lower figures (R4, R5, and R6) show responses to respective stimulation. *Large black arrows* show stimulus-response chain, while *large white arrows* show sequence of irreversible fixed actions (R6, R7, and R8). See text for labeling in (4)–(7)

The above sequences make up the chain reaction for copulation. Note that this sequence is always caused by the male and that the chain sequence proceeds in a reflexive manner. The female contributes to this sequence of events in only two

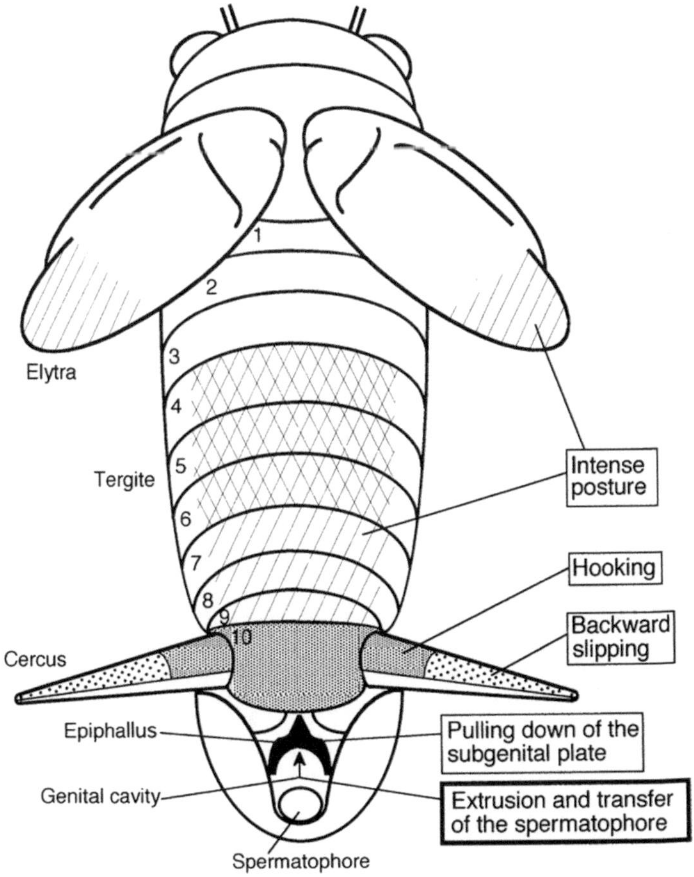

Fig. 16.5 Input and output relationships for copulatory acts in the male cricket. Each region as shown by the different patterns has mechanoreceptors to which contact stimulation elicits a specific motor response. Number (*1–10*) indicates abdominal segments. Reversible acts in early stage are shown in *thin-lined box* and irreversible fixed actions in later stage, in *thick-lined box*. See text for explanation

steps: mounting and copulatory papilla protrusion. The input and output relationships for the response chain are schematically summarized in Fig. 16.5 (Sakai and Kumashiro 2004a).

16.2.2.5 Abnormal Termination of the Mating Stage

While most of the matings result in a successful copulation, in some instances the mating stage ends prematurely due to the unusual ejection of the spermatophore (Sakai et al. 1991). There are three possible scenarios that can lead to this outcome (see their description below for details).

Pseudo-copulation

This scenario occurs when the male and the female are inadvertently separated during the final stage of copulation, genitalia coupling. The male ejects the spermatophore itself exhibiting the action of spermatophore transfer. The posture and movement are essentially the same as those of normal spermatophore transfer. Then, the male enters the sexually refractory stage. The change in the reproductive mode from the mating stage to the sexually refractory stage can be confirmed by observing the occurrence of spermatophore preparation (SP; see below).

Abortion

If the pairing female has a subgenital plate that is artificially closed with wax, the male cannot get genitalia coupling after hook hanging. In this case, the male will re-try the copulatory attempts. A similar behavior is naturally observed when the paired female is sexually reluctant and does not stay quiet on the copulating male. After a number of unsuccessful attempts, the male will eventually stop his activity. The male may recommence copulatory attempts after a long rest. Finally, the male ejects the spermatophore itself and scrapes off the substrate from the genital chamber. Then, the male enters the sexually refractory stage. In this case, however, it is uncertain when the reproductive mode was switched. Abortion certainly saves energy, as otherwise a male can spend unfruitful efforts for an extended period of time.

Spontaneous Cycle Renewal

When a male possessing a spermatophore is kept alone in an isolated condition, it keeps silent and becomes inactive. After a long time (1 or 2 days), the color of the spermatophore becomes brownish and seems to become nonfunctional in the genital chamber as it gets older. The male ejects the spermatophore without any previous sign and enters the sexually refractory stage.

16.2.3 Sexually Refractory Stage

As soon as the male extrudes the spermatophore, the sexually refractory stage begins. In this stage, the male shows a conspicuous behavior called guarding behavior (Fig. 16.1). The male also produces a new spermatophore for the next copulation. To make the spermatophore, the male first cleans the genitalia and then casts the spermatophore material into the phallic complex of the genital chamber. The latter process is called spermatophore preparation. That material soon changes into the complicated structure known as the spermatophore. The period between spermatophore extrusion and spermatophore preparation is called RS1, while the period between spermatophore preparation and the start of calling song (or mating response) is called RS2 (see Fig. 16.1).

16.2.3.1 Guarding Behavior

During the sexually refractory stage, the male is generally aggressive even when exposed to the female. When the female moves around the male, the male attacks her and exhibits body shaking. This "guarding behavior" effectively protects the transferred spermatophore from being eaten by the female (Khalifa 1950; Loher and Rence 1978). The spermatophore is made of nutritious proteins (Kaulenas 1976) and serves not only as a vessel for sperm but also as a nuptial gift. The first 20 min after spermatophore transfer is critical for the male because it takes that long for the sperm to move into the female's spermatheca. So, the male is highly aggressive in that period to prevent the female from eating the spermatophore.

16.2.3.2 Genitalia Cleaning

As illustrated in Fig. 16.6, a dramatic change occurs in the genital chamber when the spermatophore is extruded from the male genitalia (Kumashiro and Sakai 2001a, b; Kumashiro et al. 2006, 2008). The central portion (median pouch) of the genital chamber floor expands with the body fluid and goes into the empty dorsal pouch serving as the template for the attachment plate of the spermatophore (Fig. 16.6(1–4)). The median pouch wriggles at a frequency of 0.16 Hz. This peristaltic movement is composed of a large shift to right and left, and small crease like movements (Fig. 16.6(5)). These movements are involved in cleaning remnants of the previous spermatophore material or debris from the inside surface of the dorsal pouch before the spermatophore material is cast into the dorsal pouch. Every remnant or debris in the dorsal pouch is moved away by the cooperative movements of the median pouch and dorsal pouch, and this material is conveyed to the lateral pouch to be stored. This process is essential for the formation of the spermatophore (Kumashiro and Sakai 2016a, b).

16.2.3.3 Spermatophore Formation and Preparation

Spermatophore material excreted from the accessory glands and spermatozoa from the testes gradually consolidate in the dorsal pouch and the ventral lobes (Fig. 16.7 (3a-d)). The imnature spermatophore (Fig. 16.7(1)) develops into intricately organized spermatophore within 40 min (Fig. 16.7(2)). The actual mechanism of the spermatophore formation has not been inferred yet (Khalifa 1949; Gregory 1965; Mann 1984; Hall et al. 2000).

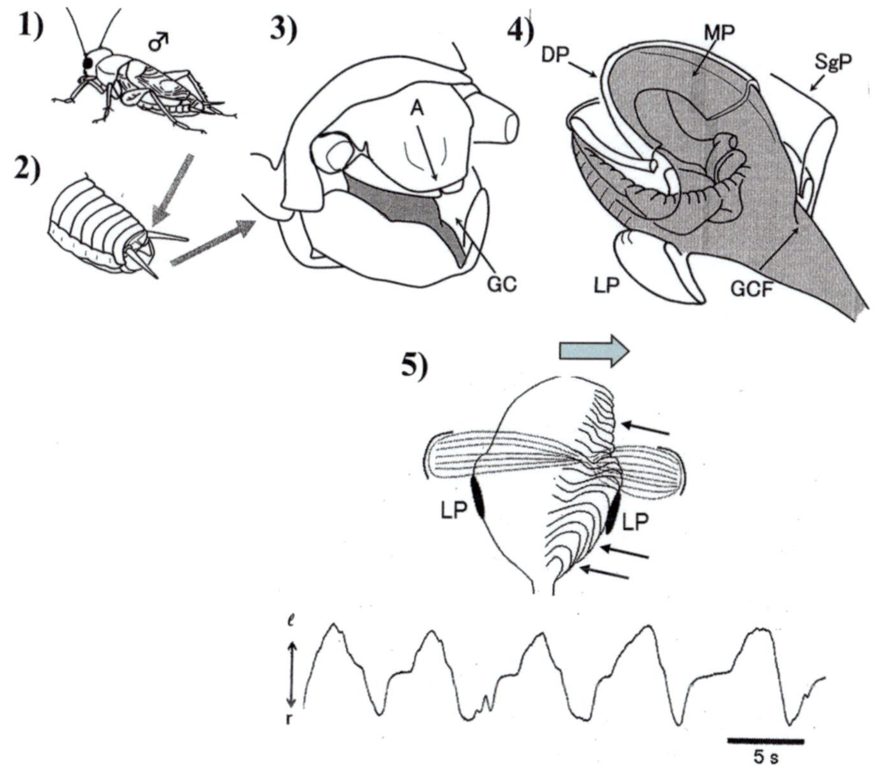

Fig. 16.6 Cleaning of the dorsal pouch following spermatophore extrusion. (*1–3*) The genital chamber (*GC*) below the anus (A) is located in the last abdominal segment of the male. (*4*) The genital membrane consists of the median pouch (*MP*) and genital chamber floor (*GCF*). The central region of the genital membrane in the last abdominal sternite (subgenital plate, *SgP*) forms a sack-like structure, the median pouch. It inflates and occupies the inside of the dorsal pouch (*gray* portion at the *bottom*) in the stage between the end of copulation and the start of spermatophore preparation. *A* anus, *DP* dorsal pouch, *GC* genital chamber, *GCF* genital chamber floor, *LP* lateral pouch, *MP* median pouch. (*5*) Movement of the median pouch. The median pouch always shows undulation which consists of a large shift (*thick arrow*) to the right and left with small crease-like movements (*thin arrows*) at about 0.16 Hz. When the pouch leans to the right, a number of crease-like movements appeared on the right side of the median pouch. The *bottom* shows the large movement of the median pouch to the *right* (r) and *left* (l)

Spermatophore preparation is initiated as described by Ootsubo and Sakai (1992). When the male antennae are stimulated by contact with a female, spermatophore preparation occurs within approximately 4 min. It is preceded 45 s by the backward protrusion of the genital pouch (Fig. 16.7(3a)). The white spermatophore material is pushed onto the ventral lobes in the genital chamber (Fig. 16.7 (3b)). If the male does not encounter the female after copulation or the male is under heavy stress, spermatophore preparation does not occur for a long time (>1 h).

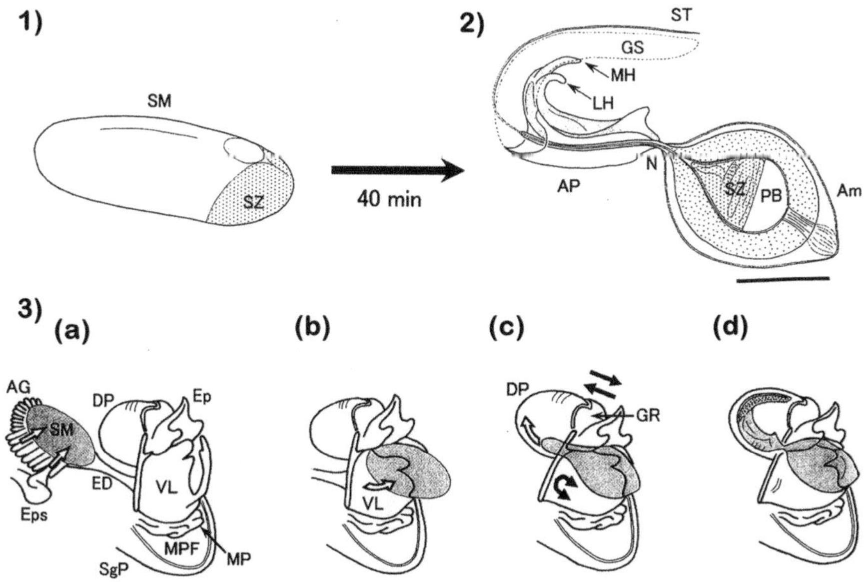

Fig. 16.7 Spermatophore formation. (*1*) Spermatophore material (SM) excreted from the accessory glands (*AG*) and spermatozoa (*SZ*) from the testis into the ejaculatory duct (*ED*). (*2*) Matured spermatophore. *Am* ampulla, *AP* attachment plate, *GS* gelatinous substance, *LH* lateral hook, *MH* median hook, *N* neck, *PE* pressure body, *ST* sperm tube, *Scale* 100 μm. (*3*) Sequence of spermatophore formation. (**a**) Excretion of SM and SZ into ED; (**b**) spermatophore preparation; (**c**) early stage of spermatophore formation; (**d**) matured spermatophore. *DP* dorsal pouch, *Ep* Epiphallus, *GR* guiding rod, *LP* lateral pouch, *MPF* median pouch floor (genital chamber floor)

16.2.3.4 Time-Fixed Interval in the Sexually Refractory Period

The male cannot recommence the calling until the spermatophore is completed. As long as the male is left alone, he does not produce the calling song (CS) even if the spermatophore is matured and the cricket is in the mating mode of the reproductive state. However, when paired with a female, the male starts calling and exhibits courtship behaviors. In the same way, when the male is presented with the mating response (MR) test (see Sect. 16.3.1) instead of a female, he begins copulatory acts to the model (see the legend for Fig. 16.1) at the same timing as CS. Thus, the time necessary to switch the reproductive mode from refractory to mating is defined as either SPMR (the time between spermatophore preparation and mating response) or SPCS (the time between spermatophore preparation and calling song).

It was found that the RS2 was time fixed when measured by SPCS (Nagao and Shimozawa 1987) differently from the first sexually refractory stage (RS1). The RS2 was measured by either SPCS or SPMR and found that it is 50–70 min long, with an average duration of 60 min (Ureshi and Sakai 2001). The RS2 varies among individuals but it is constant within an individual. This result suggested that the male may have a biological timer operating following the hourglass principle (Lees 1973), which starts at the occurrence of spermatophore preparation and stops around the start of the mating response or the calling song. The RS2 is shortened at higher temperatures and lengthened at lower temperatures. Our 24-h observation indicated that the RS2 did not change throughout 1 day as long as the temperature was kept constant. In fact, one male showed copulation 22 times a day and renewed the male cycle 22 times. Incidentally, the RS2 varies in different species; for example, the length is 116 min in *Loxoblemmus doenitzi* Stein and 136 min in *Loxoblemmus campester* Matsuura.

16.3 Physiology of Reproduction

As demonstrated in multiple different behavioral studies, cricket mating behavior comprises biologically intriguing phenomena including behavioral switching, changes in sexual excitation states, and timekeeping functions. To understand these mechanisms, it is necessary to determine the actual neuronal structure (s) responsible for each function. This has been done by using classical methods such as artificial stimuli combined with ablation and drug application.

16.3.1 Control of Copulatory Actions

When in the mating stage, the male displays the mating response (MR) consisting of body thrusts and hooking (see Sect. 16.2.2.4) in response to the application of the female model to the last abdominal tergite of the male. Sometimes, though, the male shows an evasion response to the same model. This observation suggests that the responsiveness to the model may change depending upon the male's internal state even in the mating stage. In general, the internal state can be described in terms of inhibition or excitation. For example, the MR is suppressed by inhibition when the male is under heavy stress. In contrast, the MR is elicited intesely by excitation when the male is sexually excited.

16.3.1.1 Inhibition

Classical experiments showed that while decerebrated males (lacking the brain) exhibited an evasion response to the female model, decapitated males (lacking the brain and subesophageal ganglion) showed the MR (Huber 1955). This result led to the conclusion that the inhibition center for the MR is located in the subesophageal ganglion (Huber 1955), similar to the situation in the praying mantis (Roeder 1935). More recent studies, however, exploring the effects of subesophageal ganglion ablation, did not agree to Huber's conclusion: the center for inhibition of the MR is actually located in the brain (Matsumoto and Sakai 2000a). In light of this new insight, we now recognize three different types of mating inhibition, which are described below.

Inhibition in the Mating Stage

In general, the male in the mating stage elicits the MR when stimulated with the female model. If, however, the male receives a noxious stimulus, such as a strong pinch to the leg or an antenna, he ceases to exhibit the MR and instead shows the evasion response (Matsumoto and Sakai 2000a, b). On the other hand, the male which had a leg pinching quickly recovers the ability to perform the MR when his head is cooled, chopped off, or ligated with a thread. To take advantage of such inhibitory and disinhibitory phenomena, male responsiveness was compared between decerebrated males and subesophageal ganglion-ablated males. These experiments showed that decerebrated males can quickly recover the MR after leg pinching. In contrast, the subesophageal ganglion-ablated males continued to display the evasion response after leg pinching. Hence, these results reveal that the inhibition center for the MR is located in the brain and not in the subesophageal ganglion (Fig. 16.8(1)). In addition, when one of the two connectives was cut between the head and thorax, normal mating behavior, including courting and copulation, was observed to the female. These individuals, however, were unable to inhibit the MR after leg pinching, similar to the decerebrate males. Hence, the loss of even half of the descending inhibitory input, mediated by the axons running through the connective, is enough to release the MR.

Inhibition in the RS1

In the first sexually refractory stage (RS1), the intact male never shows the MR. The mating response is elicited after the two connectives are cut at the head-thorax boundary (Matsumoto and Sakai 2000a, b). This fact suggests that brain inhibition on the MR is continuously exerted through the RS1. Furthermore, males with one of the connectives cut also had a difficulty in suppressing the MR. At first, this would appear similar to the situation in males at the mating stage whose connectives were cut following the noxious stimulation. However, males in RS1 also had a difficulty in suppressing the MR when an abdominal connective between the 6th abdominal ganglion and the terminal abdominal ganglion (TAG, 7th to 9th fused ganglia) (Fig. 16.8(2)).

These results reveal that the disinhibition that arises after cutting a connective at the head-thorax boundary is not solely dependent upon descending inhibition from the brain. This descending inhibition is commanded by an ascending signal from the TAG. The combined insight from the above experiments suggests that the reproductive mode in the TAG is switched to the refractory mode at spermatophore extrusion, and this signal is transmitted continuously to the brain through ascending connections which, in turn, inhibits the pattern generator for the MR through

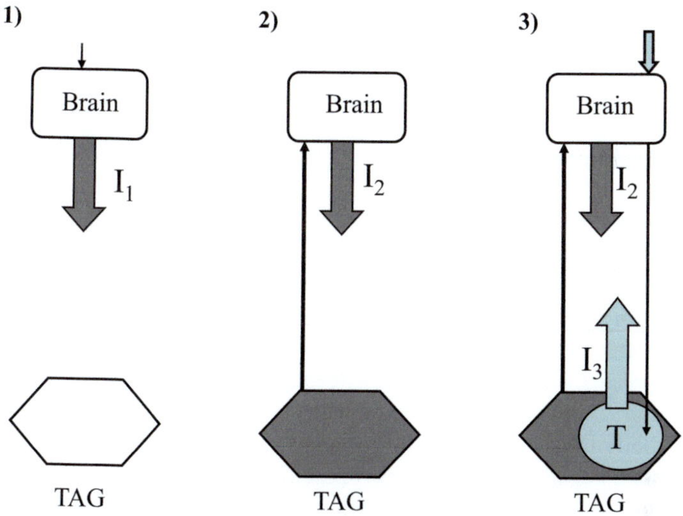

Fig. 16.8 Schematic diagrams of 3 types of inhibition. (*1*) Inhibition of mating response in the mating stage which is induced by noxious stimulation. This inhibition (I_1) of the brain is transient. (*2*) Inhibition of mating response in the sexually refractory stage (RS1) between spermatophore extrusion and spermatophore preparation. This inhibition (I_2) is exerted by the brain. The brain is tonically driven by the tonic input from the terminal abdominal ganglion (TAG) which is in the sexually refractory mode. (*3*) Inhibition of mating response in the sexually refractory stage (RS2) between spermatophore preparation and the recommencement of mating response. In this period, inhibition (I_3) commanded by the timer (T) in the TAG operates in addition to the brain inhibition (I_2). Two downward arrows to the right show the start of spermatophore preparation triggered by female stimuation to the antenna.

descending inhibition (I_2 in Fig. 16.8(2)). This scenario is consistent with the idea that the primary cause of post-copulatory inhibition resides in the TAG (Sakai et al. 1995).

Inhibition in the RS2

In the second sexually refractory stage (RS2), males do not show the MR even after decerebration, which is different from the responsiveness in the RS1. Such a difference suggests a presence of another inhibition system for the MR. Furthermore, males decerebrated just after spermatophore preparation which did not respond to the model began to respond as time elapses (>20 min after decerebration following spermatophore preparation; Sakai et al. 1995). A similar time dependency was observed even in a small isolated body region consisting of a few posterior abdominal segments which contain the TAG only. This body region which had not responded to the model began to show the movement characteristic of the MR 20 min after abdominal transection. These results suggest that the primary cause of time-dependent inhibition in RS2 is located in the TAG (I3, Fig. 16.8(3)).

16.3.1.2 Facilitation

The male responds more easily and persistently to the model when he is sexually excited. Specifically, the body thrust and hooking actions change depending upon the duration of preceding courtship (Matsumoto and Sakai 2001). The longer the male's courting of the female, the more intense the MR will be. This causal relationship can be further examined by measuring the movement parameters in decerebrate males. The duration of the MR in response to a single stimulus becomes longer, and the number of body thrusts increases as well. In addition, the time per individual body thrust (cycle length) becomes shortened. That is, the frequency of the body thrust per stimulus increases. These results reveal that the excitatory level of the nervous system underlying the MR has been facilitated during courtship.

Pharmacological experiments have elaborated on our knowledge of facilitation. Drugs injected into the abdomen of decerebrate males were tested with the model and the MR was assessed. Injection of octopamine resulted in shortened cycle length. Similarly, octopamine agonists such as synephrine and forskolin also shortened the cycle length. On the other hand, an antagonist, phentolamine, increased the cycle length. A similar facilitatory effect was induced by electrical stimulation of the connectives at the head-thorax boundary. These results suggest that some descending neurons from the head ganglia may release octopamine to facilitate the activity of the pattern generator for the MR, which is located in the thoracic and abdominal ganglia.

16.4 Neural Activity Related to Reproductive Behavior

The previous classical physiological studies indicate that the brain and the TAG are the key structures that control male behavior in the reproductive cycle. Here we review recent advancements in our understanding of how neural activity affects different aspects of this behavior.

16.4.1 Neural Activity Responsible for Spermatophore Extrusion

The spermatophore is extruded when the sensilla of the epiphallus are mechanically stimulated with the model mimicking the copulatory papilla of the female (Sakai et al. 1991). Specifically, there are relatively larger bristle sensilla on the outside of the epiphallus, and smaller sensilla (cavity hairs) are found on the inside of the epiphallus. Sensory neurons innervating both groups of sensilla project their axons to the TAG. On the motor side, the attachment plate of the spermatophore is ejected by the deformation of the dorsal pouch in the genitalia. This is caused by the contraction of the dorsal pouch muscles, each of which is innervated by a single dorsal pouch motoneuron (mDP) in the TAG (Kumashiro and Sakai 2001a, b).

Figure 16.9 shows suction electrode recordings of the genital nerve in which spikes of both sensory and motor nerves can be observed (Sakai and Kumashiro 2004b). Stimulation of the outside hairs of the epiphallus (dotted line) was not enough to activate the mDP (Fig. 16.9(1)). In contrast, stimulation of the cavity hairs activated several motor neurons including the mDP (larger spikes below solid bar) with a long latency of several seconds (Fig. 16.9(2)). When the outside hairs and cavity hairs were stimulated simultaneously, the mDP discharged with a shorter latency to elicit spermatophore extrusion (Fig. 16.9(3)). These results indicate that

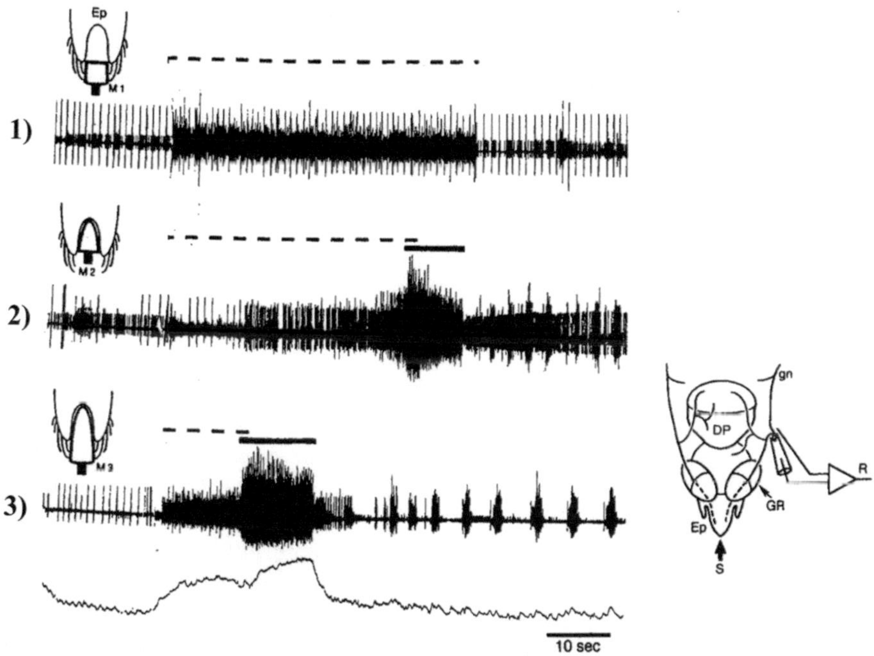

Fig. 16.9 Responses to stimulation of epiphallic sensilla with a model of the female copulatory papilla. (*1*) Response to stimulation (*broken bar*) of the outer hairs (M1) on the epiphallus (Ep) with a model. Sensory component continued to discharge for 50 s but dorsal pouch contraction was not triggered. (*2*) Responses to stimulation of cavity hairs (M2). Spermatophore extrusion and transfer (*solid bar*) was triggered after 39 s as shown by additional responses of genital motoneurons (below *solid bar*). This preparation was the same as in (*1*). (*3*) Responses to stimulation of both cavity hairs and outer hairs (M3). Spermatophore extrusion was triggered 11 s after the onset of stimulation. This preparation was different in (*1*) and (*2*). The *bottom line* is the movement of the genitalia. Right inset shows setups for recording (See Fig. 16.6 legends for abbreviations)

cavity hairs are of primary importance for spermatophore extrusion, while outside hairs play a facilitatory role.

The mDP was morphologically identified (Fig. 16.10(1)) by backfilling with cobalt and nickel (Sakai and Yamaguchi 1983), and its spike activity was singly recorded (Sakai and Kumashiro 2004b; Fig. 16.10(2)). The characteristic feature of this neuron is that the burst discharge occurs abruptly more than 10 s after the start of stimulation. This suggests that some high threshold interneuron(s) may be interposed between the sensory afferent and the mDP. This burst discharge causes the jerky contraction of the dorsal pouch muscle for spermatophore extrusion (see the deflection in the bottom trace).

16.4.2 Neural Activity Responsible for Genitalia Cleaning

After the initial bursting, the mDP firing pattern changes into spontaneous rhythmic activity (Fig. 16.10(2)). The activity continued even after the TAG was isolated by cutting the connectives between the 6th abdominal ganglion and the TAG. These results reveal that rhythmic activity is generated in the neural circuit of the TAG. The ensuing periodical dorsal pouch contractions serve for cleaning the inside of the dorsal pouch in cooperation with the rhythmic movements of the median pouch (Kumashiro and Sakai 2016b).

16.4.3 Pattern Generator for Spermatophore Preparation

Spermatophore preparation is a critical event for preparing for a new mating stage. That is, the 1-h timer starts and the RS2 begins. However, the trigger mechanism is totally unknown at the neural level. The chemosensory stimulus of contact with a female activates brain neurons, which send the command to the TAG to trigger spermatophore preparation. Even if both the connectives are cut, at any level, immediately after the pouch protrusion, spermatophore preparation occurs normally 45 s later (Ootsubo and Sakai 1992). This means that the neural circuit for spermatophore preparation is entirely located within the TAG. The brain is only needed to trigger the pattern generator for spermatophore preparation, though this can occur spontaneously, without the trigger stimulation, as well. It is interesting to note that DUM (dorsal unpaired median) neurons in the TAG are also responsible for the excretion of the spermatophore material from the accessory glands to the ejaculatory duct (Yasuyama et al. 1988; Kimura et al. 1989). Thus, the control system of DUM neurons may have a close relation to the pattern generator for spermatophore preparation.

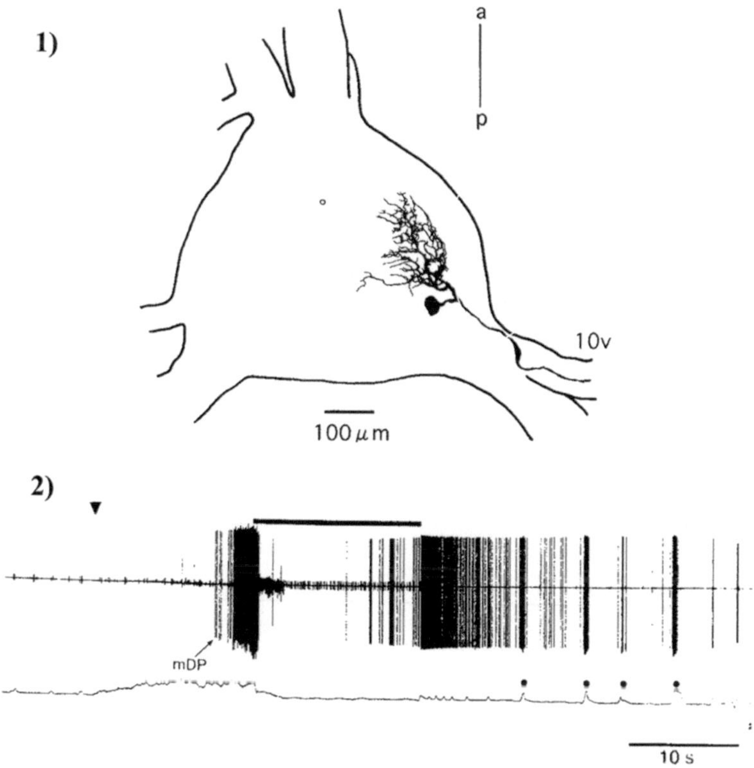

Fig. 16.10 Dorsal pouch motoneuron. (*1*) Morphology of the dorsal pouch motoneuron. *a* anterior, *p* posterior, *10v* the 10th ventral nerve root (cercal motor nerve) of the TAG. (*2*) Dorsal pouch motoneuron (mDP) discharge in normal spermatophore extrusion. The discharge pattern of the mDP (*top trace*) around spermatophore extrusion was elicited by genital stimulation. The start of stimulation is shown by *arrowhead*. The mDP did not respond in the first 13 s, but then exhibited a strong burst at spermatophore extrusion and gradually changed into rhythmic bursts after the spermatophore transfer phase (*horizontal line*). Bottom trace shows movement of the phallic complex in which each upward deflection (*dot*) indicates the rhythmic movement of the phallic complex

16.4.4 Neural Activity Associated with the 1-h Timer

For the time-fixed RS2, the previous experiments suggested that the center for time-dependent inhibition is located in the TAG. To obtain further evidence, reversible inactivation was carried out by cooling the nervous system. During the RS2, cooling either the brain, thorax, or abdomen for 30 min resulted in local inactivation of the central nervous system. Cooling at any of these locations did not lengthen the RS2. However, when the abdominal segment containing the TAG was cooled for 30 min, the RS2 was lengthened by 30 min. This demonstrates that the timer is located within the TAG (Ureshi and Sakai 2001). This is also in line with the

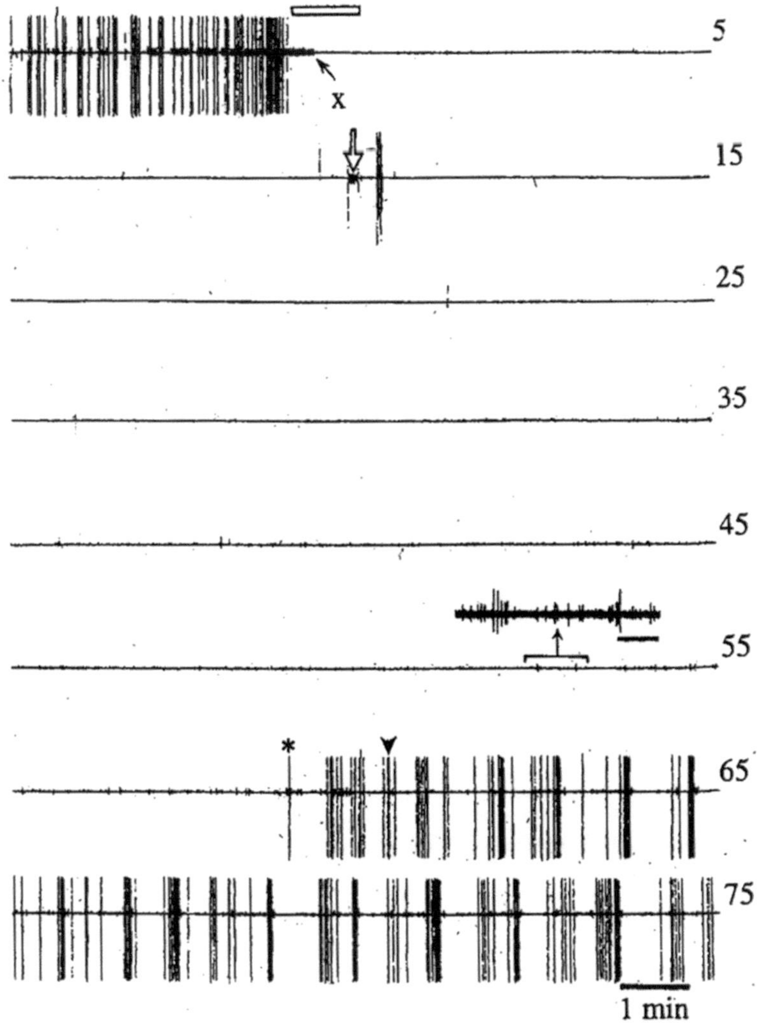

Fig. 16.11 Time-dependent activity of mDP in males with the isolated terminal abdominal ganglion. *Numbers on the right* indicate time (min) after spermatophore preparation. *White horizontal* bar on the first line indicates period between subgenital plate opening (*left end*) and spermatophore preparation (*right end*). *x* indicates small-sized spikes. The *open arrow on the second line* indicates the time of connective cut. The *asterisk* indicates the first spike of the mDP after the long silence. *Arrowhead* indicates the onset of auto-spermatophore extrusion. Inset between lines labeled 45 and 55 shows an enlarged recording of the indicated portion in which tiny spikes with at least three different amplitudes are discerned. Scale bar for *inset*, 10 s. See text for explanation

observation that the dorsal pouch motoneuron (mDP) showed time-dependent activity in the isolated TAG. Furthermore, the mDP, which is silent just after spermatophore preparation, automatically started to discharge one hour later (Fig. 16.11; Kumashiro and Sakai 2003). While it is currently unknown how the time-dependent activity of the mDP is controlled by the neuron(s) responsible for the timekeeping, there is some indication that serotonin may be involved in this process (Ureshi and Sakai 2002).

16.5 Development of Mating Behavior

While studying mating behavior was traditionally focused exclusively on adults, very little is understood about how postembryonic development might contribute to mating. In general, most of the behavioral repertoires in the larvae of hemimetabolous insects are similar to those of adults, except for flying and mating. Hence, there is a possibility that cricket larvae may already be equipped with the pattern generators for flight and mating, but they cannot exhibit such behavior because the command system is not yet operational. This scenario is supported by the classical studies in *Teleogryllus*, in which nymphs gain the ability to display mating behavior only after the ablation of the mushroom body of the brain, which is involved in inhibition (Bentley and Hoy 1970). However, as this was not the case in *Gryllus bimaculatus*, there may be some species-specific differences in this regard.

16.5.1 Mating Behavior in Nymphs

In contrast to the previous study in *Teleogryllus* (Bentley and Hoy 1970), it was reported that intact nymphs of *Gryllus bimaculatus* can exhibit mating-like actions, i.e., courtship-like behavior (CSLB) and copulation-like behavior (CPLB). The males become adults (ninth instar) after eight rounds of ecdysis. Nymphs at the eighth instar showed both CPLB and CSLB when they were paired with an adult female or male (Sakai et al. 1990). Both actions occurred at higher frequency in the middle stage of the eighth instar than in the beginning and the end. Nymphs at the seventh instar showed CSLB exclusively and much less frequently than in the eighth instar. Those at the sixth instar showed neither behavior. The movement patterns of their mating-like actions are essentially the same as those seen in adults. However, the frequency of their occurrence was less and their duration was shorter than in adults. Nymphs are very sensitive to environmental noises, and they easily stop mating-like actions and do not recommence for a long time. Since the nymph genitalia are immature, the spermatophore preparation does not occur. Thus, the male cycle or the time-fixed sexually refractory stage cannot be detected. The observations that adult-type mating actions can occur in nymphs reveal that the command system for mating behavior is already operational in nymphs. Instead, it suggests that though the mushroom bodies in the brain certainly exert inhibition, this can be canceled by the gradual increase in sexual excitation during development.

16.5.2 Mating Behavior in Fresh Adults

Adult males do not show mating-like or mating actions for a few days following the final molt. Afterward, the mating responsiveness of fresh adult males develops gradually day by day. For example, only 30 % of fresh adults exhibited some copulatory actions on the day of the final molt (day 1). In contrast, most individuals display the body thrust under the female, and hooking appears on day 2. Similarly, spermatophore extrusion and transfer becomes fully successful on day 3. Following this, mating behavior further develops to reach a mature level after 1 week. When combined with the observations in nymphs, we can now postulate that the sexual excitation gradually increases to facilitate mating behavior after the final molt.

For future studies of insect reproduction, it would be interesting to compare every aspect of male behavior between *Gryllus bimaculatus* and other cricket and orthopteran species. The nature of the sexual refractory period, especially the time-fixed RS2, should be examined for its length and consistency in a comparative context. Since the RS2 is associated with the process of spermatophore production, it should also be correlated with the structure and function of the spermatophore. Finally, the neural mechanism of the male sexual state and its switching in the reproductive cycle should be cleared.

To do so, is important to understand the nature of the communication between the brain and the terminal abdominal ganglion, which could be accomplished by recording descending and ascending spike activity. In addition, the neuron (s) responsible for the one-hour timer should be identified in the terminal abdominal ganglion. The molecular mechanism in these neurons should be elucidated, as has been done in the study of circadian clocks. With the advancement of new functional and genomic tools in *Gryllus*, such as those described in this book, these questions can soon start to be addressed.

Acknowledgments The authors thank their corroborators, Y Ogawa, M, Hiratou, T Takao, T Seno, N Umeya, R Kanazawa, Y Tsuji, M Dainobu, Y Taoda, T Katayama, F Kawasaki, and K Mori for their contributions to the study.

References

Adamo SA, Hoy RR (1995) Agonistic behavior in male and female field crickets, *Gryllus bimaculatus*, and how behavioural context influences its expression. Anim Behav 49:1491–1501

Alexander RD (1961) Aggressiveness, territoriality, and sexual behavior in field crickets (Orthopera: Gryllidae). Behaviour 17:130–223

Alexander RD, Otte D (1967) The evolution of genitalia and mating behavior in crickets (Gryllidae) and other Orthoptera. Misc Publ Zool Univ Mich 133:12–18

Beck R (1974) The neural and endocrine control of mating behavior in the male house cricket, *Acheta domesticus*. L. Ph.D. thesis University of Nottingham

Bentley DR (1969) Intracellular activity in cricket neurons during the generation of song patterns. J Insect Physiol 15:677–699

Bentley DR, Hoy RR (1970) Postembryonic development of adult motor pattern in crickets: a neural analysis. Science 170:1409–1411

Choperd L (1951) Les divisions du genre *Gryllus* basees sur l'etude de l'appareil copulateur (Orth. Gryllidae). Eos 37:267–287

Evans AR (1988) Mating systems and reproductive strategies in three Australian Gryllid crickets: *Bobilla victoriae* Otte, *Balamara gidya* Otte and *Teleogryllus commodus* (Walker)(Orthoptera: Gryllidae: Nemobiinae; Trigonidiinae; Gryllinae). Ethology 78:21–52

Gregory GE (1965) The formation and fate of the spermatophore in the African migratory locust, *Locasta maigratoria migraroides* Reiche and Fairmaire. Trans R Entomol Soc Lond 117:33–36

Hall MD, Beck R, Greenwood M (2000) Detailed developmental morphology of the spermatophore of the Mediterranean field cricket, *Gryllus bimaculatus* (De Geer)(Orthoptera: Gryllidae). Arthropod Struct Dev 29:23–32

Hörmann-Heck S (1957) Untersuchungen uber den Erbgang einiger Verhaltensweisen bei Grillenbastarden (*Gryllus campestris* L., *Gryllus bimaculatus* DeGeer). Z Tierpsychol 14:137–183

Hsu YC (1929) Crickets in China. Peking Soc Nat Hist Bull 3:5–41

Huber F (1955) Sitz und bedeutung nervöser Zentren für Instinkthandlungen beim Männchen von *Gryllus bimaculatus* DeGeer. Z Tierpsychol 12:12–48

Huber F, Moor TE, Loher W (1989) Cricket behavior and neurobiology. Cornell Univ Press, Ithaca, pp 340–363

Iwasaki M, Delago A, Nishino H, Aonuma H (2006) Effects of previous experience on the agonistic behavior of male crickets, *Gryllus bimaculatus*. Zool Sci 23:863–872

Kaulenas MS (1976) Regional specialization for export protein synthesis in the male cricket accessory gland. J Exp Zool 195:81–96

Khalifa A (1949) The mechanism of insemination and the mode of action of the spermatophore in *Gryllus domesticus*. Q J Microsc Sci 90:281–292

Khalifa A (1950) Sexual behavior in *Gryllus domesticus* L. Behaviour 2:264–274

Kimura T, Yasuyama K, Yamaguchi T (1989) Proctolinergic innervation of the accessory gland in male crickets (*Gryllus bimaculatus*): detection of proctolin and some pharmacological properties of myogenically and neurogenically evoked contractions. J Insect Physiol 35–3:251–265

Kumashiro M, Sakai M (2001a) Reproductive behavior in the male cricket *Gryllus bimaculatus* DeGeer: I structure and function of the genitalia. J Exp Biol 204:1123–1137

Kumashiro M, Sakai M (2001b) Reproductive behavior in the male cricket *Gryllus bimaculatus* DeGeer: II neural control of the genitalia. J Exp Biol 204:1139–1152

Kumashiro M, Sakai M (2003) Auto-spermatophore extrusion in male crickets. J Exp Biol 206:4507–4519

Kumashiro M, Sakai M (2016a) Genital autocleaning in the male cricket Gryllus bimaculatus (1): Structure and function of the genital membrane. Zool Sci 33 in press.

Kumashiro M, Sakai M (2016b) Genital autocleaning in the male cricket Gryllus bimaculatus (2): Rhythmic movements of the genitalia and their motor control. Zool Sci 33 in press.

Kumashiro M, Iwano M, Sakai M (2008) Genitalic autogrooming in the male cricket, *Gryllus bimaculatus* DeGeer. Acta Biol Hung 59(Suppl):137–148

Kumashiro M, Tsuji Y, Sakai M (2006) Genitalic autogrooming: a self-filling trash collection system in crickets. Naturwissenchaften 93:92–96

Lees AD (1973) Photoperiodic time measurement in the aphid *Megoura viciae*. J Insect Physiol 19:2279–2316

Loher W, Rence B (1978) The mating behavior of *Teleogryllus commodus* (Walker) and its central and peripheral control. Z Tierpsychol 46:225–259

Loher W, Weber T, Huber F (1993) The effect of mating on phonotactic behavior in *Gryllus bimaculatus* (De Geer). Physiol Entomol 18:57–66

Mann T (1984) Spermatophores. Springer-Verlag, Berlin, pp 107–111

Matsumoto Y, Sakai M (2000a) Brain control of mating behavior in the male cricket *Gryllus bimaculatus* DeGeer: the center for inhibition of copulation actions. J Insect Physiol 46:527–538

Matsumoto Y, Sakai M (2000b) Brain control of mating behavior in the male cricket *Gryllus bimaculatus* DeGeer: brain neurons responsible for inhibition of copulation actions. J Insect Physiol 46:539–552
Matsumoto Y, Sakai M (2001) Brain control of mating behavior in the male cricket *Gryllus bimaculatus* DeGeer: excitatory control of copulatory actions. Zool Sci 18:659–669
Muller M, Wehner R (1994) The hidden spiral: systematic search and path integration in desert ants, *Cataglyphis fortis*. J Comp Physiol A175:525–530
Nagao T, Shimozawa T (1987) A fixed time-interval between two behavioral elements in the mating behavior of male crickets, *Gryllus bimaculatus*. Anim Behav 35:122–130
Nishino H, Sakai M (1996) Behaviorally significant immobile state of so-called thanatosis in the cricket *Gryllus bimaculatus* DeGeer: its characterization, sensory mechanism and function. J Comp Physiol 179:613–624
Ogawa Y, Sakai M (2009) Calling and courtship behaviors initiated by male-male contact via agonistic encounters in the cricket *Gryllus bimaculatus*. Zool Sci 26:517–524
Ootsubo T, Sakai M (1992) Initiation of spermatophore protrusion behavior in the male cricket *Gryllus bimaculatus* DeGeer. Zool Sci 9:955–969
Palka J, Levine R, Schubiger M (1977) The circus-to-giant interneuron system of crickets. I. Some attributes of the sensory cells. J Comp Physiol 119:267–283
Pollack GS, Hoy RR (1989) Evasive acoustic behavior and its neurobiological basis. In: Huber F, Moor TE, Loher W (eds) Cricket behavior and neurobiology. Cornell Univ Press, Ithaca, pp 340–363
Rence R, Loher W (1977) Contact chemoreceptive sex recognition in the male cricket, *Teleogryllus commodus*. Physiol Entomol 2:225–236
Roeder DK (1935) An experimental analysis of the sexual behavior of the praying mantis (M.L.). Biol Bull 69:203–220
Sakai M, Katayama T, Taoda Y (1990) Postembryonic development of mating behavior in the male cricket *Gryllus bimaculatus* DeGeer. J Comp Physiol 166:775–784
Sakai M, Kumashiro M (2004a) Copulation in the male cricket is performed by chain reaction. Zool Sci 21:705–718
Sakai M, Kumashiro M (2004b) Auto-spermatophore extrusion reveals that the reproductive timer functions in the separated terminal abdominal ganglion in the male cricket. Acta Biol Hung 55 (1-4) 113–120
Sakai M, Matsumoto Y, Takemori N, Taoda Y (1995) Post-copulatory sexual refractoriness is maintained under the control of the terminal abdominal ganglion in the male cricket *Gryllus bimaculatus* DeGeer. J Insect Physiol 41:1055–1070
Sakai M, Ootsubo T (1988) Mechanism of execution of sequential motor acts during copulation behavior in the male cricket *Gryllus bimaculatus* DeGeer. J Comp Physiol A 162:589–600
Sakai M, Taoda Y, Mori K, Fujino M, Ohta C (1991) Copulation sequence and mating termination in the male cricket *Gryllus bimaculatus* DeGeer. J Insect Physiol 37:599–615
Sakai M, Yamaguchi Y (1983) Differential staining of insect neurons with nickel and cobalt. J Insect Physiol 29:393–394
Shimozawa T, Kanou M (1984) Varieties of filiform hairs: range fractionation by sensory afferents and cercal interneurons of a cricket. J Comp Physiol A 155:485–493
Simmons LW (1986) Female choice in the field cricket *Gryllus bimaculatus* (De Geer). Anim Behav 34:1463–1470
Stevenson PA, Hofmann HA, Schoch K, Schildberger K (2000) The fight and flight responses of crickets depleted of biogenic amines. J Neurobiol 43:107–120
Tinbergen N (1951) The study of instinct. The Clarendon Press, Oxford
Ureshi M, Sakai M (2001) Location of the reproductive timer in the male cricket *Gryllus bimaculatus* DeGeer as revealed by local cooling of the central nervous system. J Comp Physiol A 186:1159–1170

Ureshi M, Sakai M (2002) Serotonin precursor (5-hydroxytryptophan) has a profound effect on the post-copulatory time-fixed sexually refractory stage in the male cricket, *Gryllus bimaculatus* DeGeer. J Comp Physiol A 188:767–779

Weber T, Thorson J, Huber F (1981) Auditory behavior of the cricket. I. Dynamics of compensated walking and discrimination paradigms on the Kramer treadmill. J Comp Physiol 141:215–232

Yasuyama K, Kimura T, Yamaguchi T (1988) Musculature and innervation of the internal reproductive organs in the male cricket, with special reference to the projection of unpaired median neurons of the terminal abdominal ganglion. Zool Sci 5:767–780

Part III
Experimental Approaches

Chapter 17
Protocols for Olfactory Conditioning Experiments

Yukihisa Matsumoto, Chihiro Sato Matsumoto, and Makoto Mizunami

Abstract Insects have sophisticated learning abilities despite the relative simplicity of their central neural systems, which consist of small numbers of neurons as compared to vertebrates. Among insects, crickets (*Gryllus bimaculatus*) exhibit the most robust olfactory learning and memory. In this chapter, we describe protocols for classical conditioning and memory retention tests in crickets. Crickets are individually trained to associate an odor (conditioned stimulus) with water reward (appetitive unconditioned stimulus). To evaluate the effect of training, relative preference between the conditioned odor and a control odor is tested before and after training. We describe the methodology of olfactory conditioning in detail to help researchers who are interested in using crickets to study learning and memory.

Keywords Classical conditioning • Olfactory learning • Associative learning • Absolute conditioning • Differential conditioning • Memory

17.1 Introduction

Associative learning, which links two sensory stimuli, is a fundamental form of learning and is conserved among many vertebrates and invertebrates. A basic form of associative learning is classical conditioning, in which animals learn to associate a conditioned stimulus (CS) with an unconditioned stimulus (US). The US is a stimulus that innately evokes a certain behavioral response, whereas the CS is a relatively neutral stimulus that usually does not activate a specific response. After appropriate training of CS-US pairing, animals exhibit a behavioral response to the CS that is analogous to the behavioral response to the US. Thus, the animal has presumably learned that the CS is a cue for predicting the US. The response evoked by the CS is termed the "conditioned response."

Y. Matsumoto (✉) • C.S. Matsumoto
College of Liberal Arts and Science, Tokyo Medical and Dental University, 2-8-30 Kounodai, Ichikawa 272-0827, Japan
e-mail: yukihisa.las@tmd.ac.jp

M. Mizunami
Faculty of Science, Hokkaido University, Sapporo 060-0810, Japan

H.W. Horch et al. (eds.), *The Cricket as a Model Organism*,
DOI 10.1007/978-4-431-56478-2_17

Insects have good learning abilities, and since their central nervous systems are relatively simple, especially compared to invertebrates, they are an excellent model system to study learning and memory (Mizunami et al. 2004; Giurfa 2007). Several species of insects have been used as models for research on learning and memory (Mizunami et al. 2004) including honeybees, fruit flies, crickets, cockroaches, ants, and moths (Bitterman et al. 1983; Giurfa and Sandoz 2012; Tully and Quinn 1985; Matsumoto and Mizunami 2000, 2002a; Balderrama 1980; Dupuy et al. 2006; Daly and Smith 2000). Among these insects, we have shown that crickets (*Gryllus bimaculatus*) have a strong capability for olfactory learning and memory (Matsumoto and Mizunami 2002b, 2004, 2005, 2006; Matsumoto et al. 2013a; Mizunami et al. 2009; Terao et al. 2015), as do honeybees (*Apis mellifera*) (e.g., Menzel 1999; Giurfa 2007; Avarguès-Weber et al. 2011; Sandoz 2011), in experiments using a classical conditioning procedure to associate an odor (CS) with water reward (US).

In this chapter, we describe protocols for classical conditioning experiments in the cricket.

17.2 Preparation for the Experiment

17.2.1 Apparatus and Chemicals

The apparatus and chemicals used for the experiment are testing apparatus, 100 ml beakers, paper towels, filter papers, odorants (vanilla essence, peppermint essence, apple essence, etc.), cylindrical plastic containers, 1 ml hypodermic syringes, needles, gauze, distilled water, 20 % sodium chloride solution, a handheld vacuum cleaner, stopwatch, datasheets (grid sheets), and pens.

17.2.2 Animals

Adult crickets, *Gryllus bimaculatus* DeGeer, are reared in a plastic case (80 × 45 × 20 cm) under a 12 h light/12 h dark photoperiod (photophase, 08:00–20:00 h) at 27 ± 2 °C. They are fed a diet of insect food pellets (Oriental Kobo Co., Japan) and water ad libitum. Three days before the start of the experiment, young adult male crickets (about 1 week after the imaginal molt) are transferred from the laboratory colony to individual 100 ml beakers. Each beaker should be labeled to allow individual identification of the crickets throughout the experiment. A paper towel cut in a circle with a diameter of 5 cm is placed at the bottom of each beaker to provide a normal substratum for locomotion and to absorb excreta. In the beakers, crickets are fed insect food pellets, but they are deprived of drinking water to enhance their motivation to search for water.

17.2.3 Odorants

Odors are paired in these experiments. For example, in a typical experiment, one odor is used as the CS odor and the other odor is used as control odor. Mixtures of compounds (e.g., vanilla essence, peppermint essence, apple essence) are used as the odorants. The choice of a pair of odorants (e.g., vanilla–peppermint pair, apple–banana pair) depends on the purpose of the experiment. For example, naïve crickets prefer vanilla odor over peppermint odor. Pairing a less-attractive odor (peppermint odor) with reward in conditioning experiments often leads to a reversal of relative odor preference between peppermint and vanilla odors. Such an experimental design is suitable for experiments in which a high score of conditioning effect is required. Here, we describe a protocol using vanilla odor and peppermint odor.

17.2.4 Odor Ventilation

When carrying out olfactory conditioning, the odors presented should be exhausted from the experimental system as soon as the stimulation ends. After every conditioning trial, air in the beakers is vacuumed using a handheld vacuum cleaner to remove the odor left around the crickets. Alternatively, a ventilation system could be made with duct hose to chronically exhaust the air around the beakers. Ventilation should not be too strong, as unintended mechanical stimulations may interfere with conditioning.

17.2.5 Apparatus Used for Odor Preference Tests

The apparatus used for odor preference tests is made of opaque acrylic plates (black or white; 3 mm in thickness) glued using chloroform. The test apparatus consists of three chambers (Fig. 17.1b), a "test chamber" ($15.5 \times 25 \times 7$ cm), and two removable "waiting chambers" ($5 \times 7 \times 7$ cm each). The waiting chambers can be set either at the "entrance position" with an entrance (2-cm-square hole) into the test chamber or at the "waiting position" in which there is no entrance. Entry of the cricket into the testing chamber is controlled by a manual sliding door. On the floor of the test chamber, there are two circular holes (3 cm in diameter) that connect the chamber with sources of odors (Fig. 17.1b, d). For the preference test, three odor sources, one vanilla and two peppermint sources, are set beneath the test chamber on a rotating disc to exchange the position of the two odors. Each odor source consists of a cylindrical plastic container (4 cm in diameter $\times$ 4 cm in height) covered with a fine gauze net. A filter paper (15×15 mm) soaked with 3 μl of either peppermint essence (Kyoritsu-Shokuhin Co., Tokyo, Japan) or vanilla essence (Meijiya Co., Tokyo, Japan) diluted fivefold with water is placed inside each

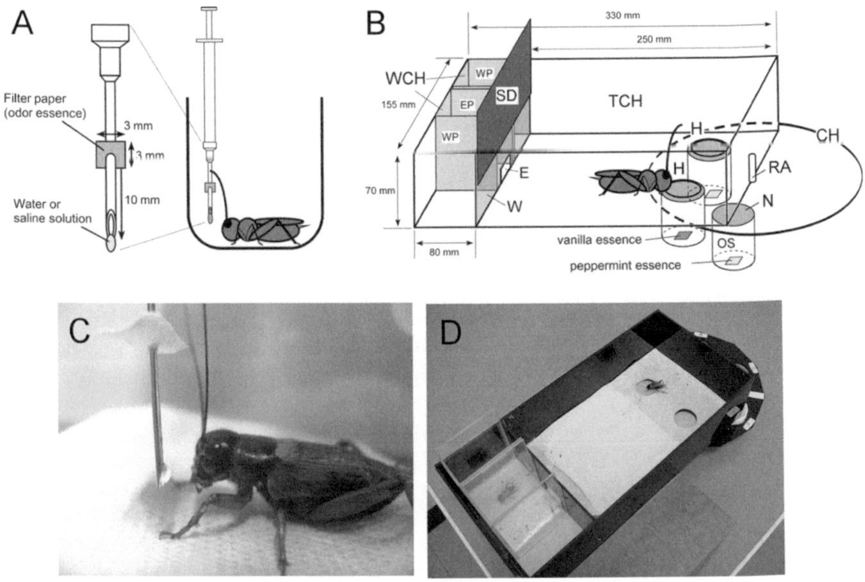

Fig. 17.1 Experimental procedure. (**a**) Experimental arrangement for conditioning. A syringe containing water or saline solution was used for conditioning. A filter paper soaked with peppermint or vanilla essence was attached to the needle of the syringe 10 mm from its tip. The filter paper was placed within 1 cm of the cricket's head so as to present a particular odor, and water or saline was then presented to the mouth. (**b**) Apparatus used for the odor preference test. *WCH* waiting chambers; *WP* waiting position, *EP* entrance position, *TCH* training chamber, *E* entrance, *W* wall, *CH* container holder, *RA* rotating axle, *OS* odor source, *N* gauze net, *SD* sliding door, *H* holes connecting the chamber with two of three odor sources. (**c**) A picture of a cricket in conditioning. (**d**) A picture of a cricket (*top right*) in an odor preference test. The cricket explores an odor source in the test chamber. Another cricket (*bottom left*) is in a waiting chamber at the "waiting position"

container. The three containers are mounted on a container holder disc (a plastic disc of 20 cm in diameter and 3 mm in thickness) with an axle (1 cm in diameter) in its center and set at the front wall of the test chamber. The test chamber is covered by a removable transparent acrylic plate (17 × 27 cm, 3 mm thick) during tests. The floor of the test chamber is covered with a paper towel with cricket scents to provide a normal substratum for locomotion.

17.2.6 Syringes Used for Conditioning

Hypodermic syringes (1 ml; TERUMO Co., Tokyo, Japan) are used for conditioning (Fig. 17.1a, c). A small filter paper (3 × 3 mm) is attached to the needle (22G × 1.5SB (0.70 × 38 mm); TERUMO Co., Tokyo, Japan) of the syringe 10 mm from its tip. The syringe used for the appetitive conditioning trial is filled with water, and the filter paper attached to the needle is soaked with 1 μl of peppermint essence. The syringe used for the aversive conditioning trial is filled with 20 % NaCl solution, and the filter paper is soaked with vanilla essence.

17.3 Protocol for Classical Conditioning and Operant Test

Crickets are subjected to an odor preference test before and after olfactory conditioning (Matsumoto and Mizunami 2002a): an innate odor preference test is followed by olfactory conditioning trials, and, after a desired interval, a retention test is performed. The experimental room is kept at 27 ± 2 °C.

17.3.1 Innate Odor Preference Test

Typically, a group of twenty crickets is subjected to an initial odor preference test. The procedures are as follows.

1. Transfer the first cricket (cricket A) from the beaker to the waiting chamber at the waiting position and wait approximately 4 min for the cricket to become accustomed to the surroundings.
2. Slide the waiting chamber containing cricket A to the entrance position and transfer the second cricket (cricket B) into another waiting chamber, and set it at the waiting position.
3. Open the door at the entrance to the test chamber to let cricket A move through the entrance into the test chamber. Push the cricket gently with a stick (e.g., a pen) if necessary.
4. When the cricket has entered the test chamber, close the sliding door and the test will begin. Start timing with a stopwatch (4 min).
5. An odor source is counted as "visited" when the cricket probes the top net of the odor source with its mouth or palpi. Register the time spent visiting each odor source cumulatively with a stopwatch and record the times on the datasheet. Use different colors to record visits to different odors.
6. Two minutes after starting the test, rotate the container holder disc to exchange the relative positions of the vanilla odor and peppermint odor sources (Fig. 17.1b). Thus, odor positions are switched once during every preference test.
7. Four minutes after starting the test, open the sliding door, push cricket A gently back into the waiting chamber, and close the door. The test session has finished.
8. Place the chamber containing cricket A at the waiting position and the chamber containing cricket B at the entrance position.
9. Remove the waiting chamber containing cricket A from the apparatus and transfer the cricket into a beaker.

10. Reposition the waiting chamber at the waiting position and place the next cricket (cricket C) inside.
11. Repeat (3–10) until the last cricket has completed the task.

Every preference test lasts for 4 min. If a cricket continues to probe an odor source 4 min after starting the test, the time of this visit is recorded until the cricket leaves the odor source. The data are discarded if the total time of a visit to either source is less than 10 s or if the cricket exhibits a locomotion defect. Renew odor essences about once an hour.

17.3.2 Olfactory Conditioning

17.3.2.1 Absolute Conditioning

Absolute conditioning is a basic form of classical conditioning in which a single stimulus (odor, CS) is paired with an appetitive reinforcer (water reward, US+) or an aversive reinforcer (saline solution punishment, US-).

In a typical absolute conditioning experiment, peppermint odor (CS) is paired with water reward (US+). Parameters such as number of trials and intertrial interval (ITI) can be changed according to the purpose of the experiment. A typical absolute conditioning procedure consists of four CS-US pairings with ITIs of 5 min (Fig. 17.2a). This procedure yields a robust and stable memory that can be retrieved several days after conditioning and is characterized as protein synthesis-dependent long-term memory (LTM) (Matsumoto et al. 2003, 2006).

1. Prepare syringes for conditioning. Set the beakers containing the crickets in rows for ease of conditioning (e.g., 5 beakers × 4 rows).
2. Start the first conditioning trial from cricket A to the last cricket T. In the conditioning trial, 4-s odor presentation (CS+) is followed by 2-s presentation of water reward (US+) (Fig. 17.1a). For odor presentation, place the filter paper within 1 cm of the first cricket's (cricket A) head. At 2 s after the onset of odor presentation, present a drop of water to the mouth of the cricket for 2 s (Fig. 17.1c).
3. Ventilate air in each beaker for 2 s using a handheld vacuum cleaner. Fill the syringe used for the conditioning with water, and soak the filter paper with 1 μl of peppermint essence.
4. Repeat (2–3) three more times (four trials per cricket in total) with an ITI of 5 min.
5. After completing the training, each cricket is fed a diet of insect pellets ad libitum but is deprived of drinking water in a beaker until the odor preference test is given.

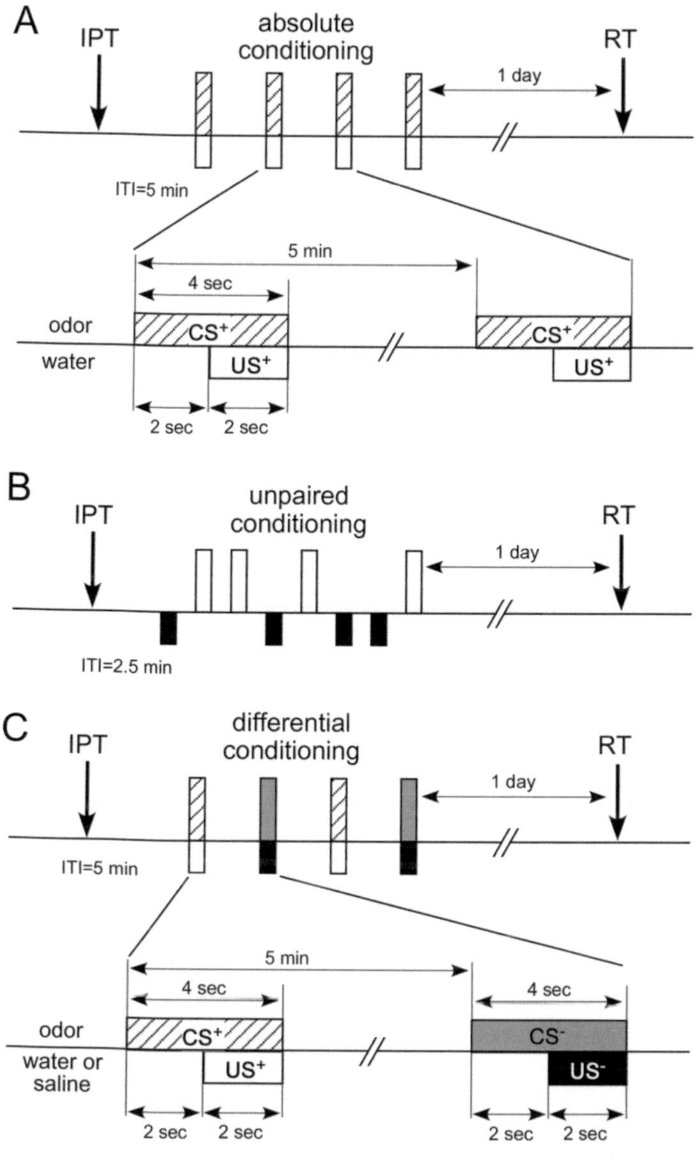

Fig. 17.2 Time schedules of three types of conditioning experiments. (**a**) Absolute conditioning. Cricket receives 4-s presentation of peppermint odor (CS+, hatched bars) and subsequent 2-s presentation of water (US+: *white squares*). Crickets are subjected to four pairing trials with an intertrial interval of 5 min, and their odor preferences are tested before conditioning (*IPT* innate preference test) and 1 day after conditioning (*RT* retention test). (**b**) Unpaired presentation of CS and US. Crickets receive unpaired presentations of the CS and of the US (four odor-only and four water-only presentations, 2.5 min apart in a pseudo-randomized sequence). Odor preferences are tested before conditioning and 1 day after conditioning. (**c**) Differential conditioning. In a set of differential conditioning trials, peppermint odor (CS+: hatched bars) is paired with water (US+: *white squares*) and vanilla odor (CS-: *shaded bars*) is paired with 20 % NaCl solution (US-: *black squares*). Crickets receive two sets of differential conditioning trials in alternate shifts, with an intertrial interval of 5 min. Odor preferences are tested before and 1 day after conditioning

17.3.2.2 Unpaired Presentations of CS and US

To ensure that crickets have acquired associative memory by absolute conditioning, several kinds of experimental controls are performed (CS only, US only, backward conditioning, etc.). Unpaired presentation of CS and US is one of them. Crickets receive unpaired presentations of the CS and of the US (four odor-only and four water-only presentations, 5 min apart in a pseudo-randomized sequence; Fig. 17.2b). Thus, both the absolute conditioning (paired) group and the unpaired group have exactly the same number of CS and US presentations (four CS and four US presentations).

17.3.2.3 Differential Conditioning

Differential conditioning is a procedure in which animals have to learn the difference between two (or more) stimuli by associating them with different USs. In a set of differential conditioning trials, one odor (CS+) is paired with water reward (US +) and another odor (CS-) is paired with 20 % NaCl solution (US-) with an ITI of 5 min (Fig. 17.2c). Typically, crickets are subjected to two sets of differential conditioning trials with an ITI of 5 min.

17.3.3 Memory Retention Test

At the desired time after conditioning, a preference test, which we refer to as a retention test, is performed again. The interval between the last training session and a retention test is called the retention interval. The length of the retention interval should be decided depending on the purpose of the experiment. During the retention interval, the crickets are placed in beakers with insect food pellets and given a drop of water on the inner wall of the beaker once daily. The procedure for the memory retention test is the same as that for the innate preference test.

17.3.4 Data Analysis

In odor preference tests, relative odor preference of each cricket is measured as the preference index (PI) for peppermint odor, defined as $tp/(tp + tv) \times 100$, where tp is the time spent exploring the peppermint odor and tv is the time spent exploring the vanilla odor. Wilcoxon's test (WCX test) is used to compare preferences before and after training, and the Mann–Whitney U-test (M-W test) is used to compare preferences between groups. For multiple comparisons, the Holm method is used to adjust the significance level.

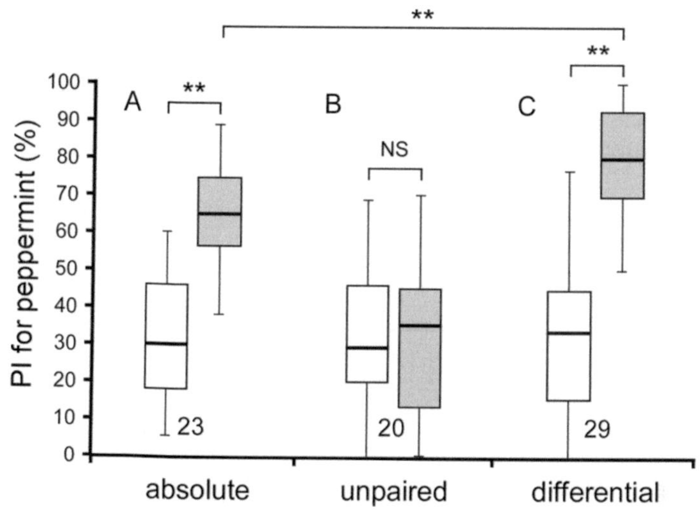

Fig. 17.3 Formation of long-term memory by olfactory conditioning in crickets. Three groups of crickets were subjected to absolute conditioning, unpaired presentations of CS and US, or differential conditioning, the procedures for which are shown in Fig. 17.2. Relative odor preferences between peppermint odor and vanilla odor were tested before and at 1 day after training. Preference indexes (PIs) for the peppermint odor before (*white bars*) and after (*gray bars*) training are shown as box and whisker diagrams. The *line* in the box is the median and the box represents the 25th–75th percentiles. Whiskers extend to extreme values as long as they are within a range of 1.5× box length. Outliers are not shown in diagrams, so as to reduce complexity of the figure and thus to facilitate visual inspection of data, but were included in data analysis. The number of animals is shown below the boxes. Wilcoxon's test was used for comparison of preference before and after conditioning and the Mann–Whitney *U*-test was used to compare among different groups. For multiple comparisons, the Holm method was used to adjust the significance level. The results of statistical comparison are shown as *asterisks* (** $p < 0.01$; NS $p > 0.05$)

17.3.5 Results

Typical results of a conditioning experiment are shown in Fig. 17.3. In the figure, PIs for peppermint (%) before and after training are shown in box plots. The line inside the box is the median and the box represents the 25–75 percentiles. Whiskers extend to extreme values as long as they are within a range of 1.5× box length. The results of statistical comparisons after adjusting by the Holm method are shown by asterisks (** $P < 0.01$; NS $P > 0.05$).

The absolute conditioning group exhibited a significant increase in preference for the rewarded odor (peppermint) 1 day after training compared to that before training (Fig. 17.3a; $p < 0.0001$, WCX test adjusted by the Holm method). The differential conditioning group also exhibited a significantly increased preference for peppermint 1 day after training compared to that before training (Fig. 17.3c; $p < 0.0001$, WCX test adjusted by the Holm method). Thus, both the absolute conditioning group and differential conditioning group exhibit 1-day memory, which is characterized as protein synthesis-dependent long-term memory (LTM)

(Matsumoto et al. 2003). On the other hand, the unpaired group did not exhibit a significantly different preference for peppermint after training compared to that before training (Fig. 17.3b; $p > 0.05$, WCX test adjusted by the Holm method). Comparisons among groups also showed that the PI after training in the differential conditioning group was significantly greater than that in the absolute conditioning group ($p = 0.00492$, M-W test adjusted by the Holm method). Thus, differential conditioning can induce a stronger, long-lasting conditioning score as compared with absolute conditioning.

17.4 Research on Cricket Learning and Memory Using Olfactory Classical Conditioning

Four-trial absolute conditioning with an ITI of 5 min can be characterized as "spaced training," in contrast to "massed training" in which trials are separated by very short (typically 30 s) intervals. Our study using different numbers of conditioning trials with different ITIs indicate the presence of at least three different memory phases after conditioning (Matsumoto and Mizunami 2002a; Matsumoto et al. 2006). The memory induced by a single-trial conditioning or massed-trial conditioning (30-s ITI) shows significant decay at 4 h after conditioning (Matsumoto et al. 2006; Matsumoto unpublished data). This memory can be dissected into two memory phases, short-term memory (STM) and medium-term memory (MTM), according to sensitivity to amnesic treatments (Matsumoto and Mizunami 2002a). Both memory phases are independent of protein synthesis (Matsumoto et al. 2003, 2006). In contrast, spaced-trial conditioning (5-min ITI) induces a stable, long-lasting memory, which can be maintained for several days (Matsumoto et al. 2003). Experiments with pharmacological or genetic intervention of biochemical cascades showed that different molecular processes underlie MTM and LTM (Matsumoto et al. 2006; Takahashi et al. 2009). LTM formation, for example, requires protein synthesis and signaling via nitric oxide (NO)–cGMP, cyclic nucleotide-gated channels, calcium–calmodulin, CaMKII, and cAMP (Matsumoto et al. 2006, 2009; Mizunami et al. 2014; see Chap. 9).

17.5 Higher-Order Forms of Conditioning in Crickets

As variants of classical conditioning in crickets, we have established aversive conditioning, in which crickets associate an odor with NaCl (or quinine) punishment (Unoki et al. 2005). In addition, we have established conditioning to associate a visual pattern (Unoki et al. 2006; Matsumoto et al. 2013b) or a color stimulus (Nakatani et al. 2009) with water reward or NaCl punishment. Our behavioral–pharmacological studies have suggested that octopaminergic and dopaminergic neurons convey information about appetitive US and aversive US, respectively, in conditioning of odors, visual pattern, or color stimulus (see Chap. 9).

By modifying the classical conditioning procedure, we have developed procedures to study some of the higher-order learning phenomena in crickets, which have been reported in many mammals but rarely in invertebrates. These procedures include second-order conditioning (Mizunami et al. 2009), sensory pre-conditioning (Matsumoto et al. 2013a), blocking (Terao et al. 2015), spontaneous recovery (Matsumoto et al. unpublished data), reversal learning (Matsumoto and Mizunami 2002a, b), and contextual learning (Matsumoto and Mizunami 2004, 2005). Pharmacological analyses of biochemical processes underlying these higher-order learning phenomena are in progress.

In conclusion, the classical conditioning paradigm and its variants in crickets may provide novel breakthroughs in the study of the neural basis of learning, memory and cognition.

References

Avarguès-Weber A, Deisig N, Giurfa M (2011) Visual cognition in social insects. Annu Rev Entomol 56:423–443

Balderrama N (1980) One trial learning in the American cockroach, Periplaneta americana. J Insect Physiol 26:499–504

Bitterman ME, Menzel R, Fietz A, Schafer S (1983) Classical conditioning of proboscis extension in honeybees (Apis mellifera). J Comp Psychol 97:107–119

Daly KC, Smith BH (2000) Associative olfactory learning in the moth Manduca sexta. J Exp Biol 203:2025–2038

Dupuy F, Sandoz JC, Giurfa M, Josens R (2006) Individual olfactory learning in Camponotus ants. Anim Behav 72:1081–1091

Giurfa M (2007) Behavioral and neural analysis of associative learning in the honeybee: a taste from the magic well. J Comp Physiol A 193:801–824

Giurfa M, Sandoz JC (2012) Invertebrate learning and memory: fifty years of olfactory conditioning of the proboscis extension response in honeybees. Learn Mem 19:54–66

Matsumoto Y, Mizunami M (2000) Olfactory learning in the cricket Gryllus bimaculatus. J Exp Biol 203:2581–2588

Matsumoto Y, Mizunami M (2002a) Temporal determinants of olfactory long-term retention in the cricket Gryllus bimaculatus. J Exp Biol 205:1429–1437

Matsumoto Y, Mizunami M (2002b) Lifetime olfactory memory in the cricket Gryllus bimaculatus. J Comp Physiol A 188:295–299

Matsumoto Y, Mizunami M (2004) Context-dependent olfactory learning in an insect. Learn Mem 11:288–293

Matsumoto Y, Mizunami M (2005) Formation of long-term olfactory memory in the cricket Gryllus bimaculatus. Chem Senses 30(suppl 1):i299–i300

Matsumoto Y, Mizunami M (2006) Olfactory memory capacity of the cricket Gryllus bimaculatus. Biol Lett 2:608–610

Matsumoto Y, Noji S, Mizunami M (2003) Time course of protein synthesis-dependent phase of olfactory memory in the cricket Gryllus bimaculatus. Zool Sci 20:409–416

Matsumoto Y, Unoki S, Aonuma H, Mizunami M (2006) Critical role of nitric oxide-cGMP cascade in the formation of cAMP-dependent long-term memory. Learn Mem 13:35–44

Matsumoto Y, Hatano A, Unoki S, Mizunami M (2009) Stimulation of the cAMP system by the nitric oxide-cGMP system underlying the formation of long-term memory in an insect. Neurosci Lett 467:81–85

Matsumoto Y, Hirashima D, Mizunami M (2013a) Analysis and modeling of neural processes underlying sensory pre-conditioning. Neurobiol Learn Mem 101:103–113

Matsumoto Y, Hirashima D, Terao K, Mizunami M (2013b) Roles of NO signaling in long-term memory formation in visual learning in an insect. PLoS One 8(7):e68538

Menzel R (1999) Memory dynamics in the honeybee. J Comp Physiol A 185:323–340

Mizunami M, Yokohari F, Takahata M (2004) Further exploration into the adaptive design of the arthropod "microbrain": I. Sensory and memory-processing systems. Zool Sci 21:1141–1151

Mizunami M, Unoki S, Mori Y, Hirashima D, Hatano A, Matsumoto Y (2009) Roles of octopaminergic and dopaminergic neurons in appetitive and aversive memory recall in an insect. BMC Biol 7:46

Mizunami M, Nemoto Y, Terao K, Hamanaka Y, Matsumoto Y (2014) Roles of calcium/calmodulin-dependent kinase II in long-term memory formation in crickets. PLoS One 9(9):e107442

Nakatani Y, Matsumoto Y, Mori Y, Hirashima D, Nishino H, Arikawa K, Mizunami M (2009) Why the carrot is more effective than the stick: different dynamics of punishment memory and reward memory and its possible biological basis. Neurobiol Learn Mem 92:370–380

Sandoz JC (2011) Behavioral and neurophysiological study of olfactory perception and learning in honeybees. Front Syst Neurosci 5:98. doi:10.3389/fnsys.2011.00098

Takahashi T, Hamada A, Miyawaki K, Matsumoto Y, Mito T, Noji S, Mizunami M (2009) Systemic RNA interference for the study of learning and memory in an insect. J Neurosci Methods 179:9–15

Terao K, Matsumoto Y, Mizunami M (2015) Conserved computational mechanisms underlying associative learning in insects and mammals. Sci Rep 5:8929

Tully T, Quinn WG (1985) Classical conditioning and retention in normal and mutant Drosophila melanogaster. J Comp Physiol A 157:263–277

Unoki S, Matsumoto Y, Mizunami M (2005) Participation of octopaminergic reward system and dopaminergic punishment system in insect olfactory learning revealed by pharmacological study. Eur J Neurosci 22:1409–1416

Unoki S, Matsumoto Y, Mizunami M (2006) Roles of octopaminergic and dopaminergic neurons in mediating reward and punishment signals in insect visual learning. Eur J Neurosci 24:2031–2038

Chapter 18
Optical Recording Methods: How to Measure Neural Activities with Calcium Imaging

Hiroto Ogawa and John P. Miller

Abstract Optical recording that provides both anatomical and physiological data has become an essential research technique for neuroscience studies. In particular, Ca^{2+} imaging is one of the most popular and useful methods for monitoring local activity at subcellular regions of single neurons and/or visualization of spatiotemporal dynamics of neuronal population activity. In neuroethological studies on the cricket, the Ca^{2+} imaging is also a powerful method for optical recording of neural activity and has yielded important information on neural mechanisms in the cricket. In this chapter, we summarize some important features for the application of Ca^{2+} imaging in the cricket nervous system. This includes the selection of an appropriate Ca^{2+} indicator and the dye loading protocols, experimental designs, and optical system configurations that are required to enable the effective use of the Ca^{2+} imaging techniques in the cricket. As an example application, we focused on Ca^{2+} imaging experiments in the cricket cercal sensory system in vivo.

Keywords Ca^{2+} imaging • Fluorescent Ca^{2+}-sensitive dye • Imaging device • In vivo imaging • Cercal sensory system • Giant interneurons

18.1 Introduction

Crickets have been used as a classical preparation for neuroethological studies. Part of the reason for their perceived value is that a wide range of powerful physiological techniques are easily applicable to the cricket. For example, electrophysiological recordings of neuronal activity have provided significant insights of neural mechanisms that underlie phonotaxis, air-current-elicited behavior, and learning and memory, as described on other chapters in this book. In particular, intracellular

H. Ogawa (✉)
Department of Biological Sciences, Faculty of Science, Hokkaido University, Sapporo 060-0810, Japan
e-mail: hogawa@sci.hokudai.ac.jp

J.P. Miller
Department of Cell Biology and Neuroscience, Montana State University, Bozeman MT 59717, USA

H.W. Horch et al. (eds.), *The Cricket as a Model Organism*,
DOI 10.1007/978-4-431-56478-2_18

recording and dye loading with sharp glass microelectrodes are fundamentally important techniques for identification and analysis of single neurons. This method elucidates details of membrane potential changes with high temporal resolution and provides information about membrane properties and synaptic connections. Extracellular recording using a multielectrode array is also an effective method for simultaneous monitoring of multiple neurons but is unable to provide anatomical information. In general, electrophysiological recording methods are not very effective for measuring neural activity of large neural ensembles simultaneously, nor local activity at multiple regions within a single cell. In contrast, optical recording techniques that have been developed since the latter part of the 1980s are powerful tools for detecting spatiotemporal dynamics of neural activity patterns ranging from whole brains to dendritic spines of a single neuron.

Optical recording techniques have been adapted to the cricket nervous system with great success and effectiveness. The earliest imaging study on the cricket was a Ca^{2+} imaging analysis of the "omega neuron": an auditory interneuron identified within the prothoracic ganglion (Sobel and Tank 1994). This study indicated that dendritic calcium accumulation induced by detection by the male song reduced the excitability of the omega neuron, via modulation of Ca^{2+}-activated K^+ channels. Ca^{2+} imaging methods have also been used in other interneurons involved in acoustic behavior within the prothoracic ganglion (Baden and Hedwig 2007, 2009). However, it is the cercal mechanosensory system of the cricket that has been the primary focus for Ca^{2+} imaging studies in the cricket nervous system. Details of the cercal sensory system are presented in Chap. 14. In one set of Ca^{2+} imaging experiments that focused on air-current-sensitive giant projecting interneurons (GIs), cytosolic Ca^{2+} concentration was shown to be elevated in dendrites by action potential generation (Ogawa et al. 2000). The presence of voltage-dependent Ca^{2+} channels on dendrites of the GIs had been suggested by whole-cell voltage-clamp experiments in cultured cells (Kloppenburg and Hörner 1998), and this prediction was verified using Ca^{2+} imaging methods (Ogawa et al. 2000). Subsequently, dendritic excitability, membrane potential dynamics, and heterogeneous distribution of synaptic inputs were demonstrated in cricket cercal GIs (Ogawa et al. 2002a, b). Furthermore, it was demonstrated that dendritic Ca^{2+} accumulation induces short-term synaptic plasticity (Ogawa et al. 2001). In vivo imaging of dendritic Ca^{2+} responses to air-current stimuli revealed synaptic integration mechanisms for the directional tuning properties of GIs (Ogawa et al. 1999, 2004, 2006, 2008). In this chapter, we will introduce the fundamental Ca^{2+} imaging techniques and illustrate some specific applications to the cricket cercal sensory system.

18.2 Principles of Calcium Imaging

18.2.1 Fluorescent Ca^{2+} Indicators

Organic Ca^{2+} indicators were developed in the 1980s by R.Y. Tsien and his colleagues (Tsien 1980). All of Tsien's fluorescent Ca^{2+} indicators are based on a fluorophore combined with BAPTA, which retains high and stable selectivity for Ca^{2+} in the neutral to weak-alkaline pH range and has a fast time constant for Ca^{2+} binding. The Ca^{2+} indicators alter their spectral properties depending on the change in cytosolic Ca^{2+} concentration ($[Ca^{2+}]_i$) associated with membrane depolarization. In general, the Ca^{2+}-sensitive fluorescent dye is loaded into nerve cells using any of a variety of staining methods, and changes in the specific fluorescent intensity of the dye are measured with an imaging device such as a camera or photomultiplier mounted on a microscope.

The primary determining factor for selection of an appropriate fluorescent probe for a particular experimental situation is its affinity for Ca^{2+}, which is indicated by its dissociation constant (K_d). A wide variety of organic Ca^{2+} probes are now available having sensitivities to $[Ca^{2+}]_i$ which range from <50 nM to >50 μM in K_d values. In a typical nerve cell, local $[Ca^{2+}]_i$ can increase by a factor of between 10 and 100 in response to action potential generation. In general, the indicators for which the K_d value is about 200 nM are used for in vivo Ca^{2+} imaging to monitor the neural activity. We used fluorescent Ca^{2+} indicators, Calcium Green 1 with 190 nM in K_d, or Oregon Green BAPTA-1 with 170 nM in K_d for detecting fast Ca^{2+} influx via voltage-dependent Ca^{2+} channels activated by action potential in the cricket GIs (Ogawa et al. 2000).

Another important point to consider in choosing an appropriate Ca^{2+} indicator is the spectral characteristics of the fluorescent dye which will bind to the Ca^{2+}. The fluorescent indicators are broadly classified into singlemetric and ratiometric dyes. The fluorescence intensity of a singlemetric dye is changed depending on $[Ca^{2+}]_i$ without a shift in its emission spectrum. Typical singlemetric Ca^{2+} indicators, including fluo-3, fluo-4, Calcium Green, and Oregon Green 488 BAPTA, show a relative increase in the emission fluorescence with an increase in $[Ca^{2+}]_i$. By monitoring changes in the total amount of light emitted from the sample within the appropriate wavelength range, the experimentor obtains a direct measurement of changes in $[Ca^{2+}]_i$. Since most of the singlemetric fluorescent dyes are excited by visible light, Ca^{2+} imaging with these dyes requires a relatively simple optical system and can be achieved with a standard confocal microscope. Further, monitoring the fluorescence at only a single wavelength enables high-speed imaging of neural activity. However, singlemetric imaging does not allow an effective cancelation of some very problematic artifactual variations in the fluorescence signals resulting from three factors: photobleaching of the dye, fluctuation of the excitation light intensity, and movement of the sample.

Techniques using ratiometric dyes can circumvent these technical limitations associated with singlemetric dyes. Ratiometric dyes do not simply change their fluorescence intensity across the entire absorption spectrum in response to changes in $[Ca^{2+}]_i$. Rather, a ratiometric dye's response to a change in $[Ca^{2+}]_i$ is a shift in its absorption spectrum. These more complicated response characteristics enable an internally calibrated measurement of $[Ca^{2+}]_i$ that circumvents the problems for singlemetric dyes. To achieve this independence from the artifacts listed above, the experimentor must record a shift in the absorption spectrum of the indicator and thus must record two independent measurements of the indicator's fluorescence for each determination of $[Ca^{2+}]_i$. Ratiometric dyes are further classified into two subtypes according to characteristic of their spectral shift. The first is the "two-excitation/one-emission" subtype, exemplified by fura-2. The second is the "one-excitation/two-emission" subtype, exemplified by indo-1. Details of their Ca^{2+}-dependent spectral shifts can be found in other reviews and textbooks (Grienberger and Konnerth 2012; Ogawa and Miller 2013). Most synthetic organic probes for ratiometric imaging are excited with UV light, which can be damaging to cells and requires a specialized laser for confocal imaging systems. Unfortunately, there are no available ratiometric Ca^{2+} sensors that have absorption within the visible range. However, it is possible to achieve effective ratiometric measurements in the visible light range using two singlemetric probes like fluo-3 (or fluo-4) and Fura Red (Lipp and Niggli 1993; Speier et al. 2008) in the same cell(s).

18.2.2 Imaging Devices and Optical Setups

Various types of fluorescent microscopes and imaging devices have been developed and are available for imaging experiments. Typically, a high-sensitivity camera mounted on a conventional fluorescent microscope is used for Ca^{2+} imaging of neural activity, because high temporal resolution is required for monitoring fast Ca^{2+} rise times corresponding to neural activity. Current implementations typically use an electron-multiplying CCD (EM-CCD) or a complementary metal-oxide-semiconductor (CMOS) sensor. These high-sensitivity cameras are capable of acquiring images at rates of up to 1,000 frames/second (fps). However, the sampling rates that are achievable depend on the staining and preparation of the tissue as well as on the specific optical configuration and imaging device characteristics. Using a digitally cooled CCD camera (ORCA-ER, Hamamatsu Photonics), rapid Ca^{2+} transients induced by single action potentials in cricket cercal giant interneurons were clearly detected with 256 × 256 pixel resolution acquired at 30 fps in our latest experiments.

All of our Ca^{2+} imaging experiments on the cricket cercal sensory system have been performed using an inverted fluorescent microscope configuration, for two reasons. First, it is essential to eliminate turbulence from the airflow stimuli applied to the cerci, which are a pair of air-current-sensitive organs in the cricket. If an upright microscope is used, the objective lens adjacent to the terminal abdominal ganglion (TAG) disturbs laminar airflow around the sensory organs. Second, simultaneous recording of optical signals and membrane potential of the GIs requires the use of electrophysiological recording electrodes as well as high-numerical-aperture microscope objectives. We need to insert a glass microelectrode into the TAG for intracellular recording and loading of Ca^{2+} dye, in addition to additional electrodes for electrical stimulation of the cercal afferent nerves and extracellular recording of ascending spikes from ventral nerve cord. The inverted microscope configuration allows total freedom of access from the top of the preparation for configuration of the necessary electrodes. Figure 18.1 shows a

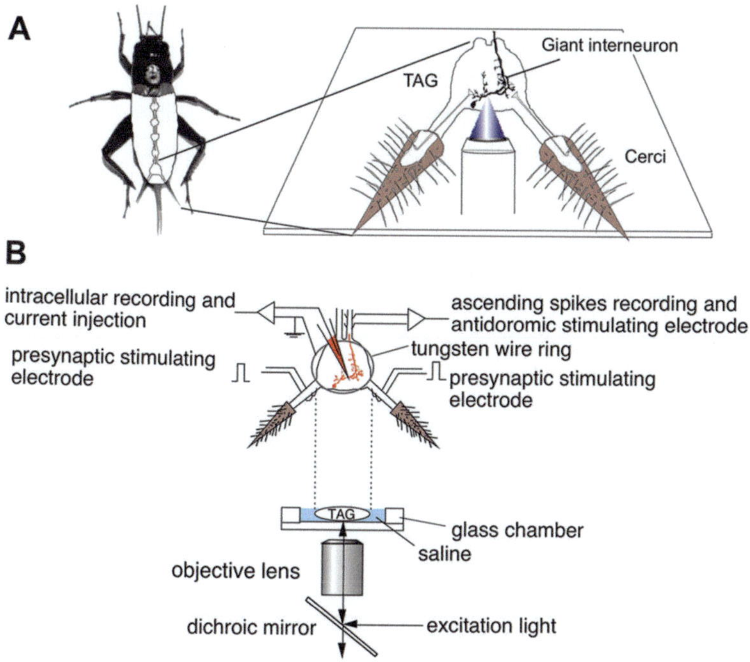

Fig. 18.1 Schematic representation of the experimental setup. (**a**) Preparation of isolated TAG, cercal nerves, and cerci in a glass chamber. (**b**) Arrangement of electrodes for intra-/extracellular recordings and for electrical stimulation (*upper*) and optical components

schematic diagram of the setup of the imaging and electrophysiological recording system for recording of Ca^{2+} responses to air-current stimulus in the cricket cercal sensory system.

For Ca^{2+} imaging experiments in the cricket cercal sensory system, we dissect the TAG away from the abdomen, along with the pair of air-current-sensitive organs called cerci that send sensory input into the ganglion. The preparation is mounted in a glass chamber made of a 0.25-mm thick cover glass (Ogawa et al. 2006, 2008). This recording chamber is mounted on the stage of an inverted microscope configured for fluorescent imaging. We can observe clear images of dye-loaded neurons in this preparation, maintaining the neural circuits and the sensory organs in conditions that are similar to their natural in vivo configuration. By removing the standard condenser tube and transmitted light source from the microscope, which normally obstruct access to the preparation from above, this setup provides excellent accessibility of the preparation to the electrodes and perfusion apparatus and has the advantage that the objective lens (placed below the stage) does not disturb the normal airflow stimulus across the sensory organs. Of course, the use of a different nervous system preparation and/or experimental arrangement, such as an isolated ganglion or in vivo preparation, might dictate the use of a different type of microscope. For in vivo Ca^{2+} imaging of brain ganglia, we used a conventional upright microscope with a water-immersion objective lens placed above the dorsal surface of the ganglia (Ogawa and Miller 2013; Ogawa and Kajita 2015).

Light scattering is one of the most serious problems that arise for in vivo brain imaging. The scattering problem is greatly reduced through the use of confocal laser scanning microscopes (CLSMs), which are currently the most popular imaging systems used for high-resolution Ca^{2+} imaging. In the cricket TAG, CLSM achieved sharper fluorescent imaging of the GIs than conventional fluorescent microscopy and enabled very effective visualization of Ca^{2+} signals induced by intracellularly recorded spike bursting without scattered fluorescence from dye-loaded glass microelectrodes (Fig. 18.2). However, CLSM yields relatively lower temporal resolution than conventional camera-based devices for capturing images of large fields, because the operation of scanning the laser over an entire field of view is more time-consuming than "snapping" an image of the same field with a conventional camera-based device. Therefore, for high-speed monitoring of Ca^{2+} signals in the nerve cells, the CLSM is usually operated in "line-scanning mode," which is typically used for measurement of Ca^{2+} transients evoked by single action potentials or EPSPs. In this mode, however, it is impossible to record the two-dimensional distribution of Ca^{2+} signals. Nipkow-type spinning-disk microscopes substantially improve the time resolution for confocal imaging over conventional scanning laser configurations. This microscope configuration provides a confocal effect by spinning a disk with many microlens-embedded pinholes in the excitation light path. The tissue sample is then scanned simultaneously with many virtual excitation beams, effectively reducing the scan time that would be required for a single scanned laser beam by a factor roughly equivalent to the number of pinholes in the disk. Recently, combining the EM-CCD camera as the image acquisition device with Nipkow-disk microscopy, we recorded Ca^{2+} responses to air-current stimulation within the cricket brain ganglion (Ogawa and Kajita 2015). The Nipkow-disk microscopy and two-photon microscopy (TPM) that are capable of observing neuronal activity located at deep regions in whole brain preparations (Helmchen and Denk 2005; Nemoto 2008; Svoboda and Yasuda 2006; Takahara et al. 2011) promise to provide remarkable and important findings on neural mechanisms underlying the operation of the cricket nervous system.

18.3 Application of Calcium Imaging to Measure Neural Activity

In the final section of this chapter, we summarize three kinds of imaging protocols that we have used for analysis of the neural circuit underlying information processing and integration in the cricket cercal system: (1) high-speed imaging of dendritic Ca^{2+} responses to air currents in single GIs, (2) visualization of ensemble activity patterns of cercal afferent terminals, and (3) simultaneous monitoring of

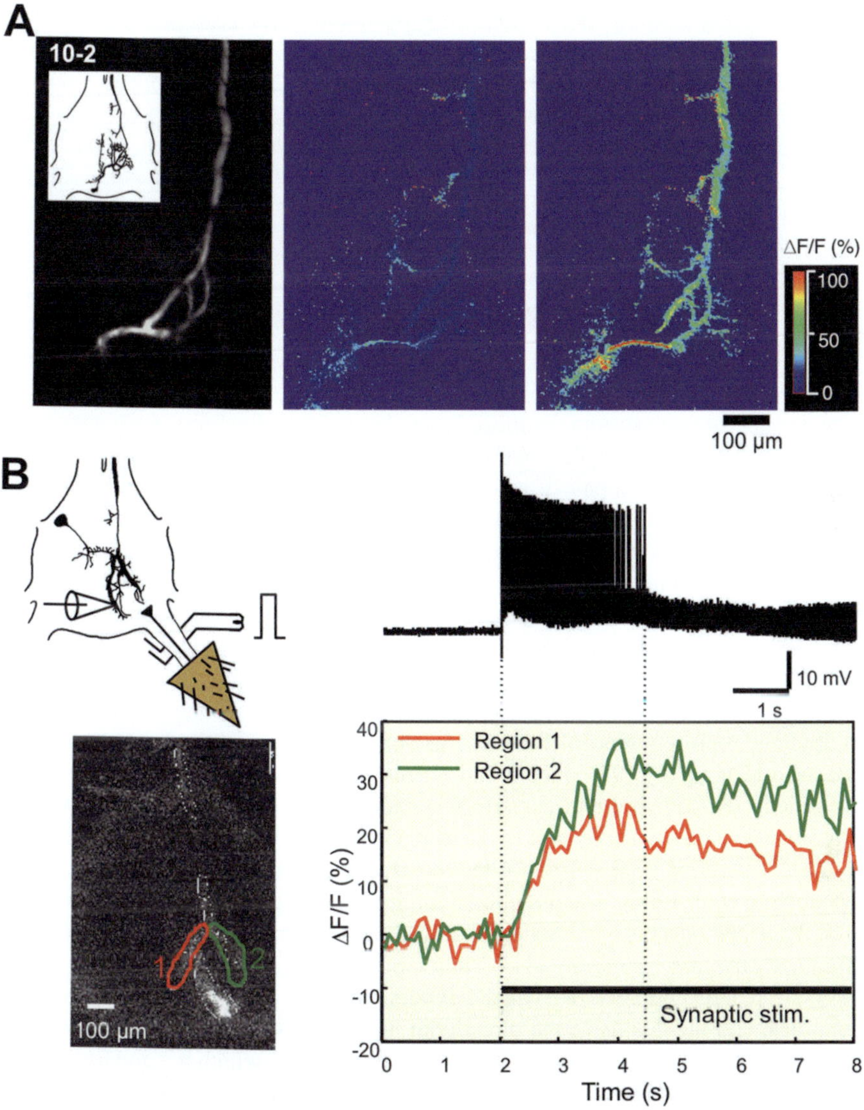

Fig. 18.2 Ca^{2+} responses of a GI to electrical stimulation of the cercal nerve. (**a**) Confocal image of the GI10-2 in dorsal view (*left image*) and pseudo-color images indicating the changes in relative fluorescence ($\Delta F/F$) (given in % values) of Calcium Green 1 in the 10-2, after 5.9 s (middle) and 29.5 s (*right*). (**b**) Simultaneous recording of membrane potentials and dendritic Ca^{2+} changes in the GI8-1. *Lower right* shows confocal images indicating the recording regions for the time-course measurement of $[Ca^{2+}]_i$ change. *Upper trace* indicates intracellular recording of action potentials, and lower (*green* and *red*) traces show time course of changes in $\Delta F/F$ in the two dendritic regions in responses to tetanic stimulation to the contralateral cercal nerve at 20 Hz (Modified from Ogawa et al. 2000)

pre- and postsynaptic activities. Our imaging studies have been performed in two species of crickets: *Acheta domesticus* and *Gryllus bimaculatus*, but the same procedures were basically applicable to both species. The microscopes, imaging devices, and software packages described here are partly outdated because they were based on our original studies reported previously (Ogawa et al. 2004, 2006, 2008).

18.3.1 High-Speed Imaging of Dendritic Ca^{2+} Responses to Air Currents

The cercal sensory system in the cricket detects the direction, frequency, and velocity of air currents with great accuracy and precision. The receptor organs of this system consist of a pair of antenna-like appendages called cerci at the rear of cricket abdomen. In *A. domesticus*, each cercus is approximately 1 cm long in a normal adult cricket and is covered with 500–750 filiform mechanosensory hairs, each of which is innervated by a single mechanoreceptor neuron (Palka et al. 1977; Miller et al. 2011). The receptor neuron is tuned to air currents from a particular direction and exhibits a change in its firing rate in response to stimuli over the entire 360-degree range of stimulus directions (Landolfa and Miller 1995). The sensory afferents of filiform sensilla project in an orderly array into the TAG to form a continuous representation of the direction of air currents in the horizontal plane (Bacon and Murphey 1984; Jacobs and Theunissen 1996, 2000; Paydar et al. 1999). Identified GIs receive direct excitatory synaptic inputs from the mechanosensory afferents. The GIs are activated by air currents and also display differential sensitivity to variations in air-current direction (Jacobs et al. 1986; Miller et al. 1991; Theunissen et al. 1996). The directional tuning curves of several of the GIs are also well described by a cosine function, that is, these GIs encode information about the stimulus direction proportional to their spiking activity. Therefore, these GIs integrate a large number of synaptic inputs from the sensory afferents on their dendrites, and the details of the directional tuning properties of any specific GI is determined largely by the subset of sensory afferents with which it is innervated. To determine the directional sensitivity in different dendritic regions of an individual GI, as well as the overall tuning property of that GI's spike response to air-current stimuli, we applied a combination of optical and intracellular recording methods to the single GIs in the cricket. Using the following procedures and imaging apparatus, we successfully recorded dendritic Ca^{2+} responses to air-current stimuli.

An air-current-evoked burst of action potentials in a GI, along with the corresponding rapid Ca^{2+} increase in two different dendritic regions of that GI, is shown in Fig. 18.3. Analysis of the dependence of Ca^{2+} influx on stimulus direction, as well as the directional selectivity of the cell as judged by the electrical signals, revealed that the preferred stimulus direction for dendritic Ca^{2+} responses corresponded to that for the overall spike response. However, the precise shapes of the directional tuning curves based on the Ca^{2+} responses for each dendritic branch were different from the overall tuning curve based on spike counts for that cell. Moreover, different dendritic branches displayed distinct directional sensitivity profiles to the air-current stimuli. These results suggest that postsynaptic activities will influence the local Ca^{2+} signals in the distal dendrites and produce differences in directional sensitivities of the different regional dendritic Ca^{2+} responses (Ogawa et al. 2004).

18.3.1.1 Preparation and Dye Loading

After removing the head, wings, and legs, an incision was made along the dorsal midline of the abdomen. The gut, internal reproductive organs, and surrounding fat were removed to expose the TAG. All peripheral nerves of the TAG were severed except for the cercal nerves. To facilitate penetration of the ganglionic sheath of the TAG, a piece of filter paper soaked in 10 % protease (Sigma, type XIV) was applied to either the dorsal or ventral side of the TAG for approximately 30 s. After the TAG was washed with cricket saline, the preparation, consisting of the sixth and terminal abdominal ganglia, abdominal connectives, cercal nerves, and cerci, was removed from the body and mounted in a glass chamber. The chamber was held on the stage of an inverted microscope (Axiovert 100, Carl Zeiss, Oberkochen, Germany). The TAG was held down within a tungsten wire ring with a diameter of 700–800 μm, which also served as the reference electrode. Fluorescent Ca^{2+} indicator (2-mM Oregon Green 488 BAPTA-1 potassium salt; OGB-1, Thermo Fisher Scientific, Waltham, MA) dissolved in 150-mM potassium acetate was iontophoretically loaded into the GI for 5 min through the glass microelectrode, using a negative current of 3 nA.

18.3.1.2 Optical Recording

An intensified CCD camera (C2400-87, Hamamatsu Photonics, Hamamatsu, Japan) attached to the inverted microscope was used to acquire the fluorescent images. Fluorescence was viewed through a 10×, 0.3 NA dry objective (Plan-Neofluar; Zeiss). For excitation of the Ca^{2+} indicator, we used a Xenon arc lamp (XBO 75 w; Zeiss) illumination with a stabilized power supply and the following filter combination: excitation, 480/20; dichroic, FT510; and emission, BP515–565. For image acquisition and analysis of fluorescence intensity of OGB-1, a software package (ARGUS-50/Ca, Hamamatsu Photonics) was used. For time-course displays of the fluorescence changes in several regions of the dendritic tree, several polygonal recording regions on a fluorescent image of each GI were selected, and the mean values of the fluorescent intensities in these regions were plotted as a function of time. A series of fluorescent images were acquired at 30 fps on 512 × 512 pixel image. Fluorescence intensities were collected from each fluorescent image, background corrected, and averaged for each selected polygonal region. $[Ca^{2+}]_i$ changes are expressed as $\Delta F/F[\Delta F/F = (F - F_0)/F_0]$, where F_0 is the background-corrected pre-stimulus fluorescent intensity of the Ca^{2+} indicator.

18.3.1.3 Intracellular Recording and Air-Current Stimulation

Intracellular recordings of the membrane potential of the GI were performed with a glass microelectrode (30–50 MΩ) used for loading of the Ca^{2+} indicator. The electrode was inserted into the GI in one of the large neurites. Action potentials of GIs were also monitored with a pair of hook electrodes positioned under the abdominal nerve cords (Fig. 18.1b). For electrical stimulation of the sensory afferents, a train of pulses (pulse duration: 100 μs) was applied to the left or right cercal nerve with two pairs of hook electrodes. All electrophysiological signals were digitized at 20 kHz through an A–D converter (Powerlab 4s, ADInstruments, Castle Hill, Australia) and analyzed with a Macintosh personal computer using data acquisition software (Chart Ver.4.2, ADInstruments).

Air-current stimuli were provided by a short puff of N_2 gas from a plastic nozzle with a diameter of 13 mm. The pressure and duration of the air puff were controlled at 20 psi and 200 ms by a pneumatic picopump (PV230, World Precision Instruments, Sarasota, FL) connected to a N_2 gas cylinder. Eight nozzles were arranged around the cerci on the same horizontal plane. The nozzle ends were positioned at 45° angle between each other at a distance of 20 mm from the center of TAG (Fig. 18.3a). For synchronization of the stimulator, electrophysiological recording apparatus, and all other peripheral equipment associated with real-time image capture, a TTL signal exported from the imaging computer was used for triggering the stimulation and electrophysiological recordings.

18.3.2 *Visualization of Activity Patterns of Cercal Afferent Terminals*

The synaptic terminals from mechanosensory afferents having similar peak directional sensitivities arborize in adjacent areas within the TAG, and the spatial segregation between afferent arbors increases as the difference in their directional tuning increases. A representation of the stimulus direction as a three-dimensional neural map of the entire filiform afferent array has been developed based on anatomical and physiological measurements taken from a large sample of individual afferents (Troyer et al. 1994; Jacobs and Theunissen 1996; Paydar et al. 1999). By combining the predicted responses of each class of afferents with this information about the spatial location of their terminal arborizations within the neural map, predictions have been made of the spatial patterns of synaptic activation that would result from sustained, unidirectional air currents (Jacobs and Theunissen 2000). To test these predictions, we tried direct visualization of the spatial patterns of

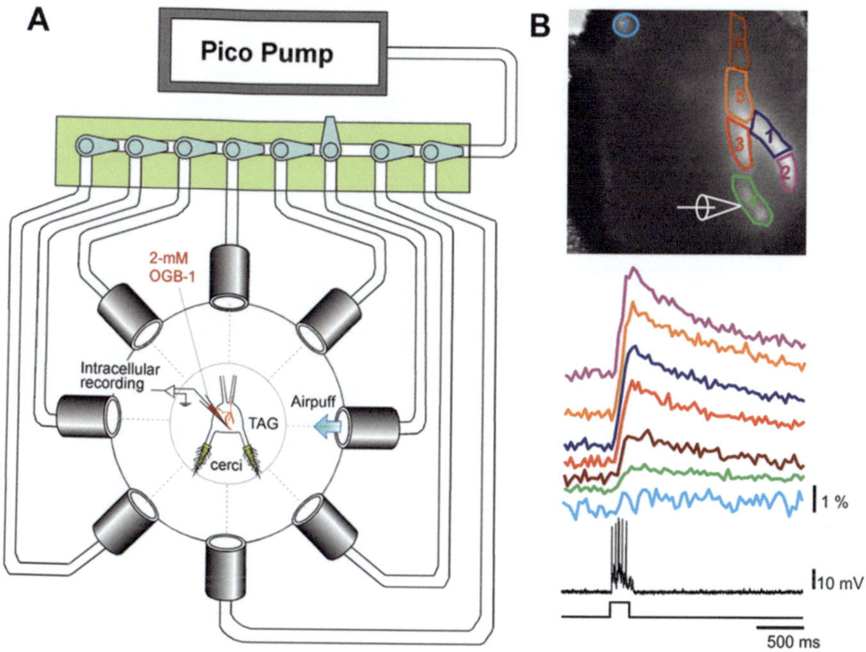

Fig. 18.3 In vivo imaging of Ca^{2+} response to air-puff stimuli on the GI. (**a**) Schematic representation of the experimental setup for in vivo Ca^{2+} imaging and air-current stimulation. (**b**) Typical responses to the single air-puff stimulus (*lowest black trace*) in membrane potential (*middle black trace*) and $[Ca^{2+}]_i$ (*upper colored traces*) in GI8-1. The air-puff stimulus was directed at 90° clockwise from the anterior of the preparation. The top fluorescent image of GI8-1 indicates the seven recording regions for time-course measurements of the changes in $\Delta F/F$ and also indicates the intracellular recording site. The recording regions corresponded to cellular components such as dendritic branches (region 4), a putative initial segment (region 5), and a soma (region 7)

air-current-evoked ensemble activity patterns of the cercal afferents using Ca^{2+} imaging techniques. In this experiment, all axon terminals of mechanosensory afferents were stained with Ca^{2+} indicator. Unidirectional air currents were applied repeatedly from eight different directions, and the optically recorded responses from each direction were averaged. The dispersion of the optical signals by the ganglion limited the spatial resolution with which these ensemble afferent activity patterns could be observed. However, resolution was adequate to demonstrate that different directional stimuli induced different spatial patterns of Ca^{2+} elevation in the terminal arbors of afferents within the TAG (Fig. 18.4). These coarsely resolved optically recorded patterns were consistent with the anatomy-based predictions (Ogawa et al. 2006).

18.3.2.1 Preparation and Dye Loading

To expose the TAG and cercal nerves, the cricket was dissected in the same procedure described in Sect. 3.1.1. A solution containing acetoxy-methyl (AM) ester

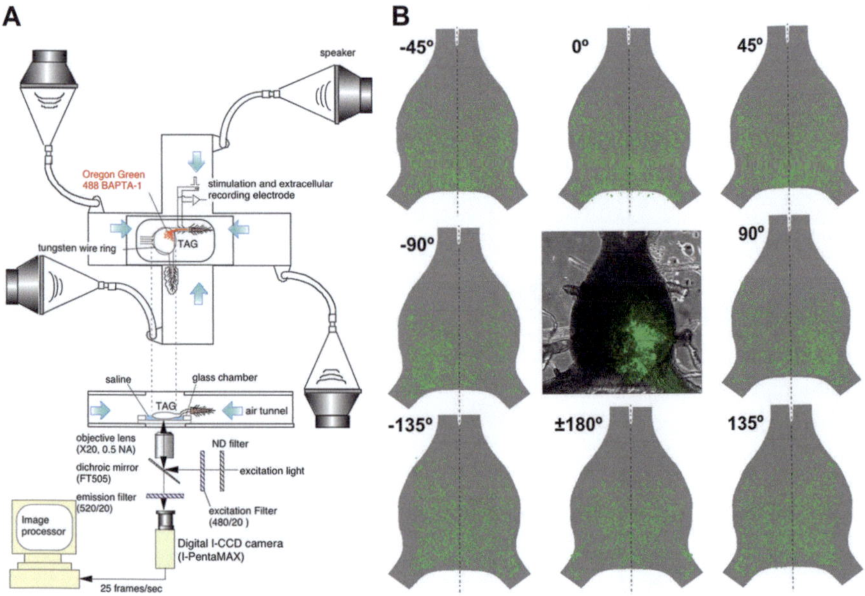

Fig. 18.4 Activity pattern map visualized using Ca^{2+} imaging of sensory afferent terminals. (**a**) Diagram of the experimental setup for air-current stimulation, electrophysiological recording, and Ca^{2+} imaging of sensory afferents in the cricket TAG. (**b**) Map of averaged afferent activity patterns elicited by eight different sets of directional air-current stimuli to one cercus. The center *inset* shows an image of TAG, produced by superimposing two views obtained with different illumination modes: a fluorescence image (obtained in confocal mode) of OGB-1 loaded into the cercal sensory afferents and a transmitted light image (Modified from Ogawa et al. 2006)

of the fluorescent Ca^{2+} indicator (OGB-1AM, Thermo Fisher Scientific) at a concentration of 0.05 % and a dispersing reagent (Pluronic F-127, Thermo Fisher Scientific) at a concentration of 1 % was pressure injected through a glass micropipette into the sheath of a cercal sensory nerve. The dye solution also contained 0.5 % tetramethylrhodamine (10,000-MW dextran-conjugated, Thermo Fisher Scientific) to enable visualization of the diffusion of the injected solution through the cercal nerve. Twelve hours after the dye injection, the axon terminals of cercal sensory neurons were stained with the Ca^{2+} indicator (Center image of Fig. 18.4b). The entire abdominal nerve cord was then dissected away from the abdomen, along with both cerci and cercal sensory nerves, and mounted on a 0.25-mm thick cover glass.

18.3.2.2 Air-Current Stimulation

Mechanosensory hairs were stimulated with a device that allowed low-velocity laminar air currents to be passed across the animal's body. The device consisted of a cross-shaped chamber milled into a 10-mm thick acrylic plastic block and sealed on the bottom surface with a 3-mm thick acrylic plate (Fig. 18.4a). The bottom plate had a hole 13 mm in diameter near the center of the chamber. Separate 15-cm diameter loud speakers were connected through Teflon tubes to the ends of each of the four arms of the chamber, so that movement of the speakers would drive air back and forth through the center of the chamber. Each pair of speakers at opposite ends of a single arm of the crossed channels was driven by identical (but inverted) voltage waveforms, to generate a unified "push-pull" displacement of the air along that chamber axis. By controlling the relative amplitude of the signals driving the speakers along each axis, air currents of any specific direction could be generated at the central crossing point of the chamber arms. The entire stimulus chamber was mounted on the stage of an inverted microscope. During experiments, the cover glass-mounted preparation was placed into the channel so that the ganglion was positioned directly over the hole in the lower plate at the point where the chamber arms crossed, allowing direct visualization from below with the objective of the inverted microscope. The top of the tunnel was then closed with an acrylic lid, which also contained a small hole directly above the prep to allow electrode access.

18.3.2.3 Optical Recording and Image Analysis

A digital intensified CCD camera (I-PentaMAX, Princeton Instruments, Trenton, NJ) was attached to the inverted microscope (Eclipse TE300, Nikon, Tokyo, Japan). Fluorescence was viewed through a 20×, 0.5 NA water-immersion objective (HCX Apo, Leica, Wetzlar, Germany). For excitation of the OGB-1, we used illumination from a 150 W xenon arc lamp (XBO 150 W/CR OFR, Osram, Regensburg, Germany) with a stabilized power supply (Model 1600, Opti-Quip, Highland Mills, NY) and the following filter combination: excitation 480/20, dichroic FT505, and emission 520/20 (Nikon). In order to avoid photodamage, the excitation light intensity was decreased to 3.1, 12.5, or 25 % through ND filters (Nikon). Tissue exposure times were limited by use of a software-controlled electromagnetic shutter. We checked the activity of afferents by extracellular recording during every experiment. Image capture and analysis software was used to acquire fluorescent images of the TAG and to measure fluorescent intensities of the afferents (MetaFluor, Molecular Devices ,Sunnyvale, CA). Fluorescent intensities were collected from each fluorescent image and corrected for the CCD's dark background noise by subtracting the image recorded in the absence of illumination. Changes in $[Ca^{2+}]_i$ were expressed as $\Delta F/F(\Delta F/F = (F - F_0)/F_0)$, where F_0 was the fluorescent image obtained in the absence of afferent stimulation.

For spatial identification of the activation area of afferents in response to air-current stimuli, we applied a total of 100 air-current stimuli: 50 from one direction and 50 from the opposite direction (switching direction between each sequential stimulus.) The TTL output signals from the stimulus waveform generator were used to synchronize the camera shutter, and one image was acquired for each of these 100 stimuli. Each exposure began 10 ms after stimulus onset and lasted for a duration of 80 ms. We then averaged all 50 fluorescent images for each of the opposing stimulus directions. In these averaged images, the values of $\Delta F/F$ in the nonstained regions of the ganglion were more variable than in the region containing the dye-loaded afferent arborizations, and the maximum intensity of the autofluorescence in the nonstained regions was lower than 25 % of the fluorescence intensity of the dye-loaded fibers. To eliminate this background fluorescence noise from the nonstained regions, we masked the regions that had less than 25 % of the maximum intensity of the initial fluorescence image. A positive fluorescence change in any region (>0 % in $\Delta F/F$) was assumed to indicate an elevation of Ca^{2+} in that region. For all image processing such as filtering, masking, and pseudo-color display, a commercial software package (Adobe Photoshop, Adobe Systems, San Jose, CA) was used.

18.3.3 Simultaneous Monitoring of Pre- and Postsynaptic Activities

The predicted neural map representing the stimulus direction leads us to the hypothesis that the directional sensitivities of the ascending interneurons, including the GIs, are primarily based on the relative positions of their dendrites within the mechanosensory afferent map (Jacobs and Theunissen 2000). To characterize the "decoding algorithm" of the directional information from the population activities of sensory afferents to individual GIs, we measured the pre- and postsynaptic local Ca^{2+} responses to the air-current stimuli, on each different GI dendrite. For this experiment, we combined single-neuron staining of a postsynaptic neuron (using microelectrode injection of one Ca^{2+} indicator) and bulk staining of presynaptic fibers with the AM ester of a second Ca^{2+} indicator. We simultaneously monitored the optical signals from both indicators in different wavelengths using a dual-view optical system. This method enabled us to measure the pre- and postsynaptic activity on specific dendrites of the identified interneuron. Using this method, we simultaneously measured the Ca^{2+} responses to air-current stimuli in the GI's dendrites and in the afferent axon terminals that make synaptic connections onto that dendritic branch (Fig. 18.5). Analysis of directional tuning properties in amplitudes of presynaptic and postsynaptic Ca^{2+} responses revealed that the individual dendrite with a distinct tuning property in its Ca^{2+} response receives synaptic inputs from different subsets of the afferents having different sensitivities to direction and that the overall directional tuning and dynamical sensitivity of each individual GI also depends on the spatial distribution of synapses onto the complex geometry of the dendritic arbors (Ogawa et al. 2008).

18.3.3.1 Preparation and Dye Loading

Two kinds of Ca^{2+} indicators with different fluorescent wavelengths were loaded to pre- and postsynaptic neurons, respectively. The sensory afferent fibers were

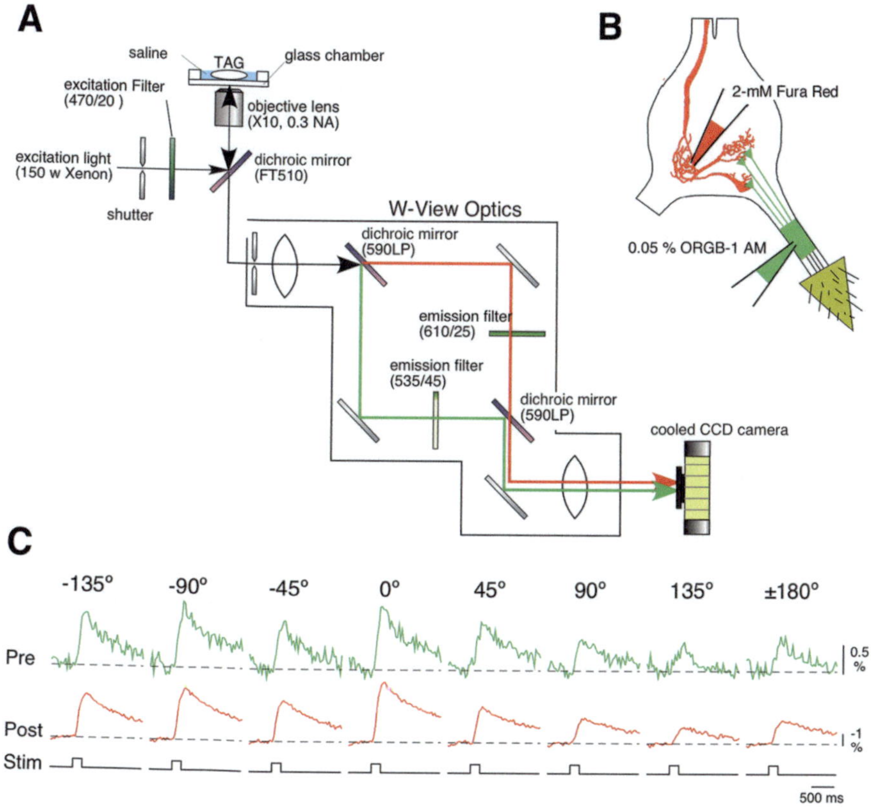

Fig. 18.5 Pre- and postsynaptic Ca^{2+} imaging of the GIs. (**a**) Diagrams of the optical splitting system for simultaneous monitoring of two fluorescent wavelengths of OGB-1 and Fura Red. A fluorescent image was divided into two images by *W-View* optics with a set of dichroic mirrors and emission filters. Both images were acquired in the same frame side by side with a cooled CCD camera at the same time. (**b**) Diagrams showing the methods for selective loading of the different Ca^{2+} indicators. (**c**) The time courses of changes in $\Delta F/F$ at 535-nm wavelength (presynaptic) Ca^{2+} signals (*green traces*) and $-\Delta F/F$ at 610-nm wavelength (postsynaptic) Ca^{2+} signals (*red traces*) in GI10-3. The air-current stimuli were applied from eight different orientations (Modified from Ogawa et al. 2008)

stained with AM ester of OGB-1, while cell-impermeant Fura Red (Thermo Fisher Scientific) was injected into the single GI (Fig. 18.5b). For staining the sensory afferents, a solution containing AM ester of OGB-1 at a concentration of 0.05 % and dispersing reagent (Pluronic F-127, Thermo Fisher Scientific) at a concentration of 1 % was pressure injected into a cercal sensory nerve through a glass micropipette. Twelve hours after dye injection, the axon terminals of cercal sensory neurons were found to be stained with OGB-1. After staining of the afferents with OGB-1, 2-mM Fura Red tetrapotassium salt was iontophoretically injected into the GI for 5 min through a glass microelectrode, using a hyperpolarizing current of 3 nA.

18.3.3.2 Optical Recording

Fluorescent signals were viewed with an inverted microscope (Axiovert100, Zeiss). A Xenon arc lamp (XBO 75 w, Zeiss) illumination with a stabilized power supply and 470/20 band-pass filter was used for excitation of the Ca^{2+} indicators, OGB-1 and Fura Red. For simultaneous measurement of pre- and postsynaptic Ca^{2+} signals, a fluorescent image passing through a FT510 dichroic mirror was divided into two images with *W-View* optics (Hamamatsu Photonics) by the following filter set: dichroic 590LP and emission 535/45 for the OGB-1 and 610/25 for Fura Red (Fig. 18.5a). The two separated images were simultaneously acquired side by side in the same frame with a digital cooled CCD camera (ORCA-ER, Hamamatsu Photonics) attached to the inverted microscope.

A series of fluorescent images were acquired at 30 fps on 160 × 256 pixel image for each wavelength. For the simultaneous measurement of two fluorescence wavelengths of OGB-1 and Fura Red, a software package (AQUACOSMOS/Ratio, Hamamatsu Photonics) was used. For time-course displays of the fluorescence changes in several regions of the dendritic tree, several polygonal recording regions on a Fura Red image of the GI were selected, and the mean values of the fluorescent intensities in these regions were plotted as a function of time. Fluorescent intensities were collected from each fluorescent image and background corrected and averaged for each selected region. Changes in $[Ca^{2+}]_i$ were expressed as $\Delta F/F[\Delta F/F = (F-F_0)/F_0]$ for OGB-1 or $-\Delta F/F$ for Fura Red, where F_0 was the background-corrected pre-stimulus fluorescent intensity.

18.4 Future Direction of Optical Recording in Cricket Neurobiology

As shown above, various types of Ca^{2+} imaging techniques have provided significant findings on information processing in the cricket cercal sensory system, including dendritic integration of the GIs, neural representation of cercal afferents, and the anatomy-based decoding algorithm for synaptic transmission between them. Genetically encoded calcium indicators (GECIs) such as GCaMP (Nakai et al. 2001) and Yellow Cameleon (YC) 3.60 (Nagai et al. 2004) enable chronic monitoring of neural activities in populations of specific cell types in the nervous system of various model animals including nematode, fruit fly, zebra fish, and mouse. In the cricket nervous system, we have recently succeeded in the expression of YC3.60 within the cricket brain using the electroporation method (Matsumoto et al. 2013). The introduction of GECI, by electroporation and other transgenic tools described in other chapters, will enable prolonged Ca^{2+} imaging in deep brain regions of the cricket. Furthermore, stable multiphoton microscopy imaging in head-fixed, awake animals allows us to monitor neuronal activity during behavioral paradigms, and *in actio* Ca^{2+} imaging in freely moving animals was achieved by novel miniature head-mounted microscopes (Grienberger and Konnerth 2012; Chen et al. 2011). Further development of in vivo Ca^{2+} imaging techniques and improvement of gene transfection methods for inducing GECIs will facilitate major advances in our understanding of the neural mechanisms underlying cricket behaviors.

References

Bacon JP, Murphey RK (1984) Receptive fields of cricket (*Acheta domesticus*) are determined by their dendritic structure. J Physiol Lond 352:601–613

Baden T, Hedwig B (2007) Neurite-specific Ca^{2+} dynamics underlying sound processing in an auditory interneurone. Dev Neurobiol 67:68–80

Baden T, Hedwig B (2009) Dynamics of free intracellular Ca^{2+} during synaptic and spike activity of cricket tibial motoneurons. Eur J Neurosci 29:1357–1368

Chen X, Leischner U, Rochefort NL, Nelken I, Konnerth A (2011) Functional mapping of single spines in cortical neurons in vivo. Nature 475:501–505

Grienberger C, Konnerth A (2012) Imaging calcium in neurons. Neuron 73:862–885

Helmchen F, Denk W (2005) Deep tissue two-photon microscopy. Nat Methods 2:932–940

Jacobs GA, Theunissen F (1996) Functional organization of a neural map in the cricket cercal sensory system. J Neurosci 16:769–784

Jacobs GA, Theunissen F (2000) Extraction of sensory parameters from a neural map by primary sensory interneurons. J Neurosci 20:2934–2943

Jacobs GA, Miller JP, Murphey RK (1986) Cellular mechanisms underlying directional sensitivity of an identified sensory interneuron. J Neurosci 6:2298–2311

Kloppenburg P, Hörner M (1998) Voltage-activated currents in identified giant interneurons isolated from adult crickets *gryllus bimaculatus*. J Exp Biol 201:2529–2541

Landolfa MA, Miller JP (1995) Stimulus-response properties of cricket cercal filiform receptors. J Comp Physiol A 177:749–757

Lipp P, Niggli E (1993) Ratiometric confocal Ca^{2+}-measurements with visible wavelength indicators in isolated cardiac myocytes. Cell Calcium 14:339–372

Matsumoto CS, Shidara H, Matsuda K, Nakamura T, Mito T, Matsumoto Y, Oka K, Ogawa H (2013) Targeted gene delivery in the cricket brain, using in vivo electroporation. J Insect Physiol 59:1235–1241

Miller JP, Jacobs GA, Theunissen FE (1991) Representation of sensory information in the cricket cercal sensory system. I. Response properties of the primary interneurons. J Neurophysiol 66:1680–1689

Miller JP, Krueger S, Heys J, Gedeon T (2011) Quantitative characterization of the filiform mechanosensory hair array on the cricket cercus. PLoS One 6(11):e27873. doi:10.1371/journal.pone.0027873

Nagai T, Yamada S, Tominaga T, Ichikawa M, Miyawaki A (2004) Expanded dynamic range of fluorescent indicators for Ca^{2+} by circularly permuted yellow fluorescent proteins. Proc Natl Acad Sci U S A 101:10554–10559

Nakai J, Ohkura M, Imoto K (2001) A high signal-to-noise Ca^{2+} probe composed of a single green fluorescent protein. Nat Biotechnol 19:137–141

Nemoto T (2008) Living cell functions and morphology revealed by two-photon microscopy in intact neural and secretory organs. Mol Cell 26:113–120

Ogawa H, Kajita Y (2015) Ca^{2+} imaging of cricket protocerebrum responses to air current stimulation. Neurosci Lett 584:282–286

Ogawa H, Miller JP (2013) In vivo Ca^{2+} imaging of neuronal activity. In: Ogawa H, Oka K (eds) Methods in neuroethological research. Springer Japan, Tokyo, pp 71–87

Ogawa H, Baba Y, Oka K (1999) Dendritic Ca^{2+} transient increase evoked by wind stimulus in the cricket giant interneuron. Neurosci Lett 275:61–64

Ogawa H, Baba Y, Oka K (2000) Spike-dependent calcium influx in dendrites of the cricket giant interneuron. J Neurobiol 44:45–56

Ogawa H, Baba Y, Oka K (2001) Dendritic calcium accumulation regulates wind sensitivity via short-term depression at cercal sensory-to-giant interneuron synapses in the cricket. J Neurobiol 46:301–313

Ogawa H, Baba Y, Oka K (2002a) Spike-triggered dendritic calcium transients depend on synaptic activity in the cricket giant interneurons. J Neurobiol 50:234–244

Ogawa H, Baba Y, Oka K (2002b) Direction of action potential propagation influences calcium increases in distal dendrites of the cricket giant interneurons. J Neurobiol 53:44–56

Ogawa H, Baba Y, Oka K (2004) Directional sensitivity of dendritic calcium responses to wind stimuli in the cricket giant interneuron. Neurosci Lett 358:185–188

Ogawa H, Cummins GI, Jacobs GA, Miller JP (2006) Visualization of ensemble activity patterns of mechanosensory afferents in the cricket cercal sensory system with calcium imaging. J Neurobiol 66:293–307

Ogawa H, Cummins GI, Jacobs GA, Oka K (2008) Dendritic design implements algorithm for extraction of sensory information. J Neurosci 28:4592–4603

Palka J, Levine R, Schubiger M (1977) The cercus-to-giant interneuron system of crickets. I. Some aspects of the sensory cells. J Comp Physiol 119:267–283

Paydar S, Doan CA, Jacobs GA (1999) Neural mapping of direction and frequency in the cricket cercal sensory system. J Neurosci 19:1771–1781

Sobel EC, Tank DW (1994) In vivo Ca^{2+} dynamics in a cricket auditory neuron: an example of chemical computation. Science 263:823–826

Speier S, Nyqvist D, Cabrera O, Yu J, Molano RD, Pileggi A, Moede T, Köhler M, Wilbertz J, Leibiger B, Ricordi C, Leibiger IB, Caicedo A, Berggren P (2008) Noninvasive in vivo imaging of pancreatic islet cell biology. Nat Med 14:574–578

Svoboda K, Yasuda R (2006) Principles of two-photon excitation microscopy and its applications to neuroscience. Neuron 50:823–839

Takahara Y, Matsuki N, Ikegaya Y (2011) Nipkow confocal imaging from deep brain tissues. J Integr Neurosci 10:121–129

Theunissen F, Roddey JC, Stufflebeam S, Clague H, Miller JP (1996) Information theoretic analysis of dynamical encoding by four primary sensory interneurons in the cricket cercal system. J Neurophysiol 75:1345–1376

Troyer TW, Levin JE, Jacobs GA (1994) Construction and analysis of a data base representing a neural map. Microsc Res Tech 29:329–343

Tsien RY (1980) New calcium indicators and buffers with high selectivity against magnesium and protons: design, synthesis, and properties of prototype structures. Biochemistry 19:2396–2404

Chapter 19
Trackball Systems for Analysing Cricket Phonotaxis

Berthold Hedwig

Abstract In order to analyse cricket phonotactic walking behaviour, different types of trackball systems have been developed. All trackball systems infer the velocity and direction of the walking insect from the movements of the trackball, however, with different degrees of resolution. Closed-loop systems compensate the animal's displacements via servomotors counter-rotating the sphere on which the cricket is freely walking and turning. In open-loop systems, the tethered cricket actively rotates the trackball but cannot change its orientation in the sound field. Trackball systems can be combined with high-speed video recordings to analyse the walking motor activity or can be incorporated into neurophysiological set-ups to explore the neural activity underlying phonotaxis.

Keywords Trackball • Open loop • Closed loop • Moment of inertia • Optical sensor • Data processing

19.1 Introduction

Measuring and analysing the locomotor behaviour of crickets in their natural environment can be quite challenging and is not always possible for practical reasons. Therefore an alternative research strategy is to bring the animals into the lab where their behaviour can be studied and quantified under controlled experimental conditions with suitable monitoring devices.

To characterise cricket phonotactic behaviour arena, observations are widely used. They aim at measuring the time taken by the insects to approach a speaker presenting acoustic stimuli and/or counting the number of animals that reach the auditory target (Tschuch 1976; Stout et al. 1983). Also Y-maze experiments, which give a more categorical indication of the animal's auditory preferences, have been employed (e.g. Popov and Shuvalov (1977); Rheinlaender and Blätgen (1982)), and steering responses of tethered flying females have been used as a measure of phonotaxis (Pollack and Hoy 1979). A technically more demanding approach is

B. Hedwig (✉)
Department of Zoology, University of Cambridge, Cambridge CB2 3EJ, UK
e-mail: bh202@cam.ac.uk

H.W. Horch et al. (eds.), *The Cricket as a Model Organism*,
DOI 10.1007/978-4-431-56478-2_19

the analysis of cricket walking behaviour by trackball systems. Different types of systems have been developed to measure and quantify female phonotactic walking and to analyse the dynamic of auditory steering. All systems keep the animal stationary and have in common that the movement of the trackball is taken as in indicator of the animal's walking speed and direction. Some approaches allow turning movements of the crickets, but they all prevent the animals from reaching the sound source located to the left or right side in front of the insect. Closed-loop systems compensate the animal's walking and turning movements, whereas in open-loop systems, the tethered animals actively rotate the trackball.

19.2 Trackball Systems for the Analysis of Phonotactic Behaviour

The first closed-loop "walking compensators" (Wendler et al. 1980; Weber et al. 1981; Schmitz et al. 1982) were based on a design by Kramer (1976). They consisted of a 30–50 cm diameter polycarbonate sphere which was supported by the driving wheels of two servomotors. Crickets were placed on top of the sphere where they were free to walk and turn in any direction. A light-reflective foil was attached to their pronotum, and an infrared video camera and illumination system monitored the insect's position. When a cricket started walking and changed its position by more than 1–2 mm, its optical displacement activated a feedback loop that controlled the servomotors and counter-rotated the sphere in a way to keep the walking cricket stationary in the centre of the infrared camera's scanning field (Fig. 19.1a). While rotating the trackball, the servomotors generated some noise and background sound levels reached up to 58 dB SPL (Schmitz et al. 1982).

These closed-loop systems captured data on the insect's walking speed and its direction as derived from either a tachometer wheel at the "south pole" of the sphere or directly from the rotation of the servomotors. Due to the properties of the servo-system, however, the trackball velocity signal, which is the indicator for the cricket walking speed, was low-pass filtered as the systems took at least 150–400 ms to adjust to the cricket's average walking speed. Velocity changes due to the insect's stepping cycle were strongly attenuated, and the dynamic of turning responses upon a change in the direction of acoustic stimulation was considerably underestimated. For example, this led to the conclusion that female crickets may evaluate a whole chirp before they start steering towards a new sound direction (Schmitz et al. 1982). These walking compensators, however, provided first quantitative data on the crickets' phonotactic walking behaviour (Schmitz et al. 1982) and valuable information on their preferences for auditory patterns based on the time and angle by which the animals tracked specific patterns (Thorson et al. 1982). Walking compensators are still successfully used for analysis of phonotactic behaviour (Verburgt et al. 2008, Hennig 2009) while considering the dynamic limitations of the systems.

Several labs developed smaller and simpler open-loop devices in which tethered crickets were positioned on a trackball and while walking stationary,

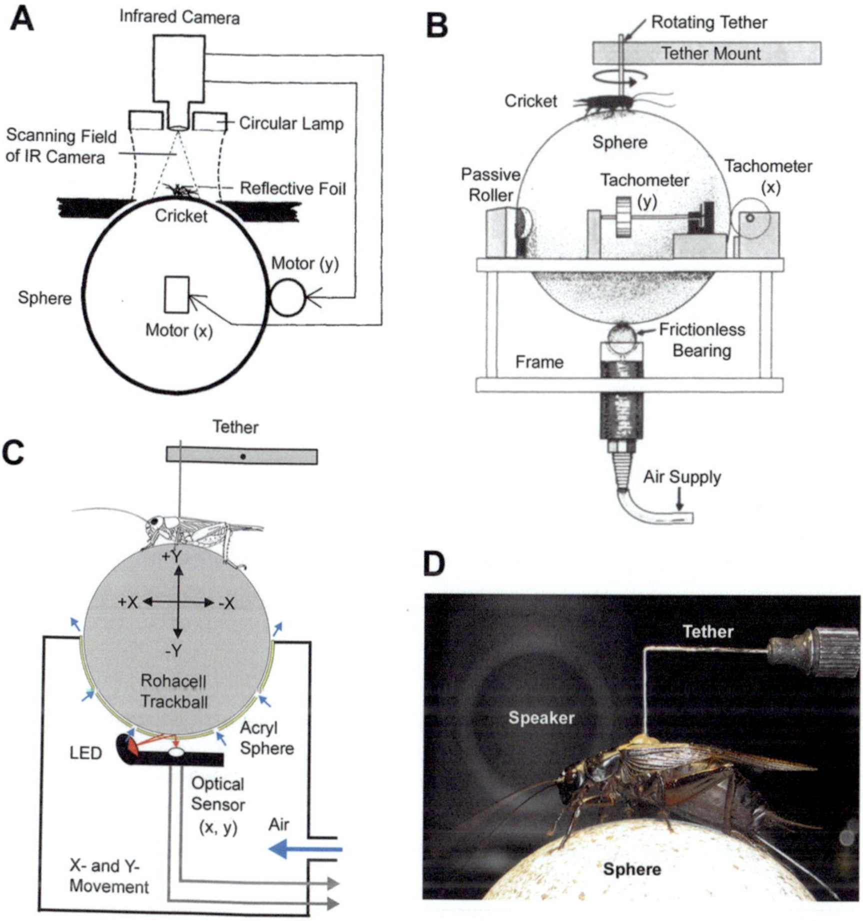

Fig. 19.1 Trackball systems for phonotaxis experiments, speakers used for acoustic stimulation are not depicted. (**a**) Closed-loop trackball system operating as a locomotion compensator. The cricket is free to walk in any direction; within the scanning field, its position is detected by an infrared camera. The camera signal is fed into an electronic servo-loop driving two servomotors which counter-rotate the sphere to continuously compensate for the cricket's movement. The diameter of the sphere is 33 cm, and its movement is calculated at 0.5 s intervals with a resolution of 1.27 mm from optical pulse generators attached to the servomotors (From Schmitz et al. 1982). (**b**) System in which the cricket walks and rotates a trackball (10 cm diameter, 8.7 g) which sits on an air-suspended Teflon ball; the cricket is free to rotate in its tether as well. Movements of the trackball are measured via two optical tachometers at the equator and fed into a microcomputer for further processing at 1 s sample intervals (From Doherty and Pires 1987). (**c**) Open-loop system in which the tethered cricket rotates a small trackball (56 mm, 3 g) that floats in an airstream. The *x* and *y* movements of the trackball are monitored with an optical flow sensor at the south pole with a resolution of 0.127 mm and 0.3 ms (From Hedwig and Poulet 2004). (**d**) Female cricket (*G. bimaculatus*) tethered at the metathorax and suspended on a trackball, speaker for acoustic stimulation in the background

they actively moved and controled the rotation of the ball (Dahmen 1980; Brunner and Labhardt 1987; Doherty and Pires 1987; Schildberger and Hoerner 1988; Walikonis et al. 1991; Gras and Hörner 1992; Staudacher and Schildberger 1998). Friction of the trackball rotation was greatly reduced by allowing the ball to float on a gentle airstream. In these systems, the insect's walking velocity and walking direction were also inferred from the trackball rotations in the forward-backward (x) and left-right (y) direction. The trackball movements were monitored mechanically via computer-mouse sensors (Fig. 19.1b) or obtained via an array of photodiodes detecting a pattern of black dots on the surface of the white trackball. As the trackballs were actively moved by the muscle power of the insect, they needed to be light-weighted. Trackballs with a diameter of 80–120 mm were made out of lightweight material (Styrofoam, Rohacell) or even hollowed out to minimise the trackball mass to 2.8–8.7 g. Temporal resolution was generally limited as signals were sampled over intervals of 0.1–1 s, corresponding to the contemporary computer technology. The most sensitive of these systems achieved a spatial resolution of 1 mm in each direction and measured walking velocities up to 3 m/s (Gras and Hörner 1992).

With the increase of computer processing power and the advent of 2-D optical flow sensors, as widely used in computer-mouse technology, temporal and spatial resolution of open-loop trackball systems could be greatly improved. Hedwig and Poulet (2004, 2005; Fig. 19.1c, d) used an Agilent ADNS-2051 with 200 counts per inch (CPI) with a spatial resolution corresponding to 127 μm and a temporal resolution of 2,300 frames/s or 38 cm/s. More recent optical flow sensors go up to 800 CPI providing even higher resolution measurements of x and y coordinates (Hedrick et al. 2007; Lott et al. 2007). Strategic positioning of the sensor at the south pole of the trackball (Fig. 19.1c) allows monitoring forward-backward (x) and left-right (y) movements with the same 2-D sensor. Such a design is sufficient when steering manoeuvers are performed during forward walking; however, it may be limited in picking up stationary rotations. Since crickets steer while they are running this is not a problem for behavioural analysis of phonotactic walking (see also Gras and Hörner 1992).

In the system of Hedwig and Poulet (2005), the output of the optic flow sensor is forwarded to a quadrature to pulse converter generating a coding pulse of 150 μs duration for each 127 μm movement in the forward-backward (x) and the left-right (y) direction. Coding pulses for (x) and (y) are sampled by a PC together with the sound stimuli and processed off-line with custom-written software to calculate the forward-backward and the left-right trackball velocity components; the velocity signals are integrated to obtain the overall length of the walked path (x) or the insect's lateral deviation (y) from a straight line or the intended walking angle. As the trackball floats (Fig. 19.1c) on an airstream generated by 24 small air inlets, only 38 dB SPL of broadband background noise occurs in this system.

The response of a female *G. bimaculatus* walking on a trackball and responding to increasing intensity levels of calling song is given in Fig. 19.2a; sound is presented from speakers at 45° to the left or right side of the animal's length axis. The lateral velocity indicating phonotactic steering clearly changed with the side of acoustic stimulation. As the sound intensity increased, the lateral velocity increased

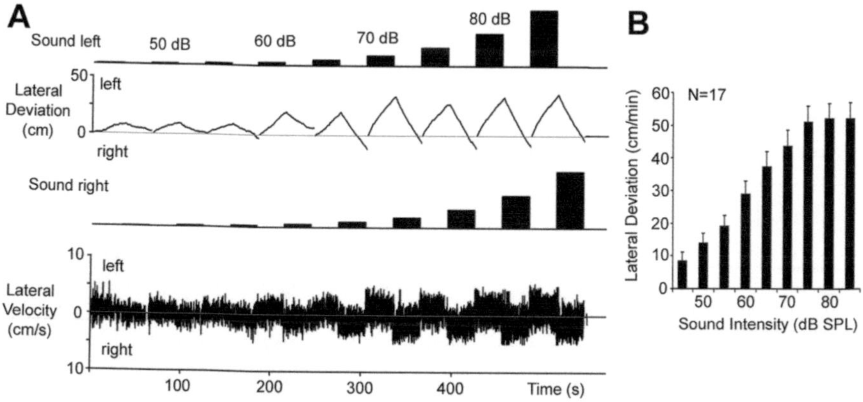

Fig. 19.2 (**a**) Phonotactic response of a female *G. bimaculatus* to calling song presented for 30 s from the left and right sides at increasing sound intensities, respectively. Steering towards the side of acoustic stimulation is indicated by the lateral velocity and the lateral deviation which both increase from 45 to 75 dB SPL and then saturate. (**b**) Lateral deviation as measured for the time interval of 1 min as a function of sound intensity. The lateral deviation towards the calling song provides a reliable measure of phonotaxis (From Hedwig and Poulet 2005)

together with the overall lateral deviation of the animal in response to each song presentation. Pooling data from many female *G. bimaculatus* crickets showed that the lateral deviation towards the sound stimulus provides a reliable measure of the phonotactic response (Fig. 19.2b); phonotaxis starts at 45 dB, increases towards 75 dB SPL, and then saturates.

19.3 Analysis of Rapid Steering Behaviour

Dynamic limitations of open-loop systems come with the mass and diameter of the trackball. The mass moment of inertia (I) of a sphere scales with the mass and the square of the radius [$I = 2/5\ mr^2$]; thus the trackball radius is decisive; a hollow sphere with a 12 cm diameter will have four times larger inertia as compared to a sphere with a 6 cm diameter given the mass of both is the same; the mass moment of inertia would increase by a factor of 32 if solid spheres of the same density are considered. Thus, mass and diameter will be the determining factor for the dynamic response properties of the system. With a small trackball diameter, the mass moment of inertia can be substantially reduced and systems sensitive enough to pick up even the cricket's tripod stepping cycle can be obtained. Hedwig and Poulet (2004) used a trackball of 56 mm in diameter and a mass of only 3 g. This corresponds to a mass moment of inertia of 1.2–1.3 g which matches the weight range of adult female *G. bimaculatus*. The high sensitivity of the system allowed analysing the dynamics of female phonotactic walking and led to a better understanding of auditory steering responses (Fig. 19.3).

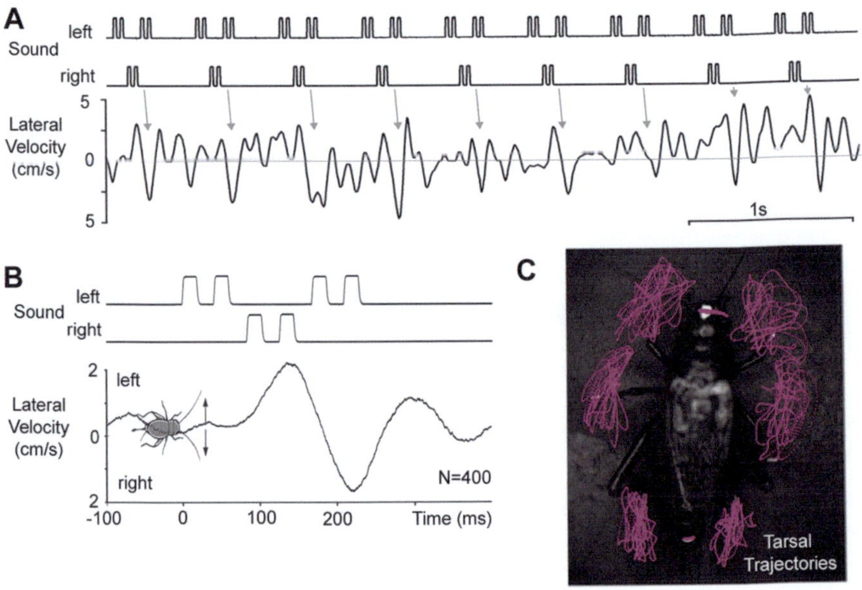

Fig. 19.3 Fast steering responses of a walking cricket towards a split-song paradigm in which pairs of pulses are presented from opposite directions. (**a**) Pairs of sound pulses elicit fast steering responses to the left and right as demonstrated by the lateral velocity signal, indicated by *arrows*. (**b**) Average of the lateral velocity signal triggered by the acoustic pulse paradigm. The female steered towards each pair of sound pulses after a latency of only 55–60 ms. (**c**) High-speed video recordings at 500 frames/s reveal the tarsal trajectories while the cricket is steering to the split-song paradigm and indicate rapid adjustments of the leg stepping pattern (Witney and Hedwig 2011 and unpublished)

Different to the walking compensator approach where the crickets can turn and alter their orientation towards the speaker, in the open-loop systems, the tethered insects stay in a constant position within the sound field. The auditory input to their hearing system is not altered by their steering movements. For quantitative analysis, steering responses towards sound stimuli can therefore be averaged over a number of repetitions. Cricket steering behaviour had been thought to depend on a sequential processing of auditory pattern recognition followed by steering commands to the motor system. However, when spilt-song paradigms (Weber and Thorson 1988) in which the sound pulses are alternating from opposing speakers were presented to females walking on an open-loop trackball system, these revealed rapid steering responses towards individual pulses (Hedwig and Poulet 2004, 2005). When exposed to a double-pulse stimulation paradigm in which pairs of calling song sound pulses were presented from opposite directions, female crickets steered to the side, from which more sound pulses were presented. However, a close-up inspection of their steering velocity revealed rapid changes in the steering velocity and demonstrated that the animals steered to each pair of sound pulses (Fig. 19.3a). Averaging the lateral velocity signal demonstrated that steering occurred with a very short latency of 55–60 ms (Fig. 19.3b). Auditory steering does not occur after the females processed the complete sequence of a chirp (Schmitz et al. 1982) but rather is a reactive response to individual sound pulses controlled by pattern recognition (Hedwig and Poulet 2004). The trackball recordings of walking crickets can easily be combined with high-speed video recordings of the insect's leg movements (Witney and Hedwig 2011). The trajectories of the tarsi indicate how the animals change their walking pattern during steering and which legs may produce the main contribution to the motor output. In the case of the split-song experiments using a double-pulse paradigm, the tarsal trajectories of all legs indicate rapid changes of the stepping movements to the left and right side in response to the acoustic stimuli. These reflect the females' rapid alternating steering responses to both sides of the sound sources as revealed by the trackball recordings (Fig. 19.3c). Thus the rapid steering movements picked up by the trackball system are also reflected and confirmed by the females' stepping pattern.

19.4 Integration with Intracellular Recordings of Brain Neurons

Due to their compact design, the open-loop trackball systems can easily be fitted into electrophysiological set-ups. Recordings of the cricket's phonotactic walking behaviour but also of wind-evoked escape running or visual steering responses have been combined with recording and labelling of individual neurons. Moreover, even a functional analysis of the neural activity is possible. Current injections to alter the spike activity of single neurons in walking crickets was applied to auditory neurons in the prothoracic ganglion (Schildberger and Hörner 1988; Hörner 1992), to giant neurons in the terminal ganglion (Kohstall-Schnell and Gras 1994) and to descending brain neurons (Böhm and Schildberger 1992; Staudacher and Schildberger 1998; Staudacher 2001; Zorovic and Hedwig 2011, 2013). This is a powerful approach to reveal causal relationships between the activity of individual neurons and the execution of motor activity or the performance of behaviour programs.

Schildberger and Hörner (1988) demonstrated that the auditory steering behaviour of walking crickets can be manipulated by altering the activity of the ascending auditory interneuron AN1. When tethered on a trackball, crickets normally turn to the side of the louder calling song; however, when AN1 is inhibited by hyperpolarising current injection while sound is presented from the same side as the dendrites and axon of the neuron, the animals will turn towards the opposite side. This indicates that the turning behaviour of the female crickets is based on the relative activity of the left and right AN1 interneurons, and it reveals their functional significance for phonotactic behaviour. Local interneurons like ON1 had only a weak effect on the direction of walking. In a similar way, Gras and Kohstall-Schnell (1998) manipulated the activity of giant interneurons and local interneurons in the terminal ganglion of crickets walking spontaneously or in response to

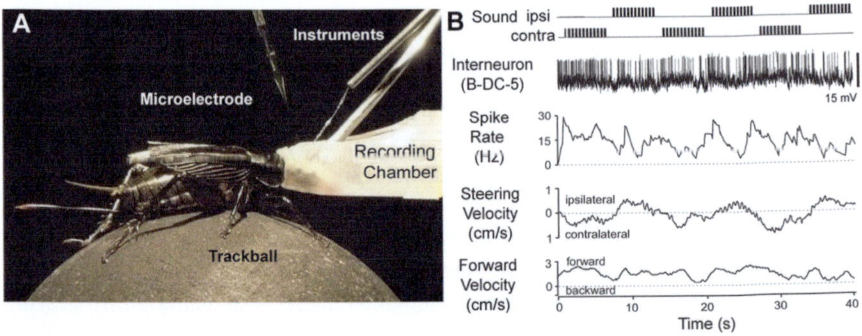

Fig. 19.4 (**a**) Cricket positioned on a trackball; its head is waxed into a recording chamber and its brain is exposed for intracellular recordings; a platform and a ring (instruments) are used to mechanically stabilise the brain. The female can freely move its legs, and it can walk and perform phonotaxis. The recording chamber is filled with saline. (**b**) Intracellular recording of a descending brain neuron while the cricket showed phonotaxis with clear changes in steering velocity towards the side of acoustic stimulation (From Zorovic and Hedwig 2013)

air-puffs and demonstrated that changes in the activity of single neurons can alter the walking speed and direction.

As the network for auditory pattern recognition is housed in the cricket brain (Kostarakos and Hedwig 2012), recording brain neurons in phonotactic walking crickets could provide a major step forward in our understanding of their behaviour. When the brain is exposed for intracellular recordings (Fig. 19.4a), crickets still perform phonotaxis, however, at a lower rate than during standard trackball tests with intact animals. Even so the auditory responses of local brain neurons and of descending interneurons can be described and the effects on the animals' walking behaviour tested (Staudacher and Schildberger 1998; Staudacher 2001; Zorovic and Hedwig 2011, 2013). In Fig. 19.4b the microelectrode recorded the activity of a descending brain neuron while the animal clearly performed phonotactic walking and steered towards the side of calling song presentation. The neuron's activity was coupled to the sound pattern; driving the neuron's spike activity by intracellular depolarisation enhanced the animals forward walking velocity and initiated turning (for details see Zorovic and Hedwig 2013). Thus recording brain neurons in phonotactic walking crickets provides us with an opportunity to analyse how auditory pattern recognition may be coupled to the generation of acoustically evoked phonotactic steering commands.

19.5 Future Prospects

Trackball systems have played a pivotal role in analysing cricket phonotactic behaviour at its behavioural and neurophysiological level. Current developments of imaging tools based on calcium-sensitive or voltage-sensitive indicators, of neural recording techniques, and of optical stimulation techniques with light-activated ion channels provide us with even more sophisticated tools. In the near future, these may be combined with trackball systems to further our understanding of the neural mechanisms underlying cricket phonotactic behaviour.

References

Böhm H, Schildberger K (1992) Brain neurones involved in the control of walking in the cricket *Gryllus bimaculatus*. J Exp Biol 166(1):113–130

Brunner D, Labhardt T (1987) Behavioural evidence for polarization vision in crickets. Physiol Entomol 12(1):1–10

Dahmen HJ (1980) A simple apparatus to investigate the orientation of walking insects. Experientia 36(6):685–687

Doherty JA, Pires A (1987) A new microcomputer-based method for measuring walking phonotaxis in field crickets (Gryllidae). J Exp Biol 130(1):425–432

Gras H, Hörner M (1992) Wind-evoked escape running of the cricket *Gryllus bimaculatus*: I. Behavioural Analysis. J Exp Biol 171(1):189–214

Gras H, Kohstall D (1998) Current injection into interneurons of the terminal ganglion modifies turning behaviour of walking crickets. J Comp Physiol A 182(3):351–361

Hedrick AV, Hisada M, Mulloney B (2007) Tama-kugel: hardware and software for measuring direction, distance, and velocity of locomotion by insects. J Neurosci Method 164(1):86–92

Hedwig B, Poulet JFA (2004) Complex auditory behaviour emerges from simple reactive steering. Nature 430:781–785

Hedwig B, Poulet J (2005) Mechanisms underlying phonotactic steering in the cricket *Gryllus bimaculatus* revealed with a fast trackball system. J Exp Biol 208(5):915–927

Hennig RM (2009) Walking in Fourier's space: algorithms for the computation of periodicities in song patterns by the cricket *Gryllus bimaculatus*. J Comp Physiol A 195(10):971–987

Hörner M (1992) Wind-evoked escape running of the cricket *Gryllus bimaculatus*: II. Neurophysiological analysis. J Exp Biol 171(1):215–245

Kohstall-Schnell D, Gras H (1994) Activity of giant interneurons and other wind-sensitive elements of the terminal ganglion in the walking cricket. J Exp Biol 193(1):157–181

Kramer E (1976) The orientation of walking honeybees in odour fields with small concentration gradients. Physiol Entomol 1:27–37

Lott GK, Rosen MJ, Hoy RR (2007) An inexpensive sub-millisecond system for walking measurements of small animals based on optical computer mouse technology. J Neurosci Method 161(1):55–61

Pollack GS, Hoy RR (1979) Temporal pattern as a cue for species-specific calling song recognition in crickets. Science 204:429–432

Popov A, Shuvalov V (1977) Phonotactic behavior of crickets. J Comp Physiol A 119(1):111–126

Rheinlaender J, Blätgen G (1982) The precision of auditory lateralization in the cricket, *Gryllus bimaculatus*. Physiol Entomol 7(2):209–218

Schildberger K, Hörner M (1988) The function of auditory neurons in cricket phonotaxis I. Influence of hyperpolarisation of identified neurons on sound localization. J Comp Physiol A 163(5):621-631

Schmitz B, Scharstein H, Wendler G (1982) Phonotaxis in *Gryllus campestris* L. (Orthoptera, Gryllidae). J Comp Physiol A 148(4):431–444

Staudacher EM (2001) Sensory responses of descending brain neurons in the walking cricket, *Gryllus bimaculatus*. J Comp Physiol A 187(1):1–17

Staudacher E, Schildberger K (1998) Gating of sensory responses of descending brain neurones during walking in crickets. J Exp Biol 201(4):559–572

Stout JF, DeHaan C, McGhee RW (1983) Attractiveness of the male *Acheta domesticus* calling song to females. J Comp Physiol A 153(4):509–521

Thorson J, Weber T, Huber F (1982) Auditory behavior of the cricket II. Simplicity of calling-song recognition in Gryllus, and anomalous phonotaxis at abnormal carrier frequencies. J Comp Physiol A 146(3):361–378

Tschuch G (1976) The influence of synthetic songs on female *Gryllus bimaculatus* de Geer. Zool J Physiol 80:383–388

Verburgt L, Ferguson JWH, Weber T (2008) Phonotactic response of female crickets on the Kramer treadmill: methodology, sensory and behavioural implications. J Comp Physiol A 194 (1):79–96

Walikonis R, Schoun D, Zacharias D, Henley J, Coburn P, Stout J (1991) Attractiveness of the male *Acheta domesticus* calling song to females III. The relation of age-correlated changes in syllable period recognition and phonotactic threshold to juvenile hormone III biosynthesis. J Comp Physiol A 169(6):751–764

Weber T, Thorson J (1988) Auditory behavior of the cricket. IV: interaction of direction of tracking with perceived temporal pattern in split-song paradigms. J Comp Physiol A 163 (1):13–22

Weber T, Thorson J, Huber F (1981) Auditory behavior of the cricket I. Dynamics of compensated walking. J Comp Physiol A 141(3):215–232

Wendler G, Dambach M, Schmitz B, Scharstein H (1980) Analysis of the acoustic orientation behavior in crickets (*Gryllus campestris* L.). Naturwissenschaften 67(2):99–101

Witney AG, Hedwig B (2011) Kinematics of phonotactic steering in the walking cricket *Gryllus bimaculatus* (de Geer). J Exp Biol 214(1):69–79

Zorovic M, Hedwig B (2011) Processing of species-specific auditory patterns in the cricket brain by ascending, local, and descending neurons during standing and walking. J Neurophysiol 105 (5):2181–2194

Zorovic M, Hedwig B (2013) Descending brain neurons in the cricket *Gryllus bimaculatus* (de Geer): auditory responses and impact on walking. J Comp Physiol A 199:25–34

Chapter 20
Synthetic Approaches for Observing and Measuring Cricket Behaviors

Hitoshi Aonuma

Abstract When we investigate animal behavior, it is necessary to quantify and to qualify the sequence of observed behaviors. It is also important to compare the behavior with the physiology of the nervous systems in order to understand underlying neuronal mechanisms. Reliable results require experimental repetition that includes consistent controls, because animals do not always respond in the same way to the same external stimuli. Instead, animals alter their behaviors in order to respond to the demands of changing environments. Engineering approaches, in particular robotics, can help us to observe and to provoke animal movements and behaviors. I describe a novel approach that provokes animal movements and behaviors in response to computer simulation and robots. These constructive approaches help us to bridge the gap between behavior and physiology. The performances of the models and robots are discussed, and the accuracies of the models are confirmed by behavior studies with animals. This approach has been named "Synthetic Neuroethology." This chapter introduces the methods used to observe and measure cricket behaviors. The aim is to understand adaptive behaviors at play in group size-dependent aggressive behavior.

Keywords Synthetic approach • Video tracking • Robot • Modeling • Simulation

20.1 Synthetic Approach for Understanding Cricket Behavior

Animals do not always respond to a particular stimulus with identical behaviors; instead, they alter their behaviors in order to respond to the demands of a changing environment. The state of the nervous systems depends on experiences as well as internal and/or external conditions. These factors are thought to mediate the threshold for releasing movements and behaviors. Biologists have performed

H. Aonuma (✉)
Research Center of Mathematics for Social Creativity, Research Institute for Electronic Science, Hokkaido University, Sapporo 060-0812, Japan

CREST, Japan Science and Technology Agency, Kawaguchi, Japan
e-mail: aon@es.hokudai.ac.jp

H.W. Horch et al. (eds.), *The Cricket as a Model Organism*,
DOI 10.1007/978-4-431-56478-2_20

many experiments at the molecular, physiological, and behavioral level that are designed to understand these sophisticated behaviors. However, even when detailed data are collected, deep gaps between physiology and behavior still exist. In order to bridge these gaps, we can use dynamic system models that are designed based on the results of biological experiments. Then the performances of the models can be analyzed by computer simulation or through the observation of the behavior of robots with embedded algorithms. The adequacy of the models can then be tested by behavioral studies using real animals. This approach is called "Synthetic Neuroethology."

20.2 Measurement of Cricket Behavior Using a Video Tracking System

When observing animal behavior, quantification and qualification of the sequence of behaviors are important. Video tracking systems are one of the powerful methods used for this purpose. In tracking a behavior, automated tracking and real-time data processing are useful functions. For the automated tracking of moving crickets, some versatile and complete commercial solutions exist (Noldus et al. 2001). Here, one of the useful tracking systems that can be customized is introduced.

This tracking system is based on SwisTrack, a simple, open-source solution (Lochmatter et al. 2008), and it is accompanied by custom Python scripts (Guerra et al. 2010). When we use video tracking systems, it is useful if we can track many crickets at once. However, crickets that overlap or crisscross in the same frame can make distinguishing individual crickets quite difficult. In order to avoid this problem, each cricket is marked a different color using colored paints or labels. The tracking of these marked crickets is performed in two passes, one for each colored cricket, and consists of image processing and computer vision stages. The image processing is performed in four steps: (1) background subtraction, (2) RGB binary threshold operation for segmentation of the color of each cricket, (3) masking off unnecessary areas, and (4) inflation and erosion to cluster together areas inappropriately separated by noise. The computer vision stage is comprised of three steps: (1) localization of blob centroids, (2) 2D calibration using a method by Tsai (1986), and (3) tracking using nearest neighbors.

For all the above steps, manual inspection is the most time-consuming and is prone to human error, serving as an obstacle for larger studies. The simple luminosity binary threshold operation is substituted by a three-channel RGB binary threshold operation calibrated so as to favor the color of either of the agents of interest. The tracking is performed in two passes, one for each agent color. This solution has proven very robust, and the processing of the whole batch of videos can be completely automatized through the use of computer scripting.

The resulting tracking data can then be processed starting from lower-level features (such as relative distance between crickets, absolute and relative velocities

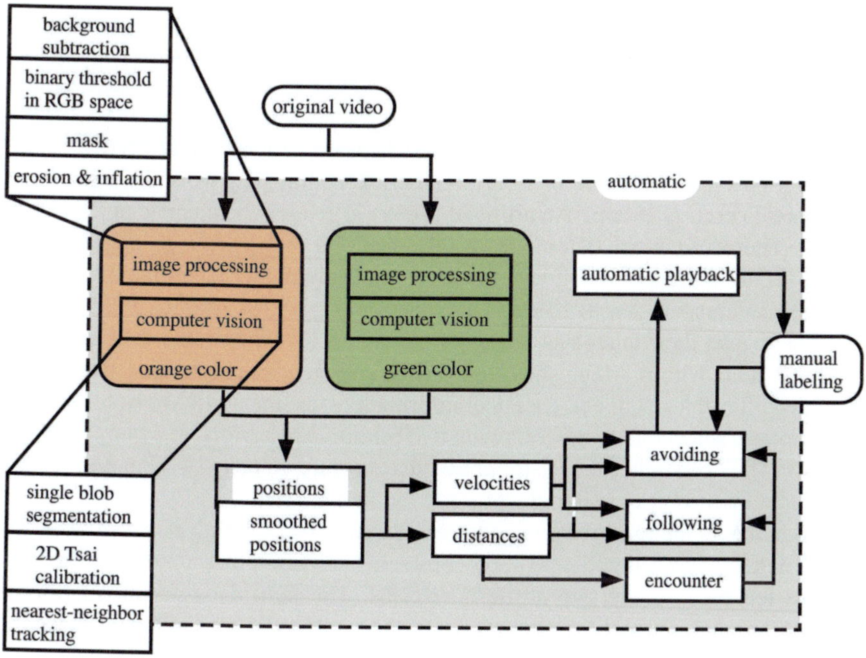

Fig. 20.1 Block diagram describing the video processing of group behavior of crickets

of the crickets, velocity cross-correlations), building all the way up to more qualitative concepts related to the interaction between the involved crickets (such as number of times they contact each other, contact durations, and escaping distances). The post-processing algorithm is implemented in Python. Tracking data can then be used for aiding with a semiautomatic qualitative analysis. The windows of time in which crickets interact are computed and catalogued automatically from the tracking data. This information is used later for playing back only those relevant windows of time in each of the videos. Users can then observe the videos and label each encounter with a few keystrokes. Automatically segmenting the videos make feasible the otherwise very tedious task of classifying each chirp, bite, antennal contact, mandible flare, etc. (Fig. 20.1).

20.3 Modeling of Group Size-Dependent Aggression in the Cricket

Using characteristics of observed behavior from experiments in modeling studies deepens our understanding of the behavior. Modeling cricket behavior has been performed based on behavioral observation and neurophysiological experiments (e.g. Ashikaga et al. 2009; Funato et al. 2011; Kawabata et al. 2012; Mizuno et al.

2012). Change of aggressiveness in male crickets depends on the size of the group in which the cricket is living. Ashikaga and colleagues (2009) established a model that describes how group size influences aggressiveness in male crickets (group size-dependent aggression).

Dynamic behavior models of artificial crickets are designed based on behavioral observations. The body size of an artificial cricket, stretching from the head to the abdominal cerci, is 25 mm. An artificial cricket is given two antennae of 28 mm in length. These values are determined by the size of a real cricket. Artificial cricket functions are assigned as follows: (1) They are individually identifiable. (2) They use their antennae to sense objects. This function is modeled after the ability of live crickets to use their antennae to survey obstacles. (3) They require 45 mm as a personal field within which they can detect opponent crickets. This function is modeled after the visual and chemical information-sensing ability of a real cricket. (4) They can select one of the following two primitive behaviors, at a sampling time of 0.1 s: They can either move straight with a constant velocity of 50 mm/s, or they can rotate themselves to an angle of $-\pi/2$ rad and $\pi/2$ rad.

Modeling of an aggressive encounter is as follows. Fighting between artificial male crickets is considered a two-body problem. The result of winning or losing the fight is fed back to the two artificial crickets. The fight does not affect a third artificial cricket. In reality, crickets select an aggressive or avoidance behavior based on past experiences. This behavioral selection cannot be explained by a known thought process. Instead, this behavioral selection requires at least one internal state. For each artificial cricket, the parameter of the internal state, α, is determined and stored, and it represents the memory of winning or losing. The internal state α expresses the level of passivity and is assigned a value of 0 through 1. In this way, artificial crickets can select behavior depending on their internal state. Artificial crickets are given three basic behavioral choices: wandering, fighting, and obstacle avoidance. The behavioral choice is determined by a finite-state machine model (Fig. 20.2) where three motion primitives are possible:

(i) Wandering: They walk randomly while going straight, and periodically, they stop and turn $\pm\pi/2$ rad.
(ii) Fighting: They select fighting if they sense another cricket in their personal field. The probability P of losing a fight is defined as $P=\alpha$. Therefore, they

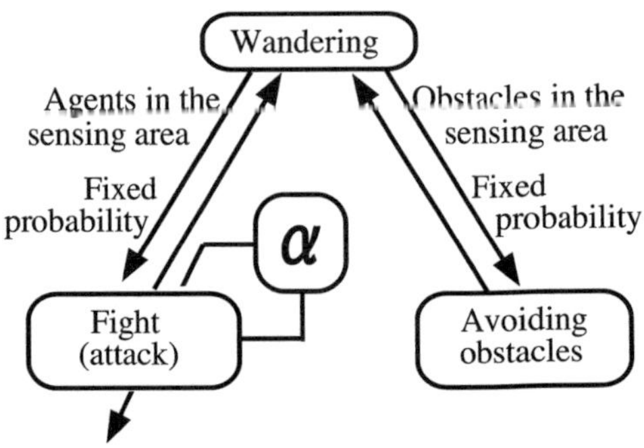

Fig. 20.2 Finite automaton model of cricket fighting behavior

lose with probability P and continue fighting at probability (1-P). This behavior continues until one of the artificial crickets retreats from the fight. The process described above is made within the sampling time 0.1 s. When the battle is settled, one of the artificial crickets turns to the opposite direction from the winner and becomes a loser. On the other hand, the other artificial cricket notices it is winning, and the winner will go on to perform wandering behavior. If the fighting continues for more than 10 s, one artificial cricket is randomly determined as the loser. This model represents actual fighting of real crickets, which typically terminates in fewer than 10 s. The possibility exists that both crickets lose a fight when they give up fighting simultaneously; this phenomenon is observed in real crickets.

(iii) Obstacle avoidance: If an artificial cricket senses an obstacle with an antenna, it will turn to the opposite side. However, if the artificial crickets sense obstacles with both antennae simultaneously, they will turn randomly right or left. This behavior will stop when the artificial crickets no longer sense obstacles. If artificial crickets sense obstacles and other crickets simultaneously, these artificial crickets respond first to obstacles and then to other crickets.

The artificial crickets update the internal state α when they interact with others as follows:

$$\alpha_{t+1} = (1-\omega)\boldsymbol{\alpha}_t + \boldsymbol{\varepsilon}_{\text{lose}}\boldsymbol{\eta}_{\text{lose}} + \boldsymbol{\varepsilon}_{\text{win}}\boldsymbol{\eta}_{\text{win}}$$

Here, α_t signifies the values of the internal states α at time t. The coefficient of a memory of loss is ω. The coefficient effects of fighting with other crickets are $\varepsilon_{\text{lose}}$ and ε_{win}. These coefficients are set by trial and error, $\omega = 10^{-2.5}$, $\varepsilon_{\text{lose}} = 0.5$, and $\varepsilon_{\text{win}} = -0.5$. Then, $\boldsymbol{\eta}_{\text{lose}}$ and $\boldsymbol{\eta}_{\text{win}}$ are defined as follows: $\boldsymbol{\eta}_{\text{lose}} = 1$ (if agent loses) and $\boldsymbol{\eta}_{\text{lose}} = 0$ (other). Likewise, $\boldsymbol{\eta}_{\text{win}} = 1$ (if agent wins) and $\eta_{\text{win}} = 0$ (other). The simulation results have shown that this proposed model reproduces group size-dependent aggressiveness in real crickets.

One of the most important things to consider when designing systematic models is to make these models as simple as possible. Simple models can tell us what is missing when we attempt to reproduce the animal or neuronal function underlying a focused behavior. These simple models can give us new insight into forming hypotheses. Indeed, the finite automaton model of cricket behavior not only describes the group size-dependent aggressive behavior but also suggests the importance of an individual's internal state in regulating group size-dependent aggression.

20.4 Reproducing a Group Size-Dependent Aggressive Behavior Using a Simple Robot

Using a robot is another approach that can be used to investigate sophisticated behaviors. Insect biological systems have provided abundant insights for a large range of biomimetic robots (Ritzmann et al. 2000). Reproduction of animal behavior using a robot helps us to understand mechanisms of locomotion and behavior. Here one of the examples using a simple robot to investigate cricket behavior is introduced (Funato et al. 2008, 2011).

Our underlying assumption is that insects temporally arrange the global structure of their neural networks to assist in the selection of the most sophisticated behavior in a given environment. Two levels of approaches have been used to elucidate the neuronal mechanisms underlying the environment-dependent behavior, i.e., (1) modeling the neuronal mechanism underlying behavior selection and (2) identification of a functional component in the brain that responds to the global environment at the individual behavioral level. To correlate these two levels, a robot controlled by a functional model of a brain has been developed. The movement of the robot is generated by an interaction between a given social interaction and the brain-function model, allowing both levels to be studied simultaneously.

Neuronal oscillation in the brain is one of the important functions necessary for processing information, and this phenomenon is central to construction of a model of brain function. The oscillations are evoked by sensory inputs in both vertebrate and invertebrate brains (Laurent and Naraghi 1994; Schadow et al. 2007; Tanaka et al. 2009). The functionality of neuronal oscillations is still disputed, and the insect olfactory system is a useful experimental model system for exploring this phenomenon. Neuronal oscillations evoked by chemical stimulation were observed as a local field potential in the brains of several insects (Laurent and Naraghi 1994; Tanaka et al. 2009). The synchrony between coupled oscillators is influenced by the network structure (Barahona and Pecora 2002). This phenomenon can be interpreted as a temporal transformation of the neural network. Imitation of the oscillation as a component of brain function makes it possible to construct an oscillator network where the output changes depending on the network transformation.

Using the oscillation as a brain function, cricket aggressive behavior can be modeled. When the battle is settled, the winner keeps aggressive status and the loser actively avoids the victorious cricket for some time (Adamo and Hoy 1995). After more than 3 h have elapsed and the crickets encounter one another again, the loser may again engage in an aggressive bout with the victor. However, if the winner and loser fight repeatedly, the defeated male fails to regain aggressiveness and avoids all opponents. As a result, the number of crickets in a population engaging in repeated fighting is limited by cricket density. The proportion of aggressive individuals declines and the frequency of aggressive encounters decreases (Funato et al. 2011). Thus, the effect of repeated encounters is to alter the behavior of the group. The speed of the emergence of this group-level behavior depends on both the density of a given group and the local results of fights. Group size-dependent aggression is known in many animals, including crayfish (Bovbjerg 1953; Patullo et al. 2009) and domestic fowl (Pagel and Dawkins 1997; Estevez et al. 2003). However, few studies have focused on the neuronal mechanisms underlying group size-dependent behavior.

Using a robot and computer modeling, the fundamental neuronal factors elicited during group size-dependent behavior in crickets can be investigated. Group size-dependent behavior must be mediated by the interaction between brain and crickets moving through an environment; so mobile robots with movements governed by a brain-function model (i.e., an oscillator network) have been developed. The question is whether the behavior of an individual cricket in a group is controlled solely by available environmental information or whether crickets need to maintain a memory of previous social interactions. To better address this question, robots either possessed an embedded memory of a previous fight or not.

The brain function required by crickets for selecting condition-dependent behavioral transitions can be investigated using the synthetic approach of building a robot and performing computer simulations. Because both environmental and internal factors are involved in the objective behavior of a cricket, the model must be designed to move and interact with the environment under the control of algorithms characteristic of a cricket brain.

Chemical cues are a major information source used in cricket decision-making. Neuronal oscillations might be involved in the mechanisms that lead to the production of motor control signals. One hypothesis is that the following three main characteristics are crucial components of the agonistic behavioral process: (1) agonistic behavior is linked to diffusive chemicals; (2) diffusive chemicals can change synaptic connections; and (3) the information governing an agonistic behavioral response is coded as an oscillation. It has been proposed that the oscillators that comprise a given network (3) can be controlled by changes in synaptic connections (2) (Funato and Kurabayashi 2008). This method allows the oscillator network to encode alternative states, such as aggression and avoidance, by considering synchrony as a behavioral state.

Figure 20.3a shows the proposed model, where circles are oscillators and the control signal of a motor unit is determined by the synchrony of oscillations. New links between oscillators, determined by the level of "capacitor" on the circuit, play a role in the switching of behavior and can change after a certain period of fighting. Here, new links mimic neuromodulation that can mediate behavioral changes. Then a robot is constructed and behavioral validation is conducted before testing the conditions that induce interaction-dependent behavior (Fig. 20.3b).

The purpose of combining a simple brain-function model with a robot is to elucidate the main factors governing environmentally dependent behaviors, in particular identifying correlations between brain mechanisms and behavior. One can now examine whether the simple robot can induce variations in the number of winning crickets and the dependence on group size. A computer simulation of a group of agents whose rules of action are consistent with the mechanisms of the constructed robot was performed. Four agents are placed in a virtual, two-dimensional, square field matching the area used in real cricket experiments (large size, 1,200 cm^2; middle size, 300 cm^2; small size, 75 cm^2). The simulated behavior of agents is observed and the number of continuously fighting agents is recorded (Fig. 20.4). Each agent moves using the rules described for the oscillator network robot. For example, during agent encounters, each agent chooses either fighting or avoidance. A change in behavior occurs if the charge level of a "capacitor" exceeds 70 %. Thus, the effect of losing, a key element of group

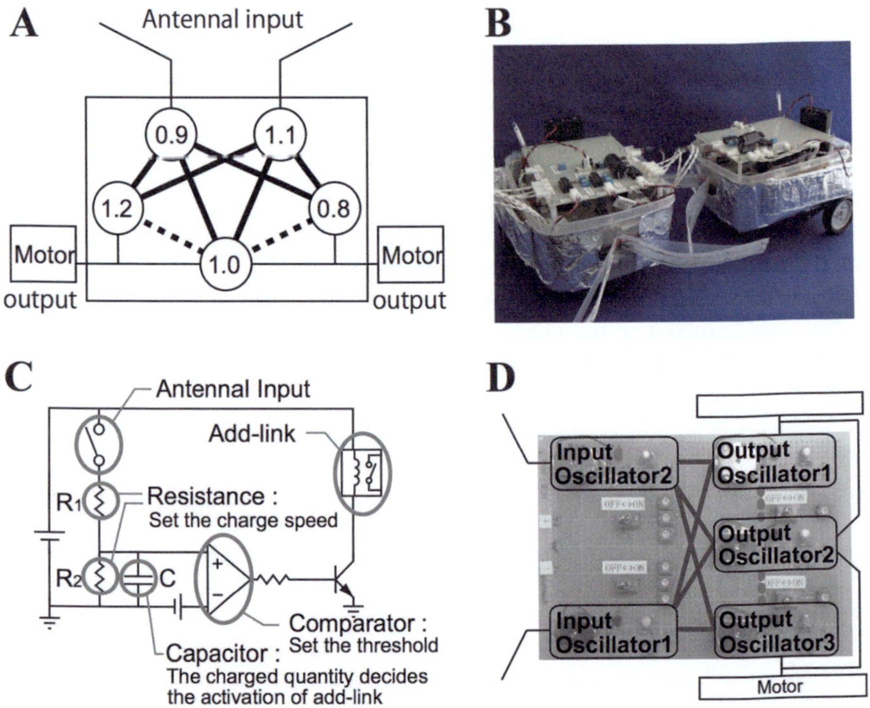

Fig. 20.3 Cricket robots. (**a**) The robot model is composed of an oscillator network that transfers the network structure to robot behavior. The circles represent oscillators and the lines represent connections between oscillators. The robot has two states depending on the existence of cross-links (*dotted lines*). The activation of motors connected to the oscillators is changed by the synchrony of the oscillators, while the synchrony of the oscillators is modified by the cross-links. Thus, the cross-link connections change the robot behavior. The connections indicated as cross-links change several minutes after the robot starts fighting, resulting in a change in the robot behavior. (**b**) Cricket robots. (**c**, **d**) Robot control circuit. The robot has two motors with wheels, two sensor inputs, and a network based on the model of the chemical processing centre. The outputs of the oscillators were connected to motors such that they activated only when they were synchronized, making the robot move in the same manner

aggression, can be closely examined. To simulate the memory of losing after avoidance behavior is selected, a mechanism is provided to accelerate the charge speed and reduce the discharge speed of the "capacitor." This operation postpones the recovery time. The behavior of a group of agents equipped with the ability to remember a losing experience (Fig. 20.4b) can now be compared to a group whose actions are not affected by memories of past fights (Fig. 20.4a). This allows us to examine how the memory of losing a fight influences group aggression.

Robot modeling of behavioral switching based on an oscillator network can also be examined. If there are no "cross-links," the robot changes its direction and heads toward the obstacle. This behavior corresponds to the cricket behavior heading toward another cricket, which is regarded as aggressive behavior. If cross-links are

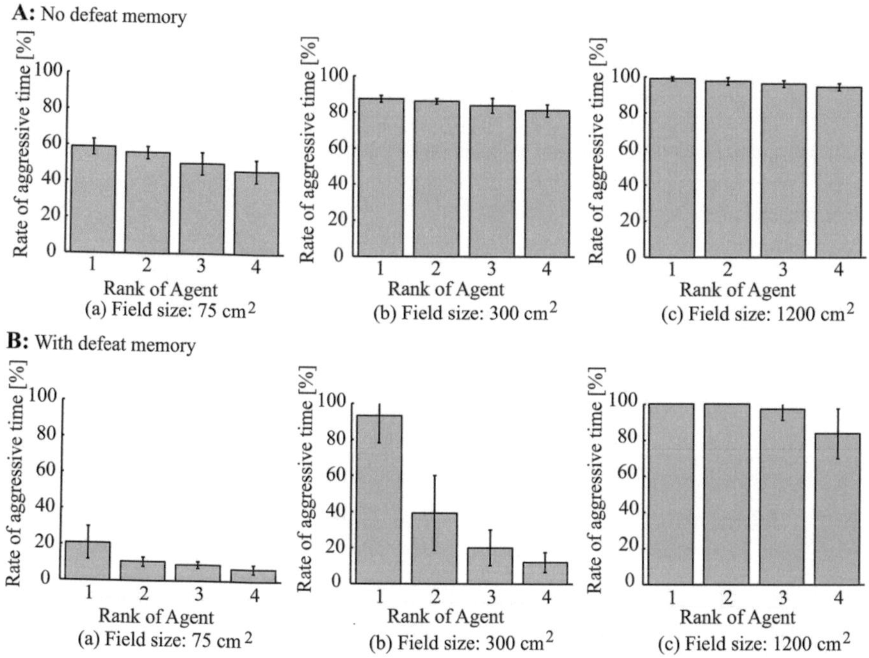

Fig. 20.4 Duration of aggressive behavior (as a percentage of total simulation time) for each agent. The duration of aggressive behavior includes the time that agents behave in an agonistic manner when they encounter other agents; this includes walking and fighting. The simulation is repeated five times with different initial conditions, and the bars are shown in order of the rank (1–4) for each trial. Values are means ± s.d. for the five trials. Results for agents that did not use the memory of their previous fights to decide their behavior. Aggression decreased at an almost equal rate according to the size of field (**a***a*–*c*). Results for agents equipped with memory, where the number of aggressive agents changed depending on the field size (**b***a*–*c*). Only one agent was continually aggressive with the medium field (**b**), every agent was aggressive with the large field (**c**), and no agents were aggressive with the small field (**a**)

present (the neural network is transformed after a fight), the robot heads away from the obstacle. Thus, the robot behaves in a submissive manner as a consequence of the structural change in the network. This indicates that we can successfully achieve a change in robot behaviors as a consequence of network structure. The delay circuit alternates the connection after a certain period of fighting, so the robot initially exhibits aggressive behavior by heading toward the obstacle, but later avoids an obstacle. Thus, the robot model simulates the individual interaction of crickets when the initial contact between two crickets results in a fight. This is followed by avoidance behavior with network transformation after a certain period.

Behavioral transitions in simulated agents with variable group densities can also be examined. Tests of group behavior using four virtual agents, which possess the mechanism described for the robots, deepen our understanding of the group behavior of live crickets, i.e., the number of aggressive crickets change depending on the arena size. Performance of the two groups can be tested using simulation to compare group behavior, i.e., the recovery time of one group is not affected by the result of the fight (without memory; Fig. 20.4a), whereas the recovery time of the other group is postponed in the defeated agent (with memory; Fig. 20.4b). Without memory, no specific agent appears dominant in the simulation, and the aggressiveness of every agent is uniformly suppressed by reducing the field size. On the other hand, with memory, the simulation results demonstrate that only the highest-ranking agent fights for most of the simulation period, whereas the other agents continuously avoid the highest ranked agent. With the large field, every agent fights continuously for most of the simulation time, whereas in the small field they stop fighting. This differentiation among agents that possess a memory of losing is similar to that observed in real cricket groups. Modeling these behaviors with a simple robot suggests to us that the loser effect plays an important role in determining group size-dependent aggression in crickets.

20.5 Manipulation of Live Cricket Behavior Using a Robot

The variety of approaches for investigating cricket behaviors can be classified into three main categories based on experimental setup: (1) one shot, (2) treadmill, and (3) free moving. In the one-shot setup, a cricket, an arena, and other apparatus are repeatedly reset into a given fixed initial condition. A stimulus is presented in a controlled fashion. Typically the analysis is focused on the behavior that immediately follows, thus allowing carefully controlled investigations of a desired behavior that can be experimentally triggered. The treadmill setup is popular for tracking crickets over a longer course. This setup is a great solution for tracking the trajectory of a single individual in a very controlled environment. However, relying on rigid controlled conditions prevents the unfolding of realistic complexity, which is typical of longer interactions. A free-moving setup is ideal for investigations of behavior. However, these experiments are difficult to control.

Behavioral manipulation using a robot is thus aimed at extending the possibilities of the free-moving setup (Guerra et al. 2010). Small attachments can be added to a controlled robot in order to convey nonstationary behavioral cues such as chemical attachments or textures. In addition, other tactile cues like antennae or visual lures can be added. For example, as shown in Fig. 20.5a-2, a robot can be fitted with the anterior portion of a male cricket, including the head and part of the thorax. Letting male crickets interact with robots rather than a live adversary or stationary cues allows the investigator to control which aspects of an opponents' movements are influential. Playing back the same sequence of movements at each trial and comparing results from trials that used robots with and without attachments allow us to study the influence of those attachments in a dynamic situation. Additionally this setup also allows the possibility of programmatically modulating the robot's movements in real time in response to the actions of the live cricket.

The micro-robot we have developed is of a size comparable to that of a real cricket (Fig. 20.5a). The robot has dimensions of $18 \times 18 \times 22$ mm and is driven by

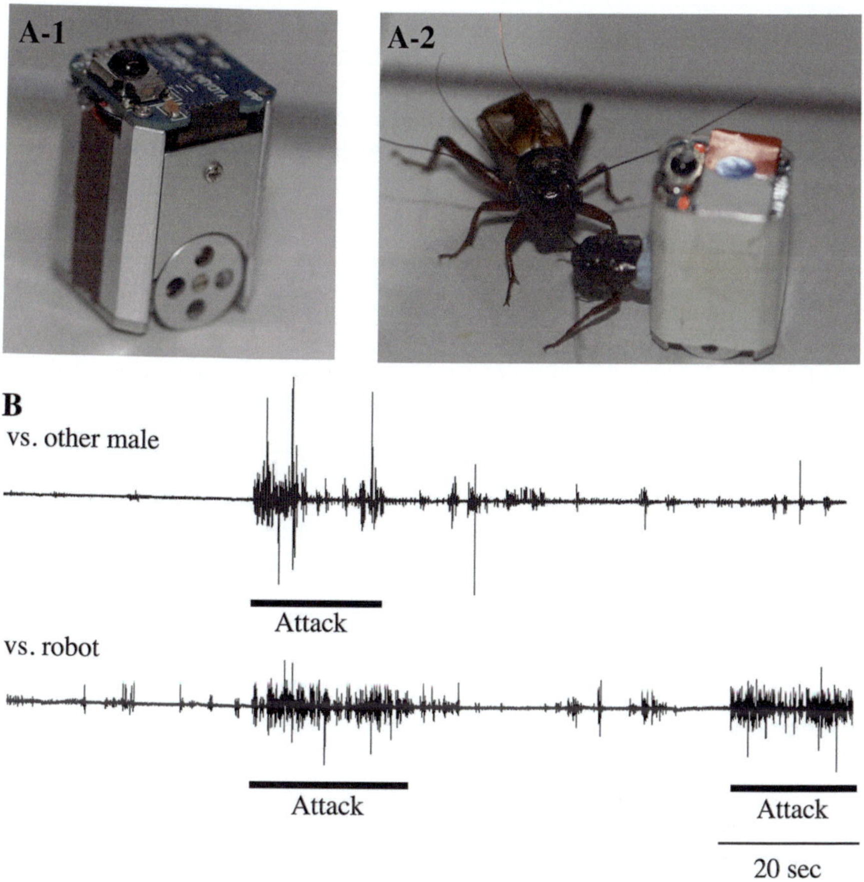

Fig. 20.5 (**a**) Manipulation of cricket behavior using a micro-robot. (**b**) Electromyograms (EMGs) of the mandibular muscles in the male cricket. A male cricket whose mandibular muscle EMG is being recorded as it attacks the robot with a male head part. *Upper* trace shows the EMG of a male as he opens his mandibles to start attacking a real cricket opponent. The *lower* trace shows the EMG of a male as he opens his mandibles to start attacking a robot with male head part

two differential wheels. It has no sensors except for an infrared receiver, which is used for receiving commands encoded in pulses of infrared light. The robot is capable of moving on smooth 2D surfaces at speeds comparable to those of a cricket when it is wandering. Movements can be controlled both in an open or closed loop with or without real-time feedback from other crickets' positions. To allow closed-loop control, an external camera gives real-time feedback of the robot's position and orientation (Guerra et al. 2007). A robot's movements can be composed of a set of prerecorded movements that are repeatedly interlaced with short random movements. Alternatively, commands playing in a loop cause the robot to move without the use of any feedback. A series of random movements can be automatically interjected between each repetition to disturb the trajectory, thus avoiding systematic preference toward specific paths. The main robot parts are a motor, customized from a wristwatch motor unit for higher torque; a battery, a miniature one-cell rechargeable 3.7 V lithium ion polymer battery with a capacity of 65 mAh; a control board, currently based on the microchip 8 bit PIC18 family of microcontrollers (each robot comes equipped with a PIC18LF1220 which features 4 kb of re-programmable flash memory); an IR sensor, used in order to listen to commands from the PC; and a body, the hard, durable body of the robot is micro-machined in aluminum.

It is possible to record physiological changes in the central nervous system, muscles, etc. from a free-moving cricket that is interacting with a robot (Fig. 20.5a-2, b). Electromyograms can be obtained using wire electrodes inserted proximally into the mandibular muscle of the free-moving cricket. Since the robot can be used to repeatedly trigger attack behavior in a free-moving cricket, such as mandible spreading, we can begin to control and measure a variety of behaviors triggered by interacting with an opponent. Given that it is typically difficult for us to predict a winner or loser of a fight, using a robot allows us to more consistently record desired responses from the free-moving cricket. Therefore, a controlled robot is a powerful approach that can be used to gain a deeper understanding of the neuronal mechanisms of cricket behavior.

References

Adamo S, Hoy R (1995) Agonistic behaviour in male and female field crickets, Gryllus bimaculatus, and how behavioural context influences its expression. Anim Behav:1491–1501

Ashikaga M, Sakura M, Kikuchi M, Hiraguchi T, Chiba R, Aonuma H, Ota J (2009) Establishment of social status without individual discrimination in the cricket. Adv Robot 23:563–578

Barahona M, Pecora LM (2002) Synchronization in small-world systems. Phys Rev Lett 89:054101

Bovbjerg RV (1953) Dominance order in the crayfish *Orconectes virilis* (Hagen). Physiol Zool 26:173–178

Estevez I, Keeling LJ, Newberry RC (2003) Decreasing aggression with increasing group size in young domestic fowl. Appl Anim Behav Sci 84:213–218

Funato T, Kurabayashi D (2008) Network structure for control of coupled multiple nonlinear oscillators. IEEE Trans Syst Man Cybern B Cybern 38:675–681

Funato T, Kurabayashi D, Nara M, Aonuma H (2008) Switching mechanism of sensor-motor coordination through an oscillator network model. IEEE Trans Syst Man Cybern B Cybern 38:764–770

Funato T, Nara M, Kurabayashi D, Ashikaga M, Aonuma H (2011) A model for group-size-dependent behaviour decisions in insects using an oscillator network. J Exp Biol 214:2426–2434

Guerra RS, Boedecker J, Yanagimachi S, Asada M (2007) Introducing a new minirobotics platform for research and edutainment. In: Proceedings of the 4th international symposium on autonomous minirobots for research and edutainment. p 216

Guerra RDS, Aonuma H, Hosoda K, Asada M (2010) Semi-automatic behavior analysis using robot/insect mixed society and video tracking. J Neurosci Meth 191:138–144

Kawabata K, Fujii T, Aonuma H, Suzuki T, Ashikaga M, Ota J, Asama H (2012) A neuromodulation model of behavior selection in the fighting behavior of male crickets. Robot Auton Sys 60:707–713

Laurent G, Naraghi M (1994) Odorant-induced oscillations in the mushroom bodies of the locust. J Neurosci 14:2993–3004

Lochmatter T, Rodult P, Clanci C, Correll N, Jacot J, Martinoli A (2008) A flexible open source tracking software for multi-agent systems. In: Proceedings of the international conference on intelligent robots and systems (IROS), IEEE, pp 4004–4010

Mizuno T, Sakura M, Ashikaga M, Aonuma H, Chiba R, Ota J (2012) Model of a sensory-behavioral relation mechanism for aggressive behavior in crickets. Robot Auton Syst 60:700–706

Noldus LP, Spink AJ, Tegelenbosch RA (2001) Ethovision: a versatile video tracking system for automation of behavioral experiments. Behav Res Methods Instrum Comput 3:398–414

Pagel M, Dawkins MS (1997) Peck orders and group size in laying hens: 'futures contracts' for non-aggression. Behav Proc 40:13–15

Patullo BW, Baird HP, Macmillan DL (2009) Altered aggression in different sized groups of crayfish supports a dynamic social behaviour model. Appl Anim Behav Sci 120:231–237

Ritzmann RE, Quinn RD, Watson JT, Zill SN (2000) Insect walking and biorobotics: a relationship with mutual benefits. Bioscience 50:23–33

Schadow J, Lenz D, Thaerig S, Busch NA, Frund I, Herrmann CS (2007) Stimulus intensity affects early sensory processing: sound intensity modulates auditory evoked gamma-band activity in human EEG. Intern J Psychophys 65:152–161

Tanaka NK, Ito K, Stopfer M (2009) Odor-evoked neural oscillations in Drosophila are mediated by widely branching interneurons. J Neurosci 29:8595–8603

Tsai RY (1986) An efficient and accurate camera calibration technique for 3d machine vision. In: Proceedings of the IEEE conference on computer vision and pattern recognition, pp 364–374

Chapter 21
Protocols in the Cricket

Hadley Horch, Jin Liu, Taro Mito, Aleksandar Popadić, and Takahito Watanabe

Abstract The last decade has witnessed the rapid growth of techniques that can be used to examine a variety of questions in the cricket *Gryllus bimaculatus*. In this chapter, we provide detailed and well-established protocols for injections (21.1), RNAi approaches (21.2), genome editing (21.3), and gene expression analysis (21.4). These protocols can be used not only in other Orthopteran species, but also as a starting point for functional analyses in other hemimetabolous insects from Odonatans to Hemipterans.

Keywords Methods • Injection • RNA interference • dsRNA • Gene editing • Transgenic cricket • CRISPR • In situ hybridization • Immunohistochemistry

21.1 Injection Methods

For several of the techniques described in this chapter, efficient delivery of material into the cricket is a key step. Whether one is attempting RNA interference experiments or attempting genome modification, injections that effectively deliver materials without compromising the health of the cricket are critical to experimental success. Depending on the experimental design, it may be imperative to introduce this material into eggs, larval nymphs, or adults. Techniques for all three are outlined below.

These protocols were adapted from those developed in several labs predominantly the labs of Sumihare Noji, Aleksandar Popadić, and Cassandra Extavour.

H. Horch
Departments of Biology and Neuroscience, Bowdoin College, 6500 College Station, Brunswick, ME 04011, USA

J. Liu • A. Popadić
Biological Sciences Department, Wayne State University, Detroit, MI 48202, USA

T. Mito (✉) • T. Watanabe
Graduate school of Bioscience and Bioindustry, Tokushima University, Tokushima 770-8513, Japan
e-mail: mito.taro@tokushima-u.ac.jp

H.W. Horch et al. (eds.), *The Cricket as a Model Organism*,
DOI 10.1007/978-4-431-56478-2_21

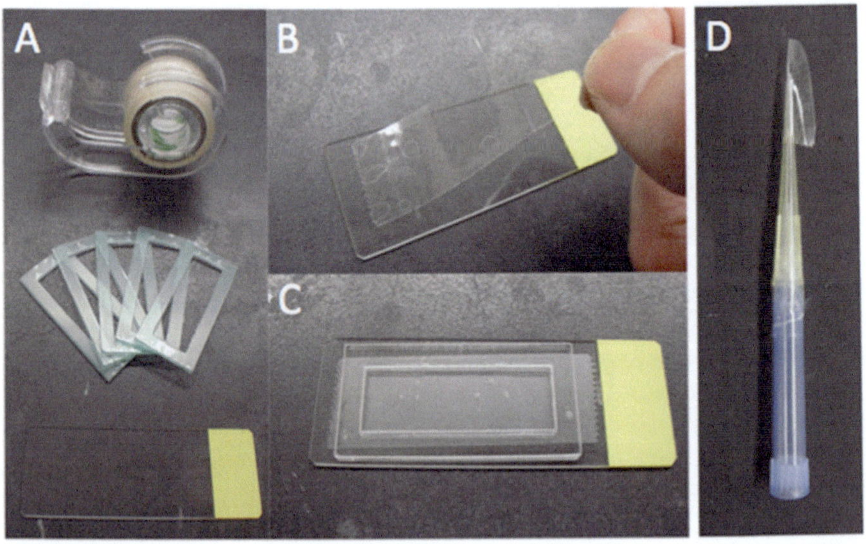

Fig. 21.1 Egg chambers for egg injection "Method 1" can be prepared using (**a**) double-sided tape, Watson chambers, and microscope slides. (**b**) Double-sided slides applied to the clear part of the slide. (**c**) Remove the backing of the Watson chamber to expose sticky side and stick to the slide. Make sure there are no gaps in between the double-sided tape and the Watson chamber. (**d**) "Spatula" made from pipette tips and double-sided tape can be used to transfer eggs

21.1.1 Egg Injections

Two different injection procedures will be described. The original method was developed at Tokushima University in the lab of Dr. Sumihare Noji. This method will be referred to as Egg Injection Method 1. The second method was developed at Harvard University in the lab of Dr. Cassandra Extavour and will be referred to as Egg Injection Method 2.

21.1.1.1 Method 1

1. Preparing egg injection chambers.
   (a) Chambers can be made by applying double-sided tape to a glass slide and mounting a Watson chamber onto the tape. Pressure from a table edge can be applied to remove any bubbles (Fig. 21.1a).
2. Preparing egg spatula
   (a) Folding double stick tape over the end of a plastic pipette tip. Trim to shape spatula into a flat structure with a tiny roll at the crease. This can be taped to a larger (p1000) pipette tip for ease of handling (Fig. 21.1b).

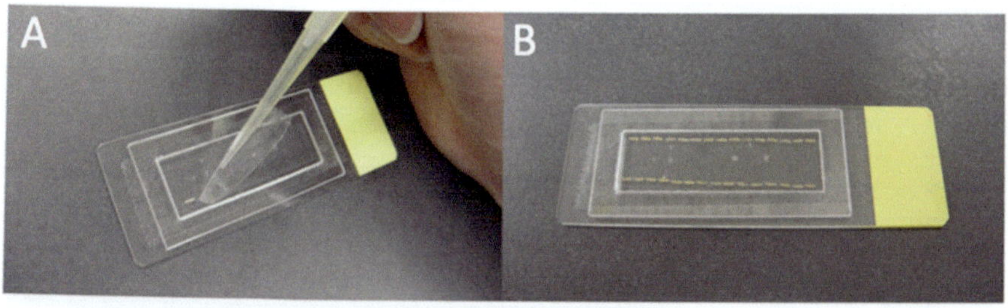

Fig. 21.2 (**a**) Use the spatula to pick up an individual egg and transfer to the edge of the well. (**b**) Eggs can be arranged end-to-end along each long edge of the chamber

3. Collect timed eggs.

   (a) In order to collect eggs for injections, adult cricket should be deprived of any laying material, such as cotton-stopped water vials, for at least 8–10 h. A number of methods for egg collection are possible, including layering moist paper towels into Petri dishes or offering dishes of fine sand. These materials are offered for only 1–2 h so that egg age can be closely timed.

4. Remove the egg dish and place in a 28 °C incubator for 1 h.
5. The injection room should be warm (25–28 °C) and humid (50–60 % RH) with no blowing, dry air.
6. Eggs can be collected from the egg dish of any type by rinsing eggs from the laying material and collecting in a sieve or mesh basket.
7. Soak collected eggs (in a strainer) in 70 % EtOH for 3–5 s, and then dip the eggs through freshwater to rinse.
8. Spread the eggs on a lightly moistened paper towels.

   (a) Pick up individual eggs with the "egg spatula" (a paintbrush may work as well) and place into chamber dorsal side facing the center of the slide (Fig. 21.2a). For ease of injection arrange end to end along each long edge of the Watson chamber as shown. Gently push the eggs into the tape to make sure they won't roll when injected. Try to get eggs on the slide within about 10 min (Fig. 21.2b).

9. Cover eggs with 500–600 uL of mineral oil and place in moist chamber in a 28 °C incubator.
10. Prepare injection needles.

    (a) Pull Drummond capillary glass (#21) needles using a Sutter Micropipette Puller (P-1000IVF).

      (i) HEAT 743, PULL 0, VEL 20, TIME 250
      (ii) HEAT 743, PULL 0, VEL 20, TIME 250
      (iii) HEAT 743, PULL 0, VEL 20, TIME 250
      (iv) HEAT 743, PULL 70, VEL 25, TIME 150

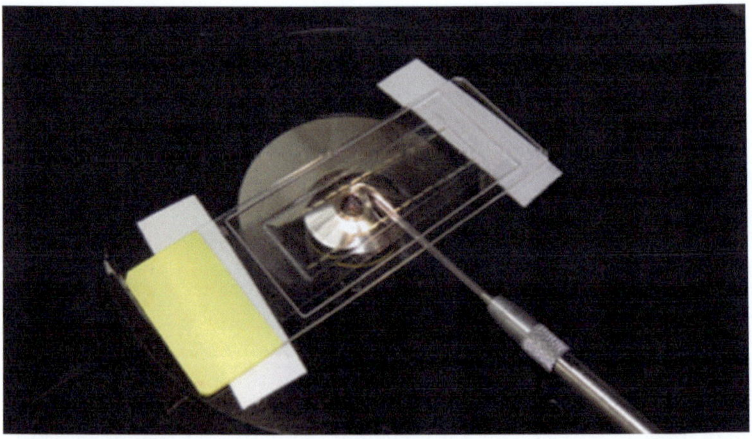

Fig. 21.3 Injection of eggs. Once egg chamber is placed on microscope slide, a micromanipulator can be used to center the needle in the microscope visual field. The microscope stage can be used to move the eggs toward and away from the needle for injection

   (b) The needle shape should be fairly blunt and not too long and bendy. Different pullers may require different program optimization.
   (c) Carefully, under the microscope, break the tip of the needle against the blunt end of another so that the opening is approximately 5–7 μm in diameter.

11. Load needles with injection solution by adding liquid via the blunt end.
12. Inject eggs (Fig. 21.3)

   (a) Place needle on a micromanipulator setup next to a dissecting or compound microscope.
   (b) Attach a pressure-injection device (60 cc syringe) via tubing to the back of the needle.
   (c) Visualize eggs and needle through the microscope at low power.
   (d) Align needle near the posterior end of the first egg (about 20 % egg length from the posterior end). Move the microscope stage to push the egg onto the needle.
   (e) Inject a small amount, typically 2–3 nL
   (f) Move stage to retract needles from egg, and move to the next egg and repeat.

13. After injections, keep eggs on slide in a moist chamber and incubate at 28 °C for 2–3 days.
14. After 2–3 days check eggs. Record the total number of embryos that are alive and that appear transfected (if possible).
15. Make a moist chamber to incubate the eggs by cutting paper towels in ½, four times. Place small squares in a Petri dish and moisten. Press most of excess water out. The squares should be moist but not soaking wet.

16. Very carefully remove eggs from mineral oil and tape using fine forceps.
    (a) Slowly slip the forceps under the egg, moving gently back and forth to loosen the egg from the tape. Use kimwipes or paper towels to remove as much mineral oil as possible. It is best to roll the egg gently on the kimwipe.
17. Place cleaned eggs onto paper towel square in the moist chamber. Lay several eggs every few layers. Cover each layer with moist squares. Once all eggs have been transferred, cover with the lid and place container in small bin in the cricket room.
18. Crickets should hatch in 10–15 days.

21.1.1.2 Method 2

This method is similar in many ways to Method 1, but includes a few key differences, highlighted here. This also requires a few pieces of additional equipment, including a micropipette grinder, a Picospritzer, and a fluorescent dissecting microscope.

1. Needle preparation
    (a) Needles (borosilicate glass capillaries with filament from WPI, cat # 1B100F-4) are pulled in the "Bee Stinger" configuration (#60) with a Sutter Instrument P-97 Flaming/Brown Micropipette Puller using the following program:
        (i) HEAT 621, PULL 100, VEL 12, TIME 250
        (ii) HEAT 621, PULL 100, VEL 12, TIME 250
        (iii) HEAT 621, PULL 100, VEL 33, TIME 250
    (b) If you are using a Sutter Micropipette Puller (P-1000IVF), use the following program:
        (i) HEAT 816, PULL 0, SPEED 15, TIME 250, PRESSURE 500
        (ii) HEAT 816, PULL 0, SPEED 15, TIME 250, PRESSURE 500
        (iii) HEAT 816, PULL 0, SPEED 15, TIME 250, PRESSURE 500
        (iv) HEAT 816, PULL 80, SPEED 15, TIME 200, PRESSURE 0.
    (c) Bevel the tip to 20° using a Narishige Micropipette Grinder (model EG-44) with dropping water. The opening should be 9–13 μm in diameter. Sharpness and fineness of capillary tip is quite important for viability after injection.
2. Preparation of egg trough (Fig. 21.4a)
    (a) Eggs are held in rectangular-shaped troughs made with a plastic mold set in 1.0 % agarose in water.
    (b) Pour approximately 20 ml of hot 0.1 % agarose into a 9 cm Petri dish.
    (c) Set the mold groove side down, tapping to eliminate bubbles. Grooves on mold are 0.7 mm wide × 0.5 mm deep (Fig. 21.4a).
    (d) When solidified, remove the mold and wrap dish in parafilm, and store at 4 °C.
    (e) Add 1X PBS with 50 U/ml penicillin and 50 μg/ml streptomycin.
    (f) Transfer eggs into troughs. Each trough can hold ~35 embryos.

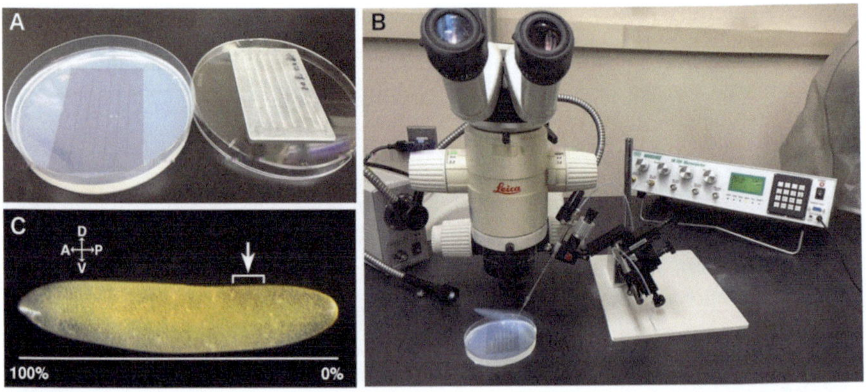

Fig. 21.4 Injection system for cricket eggs. (**a**) The injection gel-board and resin form for making groove. (**b**) Injection apparatus for cricket eggs. This apparatus consists of the microscope and microinjector (NARISHIGE IM300). (**c**) Picture of a cricket egg. Anterior end is *left*; dorsal side is *up*. The injection needle was inserted in the dorsal side of the egg near the posterior end (20–25 % of the egg length from the posterior end; *arrow*)

3. Injection of eggs (Fig. 21.4b)
   (a) Transfer Petri dish to a dissecting microscope equipped with fluorescence.
   (b) Fill the injection capillaries from the blunt end with the injection solution. (Use the typical Tribolium injection solution. 10× recipe is14 mM NaCl, 0.7 mM Na_2HPO_4, 0.3 mM KH_2PO_4, 40 mM KCl.) Rhodamine dextran can be added to the injection solution in order to visualize the injection.
   (c) Mount needle onto a micromanipulator (Fig. 21.4b).
   (d) Connect to a Picospritzer, such as a Narishige, IM-300.
   (e) Set injection pressure to 5–10 psi and injection time to 0.1 s.
   (f) Tribolium injection buffer is used and Rhodamine dextran is added in order to visualize injection.
   (g) Inject the posterior end of the egg (20–25 % of egg length from the posterior end; Fig. 21.4c) with approximately 5 nL of solution.
4. Use forceps to slide the egg off the needle.
5. Using a P1000 micropipette, transfer injected eggs to a 6 cm plastic dish filled with PBS^+ buffer.

21.1.2 Nymphal/Adult Injections

Experiments designed to explore the effects of dsRNA on nymphal development or regeneration may require injections of small cricket larvae. It may also be desirable to inject adults to explore the effects of dsRNA on adult behavior or to deliver dsRNA into embryos via gravid females ("parental" RNAi). The protocol below recommends using a Drummond Nanoject for the injections. However, you can also place needles onto a Hamilton syringe and inject by hand.

1. Prepare injection needles.

2. Use borosilicate capillary tubes that are packaged with the Drummond Nanoject (cat# 3-000-203-G/X) and pull needles using a Sutter Micropipette Puller (P-1000IVF) using the following program (programs may need to be adapted for different pullers):

   (i) 1 Heat = 743, Pull =, vel = 30, Time = 250
   (ii) 1 Heat = 745, Pull =, vel = 30, Time = 250
   (iii) 1 Heat = 743, Pull = 70, vel = 35, Time = 150

3. Collect crickets of the appropriate age and immobilize by cooling.
4. Load injection solution into injection needle.

   (a) Use a small-gauge needle to fill entire needle with mineral oil.
   (b) Mount needle on Nanoject and advance plunger to fully extend.
   (c) Place 3 μl injection solution onto parafilm.
   (d) Lower the needle into the solution and draw into the tip of the needle.

5. Holding the immobilized cricket, advance the needle and inject larval nymphs between abdominal segments A2 or A3. Alternatively adults can be injected in the soft tissue of the joint between the T3 coxa and thorax.
6. To ensure that the needle isn't too deep or pressed against internal structures such as a muscle, retract the needle slightly before beginning to inject.
7. Set the Nanoject such that each push will inject approximately 70 nL.
8. Plan on approximately 70 nL per instar age. For example, 200 nL (three pushes) for third instar larvae, seven pushes for seventh instar larvae, etc. Alternatively for adults, a total of 1 uL is injected (approximately 14 pushes).
9. After injections are completed, slowly retract the needle.
10. Once the needle has been removed, hold the cricket and examine the injection site for leakage. Though hemolymph may leak out, excessive leaking may lead to lower concentrations of injected material and variable results.
11. If performing a "parental" injection, begin collecting all eggs laid by injected females immediately. Phenotypes are typically evident in eggs laid starting 3 days after injection.

21.2 RNAi Approaches

RNA interference (RNAi) results in the near-complete depletion of the targeted mRNA. More than two dozen studies in the past decade have established that RNAi works effectively in *Gryllus bimaculatus*, making this technique an easy and reliable way for performing loss-of-function analyses. There are several important points, though, that have to be kept in mind to assure a successful RNAi experiment:

1. If possible, the targeted region should be close to the 3′ end of the gene so it can encompass at least a part of the UTR sequence. This will reduce the chance of off-target effects.

2. While using a single dsRNA sequence is sufficient in most instances, it is preferable to design and use another dsRNA sequence that spans another region of the gene of interest. If both dsRNAs produce similar phenotypes, then the likelihood of off-target effects is quite low. In addition, some regions simply do not result in RNA depletion and have to be replaced by alternative target sequences.
3. The optimal size of the resulting double-stranded RNA (dsRNA) is 300–500 bp.

21.2.1 Protocol

21.2.1.1 Double-Stranded RNA Preparation

PCR Amplification of Template DNAs for Transcription

1. Design gene-specific primers, and amplify target region (optimal size is 300–500 bp and include UTRs as mentioned above).
2. Clone the generated amplicon into pGEM-T Easy (Promega) or similar vector that has T7 and SP6 polymerase sites.
3. Use this clone with an insert as template for an 80 μl PCR.

2.5 mM dNTP mix	8 μl
10X PCR buffer	8 μl
DNA vector template	4 μl
10 μM T7 primer	4 μl
10 μM SP6 + T7 primer	4 μl
Taq polymerase	0.4 μl
ddH_2O	up to 80 μl

PCR cycle parameters: 95 °C for 1 min; 95 °C for 30 s; 56 °C for 30 s; 72 °C for 60 s (this can change depending on size of product); repeat cycle 35 times; 72 °C for 4 min; hold for 4 °C.

4. Run 2 μl of product on a 2 % agarose gel to visually inspect the quality of DNA. A single, strong band should be evident.

Purification of PCR Product

1. Use MQ water to bring volume up to 200 μl.
2. Add 200 μl PCI (phenol/chloroform/isoamyl alcohol (25:24:1, v/v)); mix and centrifuge for 5 min at maximum speed at RT.
3. Transfer top phase to a fresh tube.
4. Add 2.5× volume (~500 μl) EtOH and 0.01 volume 3M NaOAC (20 μl).
5. Incubate on ice for 10–20 min.

6. Mix and centrifuge at 4 °C, at 15,000 RPM for 15 min to pellet.
7. Rinse pellet in 1 ml 70 % EtOH. Vortex and centrifuge for 5 min at 4 °C, at 15,000 RPM.
8. Remove EtOH, dry pellet, and resuspend in 11 μl TE (depending on size of pellet).
9. Check concentration on a NanoDrop and dilute to 0.5 μg/μl TE.

In Vitro Transcription and RNA Purification

1. Using the Megascript T7 kit (Ambion), set up a transcription reaction as follows, and incubate at 37 °C for about 3 h.

PCR template	(1 μg – probably ~2 μl)
T7 10× reaction buffer	2 μl
CTP solution	2 μl
ATP solution	2 μl
GTP solution	2 μl
UTP solution	2 μl
Enzyme mix	2 μl
RNase-free water	up to 20 μl

2. Add 1 μl Turbo DNase to the transcription reaction.
3. Incubate at 37 °C for 15 min.
4. Add 115 μl RNase-free water.
5. Add 15 μl ammonium acetate (from kit).
6. Extract with equal volume PCI (~150 μl).
7. Vortex and centrifuge at 15,000 RPM for 5 min at RT.
8. Transfer top phase to new tube and add 1 volume isopropanol.
9. Chill at −20 °C for 15 min.
10. Centrifuge at 15,000 for 15 min at 4 °C.
11. Remove EtOH and avoid disturbing pellet.
12. Rinse in 70–80 % EtOH.
13. Centrifuge at 15,000 for 5 min at 4 °C.
14. Redissolve pellet in 50 μl TE.

Annealing RNAs

1. Use a beaker about half full with water, and heat to boiling in a microwave.
2. Place the beaker on stand over Bunsen burner or hot plate.
3. Wrap tubes with parafilm, and then cap to make sure Eppendorf tubes stay closed.
4. Place tubes in a floater and place in boiling water.
5. Boil for 5 min.

6. Remove beaker from heat source and let cool O/N on counter.
7. Add 100 μl EtOH (two volumes).
8. Add 5 μl NaOAc (0.1 volume)
9. Incubate at −20 °C for 10 min
10. Spin for 15 min at 15,000 rpm at 4 °C
11. Remove EtOH
12. Rinse with 500 μl 70 % EtOH
13. Centrifuge for 5 min at 15,000 RPM at 4 °C
14. Remove EtOH and let pellet dry for 3–4 min only
15. Depending on size of pellet, dissolve in about 20 μl MQ H_20 or vehicle.

Preparation of Injection Solution

1. Dilute 1 μl of your dsRNA into 9 μl water.
2. Check concentration using a NanoDrop.
3. Multiply your measured amount by 10 (to make up for the 1:10 dilution you did).
4. Calculate concentration using the following formula:

$$\frac{X \mu g/\mu l}{310 \times 2 \times \text{bp size of inset}}$$

   where 310 is the average molecular weight of a base and 2 is the number of strands you have. The generated result will be in μM.
5. Adjust concentration to 20 μM by adding the necessary amount of water or vehicle.
6. Check 1.5 μl of remaining diluted sample on gel. Ideally only one large band, a bit bigger than the expected size, should be observed. Alternatively, a smear above the expected band can be observed, which is also OK. [Note: a smear below the expected band means that the sample is likely degraded.] Larger bands may be evident as well.
7. Aliquot 5 μl per tube, label, and freeze.

21.2.1.2 RNAi Injections

Double-stranded RNA solution is injected into early embryos (embryonic RNAi), nymphs (nymphal RNAi), or adults (parental RNAi). For basic injection techniques for each case, refer to the appropriate part of Sect. 21.1. Supplemental information below.

Embryonic RNAi:

1. Allow females to lay in egg dish for 2 h and use the eggs for injection within 1 h of collection.

Regeneration-Dependent RNAi:

1. To analyze gene function during the process of leg regeneration, amputate the leg at the distal tibia after dsRNA injection into a nymph at the third instar stage. You can observe effects of a specific RNAi in subsequent molts.

Parental RNAi:

1. The day after injection, add male adults into the cage of the injected females to breed them (you should not add males immediately after injection, because the injected females require time to recover from injection damage).
2. Collect eggs starting from 3 days after injection and incubate them at 28 °C until an appropriate stage for analysis. The ratio of affected animals will reach a maximum within a few weeks after the start of egg collection.

21.3 Gene Editing

21.3.1 Introduction

The two-spotted cricket, *Gryllus bimaculatus* (Orthoptera: Gryllidae), is one of the most abundant cricket species. It can be easily bred in the laboratory and has been widely used to study insect physiology and neurobiology (Niwa et al. 2000). This species was recently established as a model system for studies on molecular mechanisms of development and regeneration. To analyze gene functions during embryogenesis and regeneration in *G. bimaculatus*, the RNA interference (RNAi) technique can be used (Mito et al. 2005; Nakamura et al. 2008). Furthermore, an approach to generate transgenic crickets using the *piggyBac* transposase has recently been established (Shinmyo et al. 2004; Nakamura et al. 2010). *eGFP*-expressing transgenic cricket lines allowed us to perform live imaging of fluorescently labeled embryonic cells and nuclei. Blastoderm cells were found to move dynamically, retaining their positional information to form the posteriorly localized germ anlage (Shinmyo et al. 2004).

Although the above RNAi and transgenic systems are effective for analyzing gene function, each system has some shortcomings. For example, RNAi can easily and efficiently inhibit gene functions, but inhibition is not complete, because some RNA remains intact. In the case of *piggyBac*-based transgenesis, transgenes are randomly integrated into the genome, meaning that one cannot control either the copy number or the genomic locus of a transgene. Thus, in order to conduct more sophisticated functional analyses of individual genes, a technique for modifying the cricket genome at a specific site is needed.

Genome editing tools such as zinc-finger nucleases (ZFNs), transcription activator-like (TAL) effector nucleases (TALENs), and clustered regularly interspaced palindromic repeat (CRISPR)/CRISPR-associated nuclease 9 (Cas9) (known as the CRISPR/Cas9 system) can be used to induce targeted DNA double-stranded breaks (DSBs) into specific regions of the genome (Miller et al. 2011; Jinek et al. 2012). ZFNs and TALENs consist of an engineered array of zinc-finger or TAL effector repeats, for binding specific target sequences, fused to the *FokI* cleavage domain (Porteus and Carroll 2005; Moscou and Bogdanove 2009). In order for *FokI* dimers to cleave DNA, two ZFN or TALEN molecules are constructed to target contiguous sequences in each DNA strand, separated by a spacer sequence (Rémy et al. 2010). On the other hand, the CRISPR/Cas9 system is capable of recognizing and cleaving specific target sequences by Cas9 nuclease and a guide RNA (gRNA). A gRNA, which is a synthetic fusion of two RNAs, crRNA and tracrRNA, is incorporated into the Cas9 protein, which contains two nuclease domains. The gRNA guides the Cas9 nuclease to the targeted location on the DNA, allowing Cas9 to cut the DNA at that specific location. The Cas9 nuclease from *Streptococcus pyogenes*, which is most commonly used in genome editing studies, can be programmed to cleave any DNA sequence three base pairs upstream of a 5′-NGG-3′ PAM sequence (Mojica et al. 2009; Cong et al. 2013). Because of this flexibility, the CRISPR/Cas9 system has been more broadly used in genome editing studies than ZFNs and TALENs.

Induction of DSBs by any of these genome editing tools activates the DNA damage response (Rémy et al. 2010). A DSB can be repaired by nonhomologous end joining (NHEJ), in which short insertions or deletions are generated at the cleavage site (Lieber et al. 2003), or by homology-directed repair (HDR) with a DNA template, in which gene knock-in results in a perfect repair or a sequence replacement, if a modified template is used (Weiner et al. 2009; Fig. 21.5). HDR

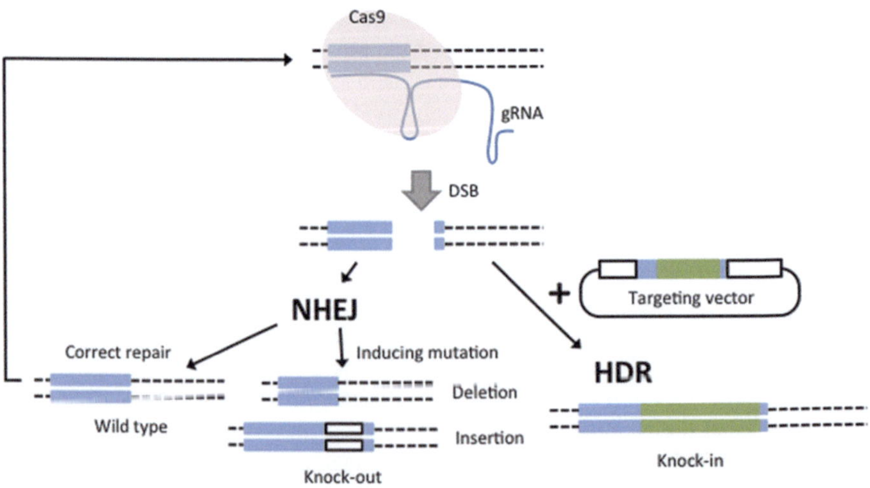

Fig. 21.5 Summary of genome editing using CRISPR/Cas9 system. The gRNA incorporated into the Cas9 nuclease recognized a target sequence on the genome. DSBs were introduced into the targeted site by the CRISPR/Cas9 system. DSBs are repaired via NHEJ or HDR (if donor plasmids are provided). In the NHEJ repair pathway, these breaks are either repaired correctly or In/Del mutations are introduced

can be much more precise, since DSBs are repaired through a specific homologous template. Although gene knock-in is indispensable for further detailed analysis of genome function, establishment of the knock-in method through an HDR pathway is far more difficult than gene knock-out, because of the general low efficiency of HDR in eukaryotes (Hagmann et al. 1998).

Recently, efficient methods for gene knock-in through NHEJ have been developed in zebra fish (Auer et al. 2014). In this method, both genome and donor vectors are cleaved in vivo and then some percentage of the time the donor is added at the genomic cut site through NHEJ (Fig. 21.6). This method has advantages in that the knock-in efficiency and the length of integrated sequences are greater compared to HDR-mediated knock-in.

In the cricket, we reported the first example of genome editing in the cricket, demonstrating the effectiveness of ZFNs and TALENs for generating knock-out animals (Watanabe et al. 2012). We have also demonstrated gene knock-outs using the CRISPR/Cas9 system and explored the role of dopamine in learning and memory in the cricket (Awata et al. 2015). In addition, we succeeded in performing functional analysis of developmental genes by generating knock-out lines using the CRISPR/Cas9 system. Recently, we also succeeded in generating knock-in crickets by utilizing a homology-independent gene knock-in method. We generated a donor vector containing gRNA target sequence from the DsRed gene, which does not exist

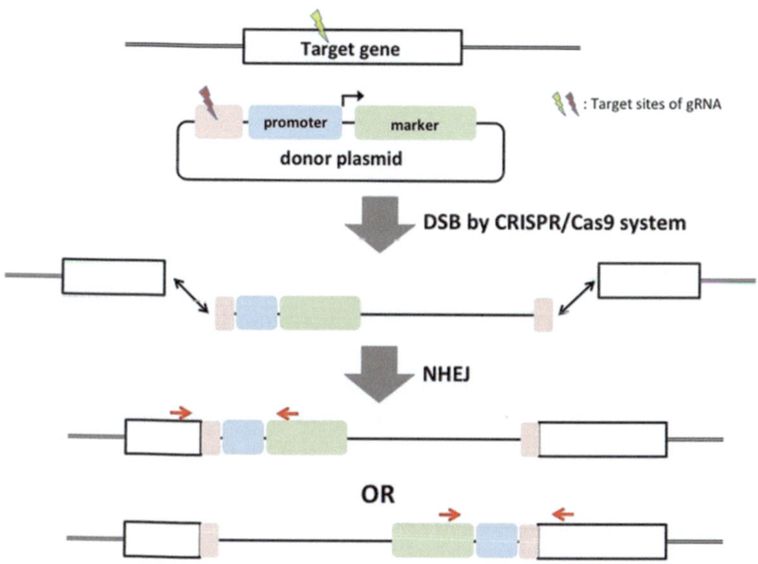

Fig. 21.6 Outline of NHEJ-mediated knock-in. Two gRNAs for a target gene and a donor plasmid were designed and co-injected with Cas9 nuclease and donor plasmids. The donor plasmid (in our case) contained a target site for gRNA and an expression cassette for a marker or gene of interest. Induced DSBs in genome and donor plasmids were ligated via the NHEJ repair pathway. For detection of knock-in events, inside-out PCRs were performed using primer pairs shown by *red arrows*

in the cricket genome, and targeted this sequence upstream of autonomous GFP expression cassette. Cas9 mRNA, gRNAs targeted against the genome sequence, and a donor vector were co-injected into fertilized eggs. To check for the presence of the knock-in event, inside-out PCR was performed at 7 days postinjection using specific primers for each 5′ and 3′ junction point (Fig. 21.6). For some genomic loci, we were able to isolate knock-in lines and observe GFP expression in embryos (unpublished). In this chapter, we provide the protocols for using CRISPR/Cas to generate knock-out and knock-in cricket lines.

21.3.2 CRISPR/Cas

21.3.2.1 Preparation of Plasmid for Cas9 mRNA In Vitro Transcription

There are several plasmids available from Addgene for the in vitro transcription of Cas9 mRNAs. We usually use the pMLM3613 plasmid containing wild-type *Streptococcus pyogenes* Cas9 (*SpCas9*) under the T7 promoter and without codon optimizing (Hwang et al. 2013). Currently, although plasmids containing Cas9 variants from other bacteria species (e.g., *Staphylococcus aureus*, *Neisseria meningitidis*) are available, cleavage activities of these variants have not been evaluated in the cricket. Cas9 protein is also available from PNA Bio and Thermo Fisher Scientific, and they work well in cricket eggs.

21.3.2.2 Preparation of Plasmid for gRNA In Vitro Transcription

For designing gRNA spacer sequences against each target genome, many web-based or stand-alone tools are available. We usually combine the CRISPR/Cas designer tool from the ZiFiT web site with the Cas-OT stand-alone program (Sander et al. 2010; Xiao et al. 2014). For in vitro transcription of designed gRNAs, we usually use the pDR274 plasmid, obtained from Addgene, containing an empty gRNA scaffold under control of the T7 promoter (Hwang et al. 2013). To clone designed target sequences into pDR274, each target must have a GG at the 5′ end, because transcription is started from the GG of the T7 promoter 3′ end.

1. For cloning designed target sequence into pDR274, prepare two oligo DNAs.

   Forward; TA-GGNNNNNNNNNNNNNNNNNN
   Reverse; AAAC-NNNNNNNNNNNNNNNNNN

2. Mix 1 μl of each DNA oligo (100 μM concentration) and 8 μl water, and anneal using decreasing temperature steps as follows:

   95 °C, 1 min –> (−2.0 °C/s) –>85 °C, 10s –> (−0.3 °C/s) –>25 °C, Hold

3. Dilute annealed oligo DNA to 1 µM.
4. Ligate pDR274 and annealed oligo DNA using the Golden Gate cloning method with BsaI and T4 DNA Ligase (NEB).

10× T4 DNA Ligase buffer	2 µL
pDR274 plasmid (100 ng/µl)	1 µl
Annealed oligo DNA (1 µM)	3 µl
BsaI	1 µl
T4 DNA Ligase	1 µl
ddH$_2$O	up to 20 µl
[37 °C, 5 min/16 °C, 10 min] × 10 cycles –>12 °C	

5. Add 0.5 µl BsaI to the Golden Gate cloning mix for complete digestion of pDR274.

Golden Gate cloning mix	20 µl
BsaI	0.5 µl
50 °C, 60 min –>80 °C, 15 min –>12 °C	

6. Transform competent cells using 10 µl of the Golden Gate cloning mix.
7. After plasmid preparation, check the cloned oligo DNA sequence.

21.3.2.3 In Vitro Transcription of Cas9 mRNA and gRNA

1. Purify plasmids for the Cas9 mRNA and gRNA using a plasmid purification kit.
2. Suspend plasmids in 50 µl of TE buffer.
3. Linearize the plasmids for mRNA synthesis with a restriction enzyme (*Pme*I for pMLM3613, *Dra*I for pDR274).
4. Extract linearized DNAs using phenol:chloroform:isoamyl alcohol.
5. Perform ethanol precipitation, and resuspend DNA in TE buffer at 0.5 µg/µl.
6. Synthesize Cas9 mRNA and gRNA with the mMessage mMachine T7 Ultra Kit, according to the manufacturer's protocol.
7. Resuspend RNA pellets in nuclease-free water, and adjust to a final concentration of 2 µg/µl.
8. Aliquot the solution into a 1.5-ml tube, and store at −80 °C.

21.3.2.4 Injection of Cas9 mRNA and gRNA Cocktail into Cricket Eggs

Follow "Method 2" directions for egg injections (Sect. 21.1.1.2).

1. Make injection solution.

    (a) Mix Cas9 mRNA and gRNA for the injection solution (final conc. 1 µg/µl and 0.5 µg/µl, respectively). To obtain maximal efficiency for each of target, one may need to adjust the ratio of Cas9 mRNA and gRNA conc.).

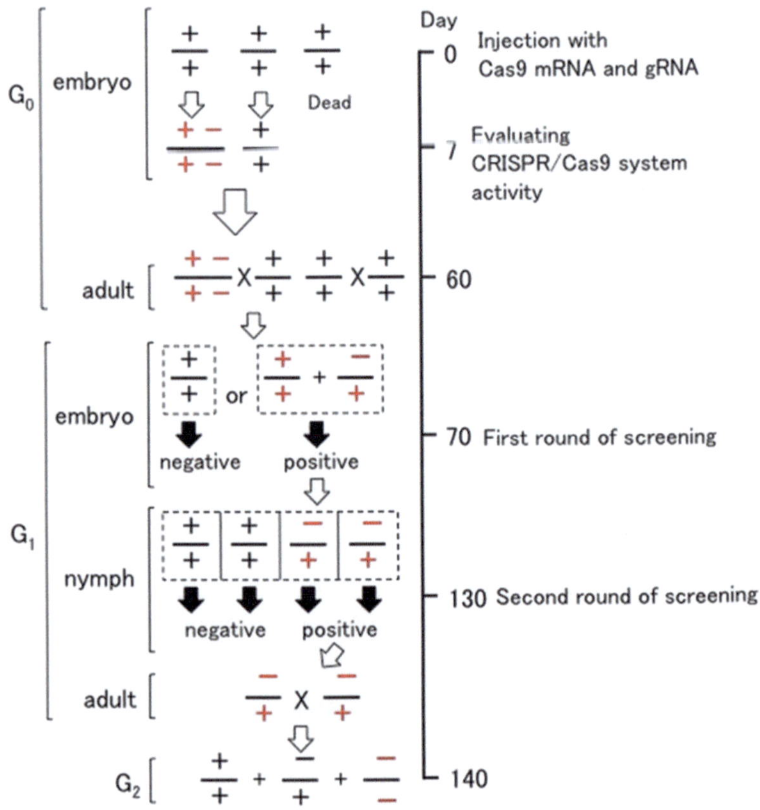

Fig. 21.7 Illustration of the scheme to isolate homozygous mutations. Mutagenized G_0 adults were crossed to wild-type adults. G_1 embryos and nymphs were checked for heterozygous mutations in two rounds of screening using the Surveyor nuclease assay. Positive G_1 adults were then crossed for each strain to obtain homozygous mutants in G_2 generation

21.3.2.5 Generation of Homozygous Knock-Out Cricket Lines via Two Selection Stages (Fig. 21.7)

1. On day 0, inject RNAs into the posterior end of ~150 eggs.
2. On day 7, evaluate the CRISPR/Cas9 system activity using ten eggs (Fig. 21.7 Day 7).

   (a) Select ten CRISPR-treated eggs, and extract genomic DNA from each egg individually, using a genomic DNA purification kit.

   (b) Amplify the 150- to 300-bp fragment, including the gRNA target site, using a 20 μl PCR. Divide the PCR products into halves for SURVEYOR nuclease (Integrated DNA Technologies) treatment or nontreatment.

(c) To allow complementary but mismatched strands to anneal, incubate 10 μl of the PCR products at 95 °C for 5 min, ramp the temperature from 95 to 85 °C at −2 °C/s, and ramp the temperature from 85 to 25 °C at −0.1 °C/s.
(d) Add 0.5 μl Nuclease S and 0.5 μl Enhancer S to the reannealed PCR products.
(e) Incubate the mixture at 42 °C for 45 min (if the digested bands appear smeared during electrophoresis, reduce the reaction time to 15 min or 30 min).
(f) Visualize the SURVEYOR nuclease-treated and nuclease-untreated products by agarose gel electrophoresis immediately after incubation. (The digested bands are often broad because the size of the insertions/deletions can vary among the mutagenized sequences. By using gels that have 6 or 7 mm wells, a clearer band pattern may be achieved.)
(g) Calculate the percentage of eggs for which mutations have been introduced into a gRNA target site. (To obtain G_1 mutants effectively, we recommend the use of gRNAs which introduce mutations into over 50 % eggs.)

3. Approximately 70–80 % of nymphs will hatch on day 13. This hatching rate will be lower if genes related to embryonic development have been targeted. If the knock-out phenotype is visible (as is the case in *laccase2*), then nymphs will exhibit a partial knock-out phenotype because G_0 nymphs are chimeric insects consisting of CRISPR-induced mutant cells and wild-type cells.
4. From days 14 to 50, rear the G_0 nymphs until the eighth instar.
5. To avoid unintentional mating, divide the G_0 nymphs into males and females once they become eighth instar nymphs, and continue to rear separately (about day 50).
6. Three to 7 days after the nymphs become adults (about day 60), mate the G_0 adults to wild-type adults, and collect 200–300 G_1 eggs.
7. One week after the eggs have been laid (approximately 70 days from the original injection), perform the first round of screening with nuclease, to identify cricket lines containing CRISPR-induced mutants (Fig. 21.7 Day 70).

   (a) Combine 25 embryos from each line and extract genomic DNA, using a genomic DNA purification kit. (By using a larger number of embryos, the sensitivity of mutation detection may be reduced.)
   (b) Perform SURVEYOR nuclease treatment (see Sect. 21.3.2.5b–f).
   (c) Isolate SURVEYOR nuclease-sensitive (positive) lines, which contain the mutants.

8. Keep the G_1 eggs from the positive lines of the first screening, rearing the hatched nymphs until the eighth instar stage (day 70–120).
9. A few days after the nymphs from step 8 become eighth instar nymphs (about day 120–130), perform a second round of screening with SURVEYOR nuclease to identify the G_1 heterozygous mutants (Fig. 21.7 Day 130).

   (a) Separately extract genomic DNA from the T3 leg tips of 24 nymphs (12 males and 12 females) in each line, using a genomic DNA purification kit.

(b) Perform SURVEYOR nuclease treatment for each genomic DNA (see Sect. 21.3.2.5b–f).
(c) Isolate the SURVEYOR nuclease-sensitive (positive) crickets, which correspond to the G_1 heterozygous mutants.

10. Rear the G_1 heterozygous mutants until the adult stage.
11. At 3–7 days after the G_1 heterozygous mutants become adults (about day 140), mate the G_1 heterozygous mutant adults individually, and collect the G_2 eggs.
12. Based on Mendelian inheritance, G_2 eggs will include homozygous mutants, heterozygous mutants, and wild types.
13. Genotype the eggs using the SURVEYOR nuclease system or by direct sequencing of the target region.

21.3.2.6 Construction of the Donor Vector for Knock-in Experiment and Checking the Knock-in Event

Donor Vector The summary of NHEJ-mediated knock-in is shown in Fig. 21.6. The donor vector includes a gRNA target sequence, which does not exist in the cricket genome (we usually use partial sequence of *DsRed* gene). We constructed the donor vector containing an *eGFP* expression cassette driven by *Gryllus* actin (G'act) promoter.

Microinjection Microinject as described in Sect. 21.1.1. The injection solution contains Cas9 mRNA (100 ng/μl), the donor vector (100 ng/μl), gRNA for the genome target (50 ng/μl), and gRNA for the donor target (50 ng/μl).

PCR Check of the Knock-In Events Check the knock-in event with inside-out PCR (Fig. 21.6; Auer et al. 2014).

1. Extract genomic DNA from G_0 eggs 7 days after injection.
2. Amplify and determine the junction sequence between the integrated donor vector and flanking genome region using donor vector-specific and target genomic region-specific primers.

Observation of the Fluorescent Marker Expression Check for the presence of marker fluorescence in the embryonic body 5 days after injection.

(Note: Because activity of the G'act promoter would depend on integrated loci, the marker expression may not be detectable as fluorescence.)

Obtaining Knock-In Lines

1. Obtain G_1 eggs by crossing G_0 adults and wild type as described in Sect. 21.3.2.5, steps 3–6.
2. Perform inside-out PCR using genomic DNA extracted from batches of 25 G_1 eggs at 7 days after egg-laying for each cross. (If fluorescence is detectable in embryos, a positive line can be identified without inside-out PCR.)
3. To identify knock-in individuals, perform an additional round of inside-out PCR using genomic DNA separately extracted from the T3 leg tips of 25 nymphs.

21.4 Gene Expression Analysis

The detection of the expression pattern of a gene of interest is an indispensible part of understanding a gene's function. Here we describe protocols for detecting both the mRNA (*by* in situ *hybridization, ISH*) and protein expression (*by immunohistochemistry using Antibody staining, IAb*). The *ISH* protocol was originally developed for the analysis of gene expression in whole-mount embryos and represents the primary technique for studying the localized accumulation of mRNA. We also provide three modified versions of this method, optimized for using a specific tissue source (nymphal legs, nymphal wings, and neuronal tissue). Similarly, the *IAb* protocol was developed for analyzing protein expression in whole-mount embryos and can be used as a source for additional optimization when applied to other tissues.

21.4.1 ISH in Whole-Mount Embryos (Embryonic ISH)

The most critical aspect of embryonic ISH is to maintain an RNAse-free environment, both while fixing the embryos and during day 1 of the procedure. Another important consideration is the duration of the Proteinase K step, which can range between 5 and 10 min depending on the age of the embryo. Finally, while the entire protocol is typically performed in round-bottomed 2 ml tubes (referred to throughout as "Eppendorf tubes"), one can also use 24-well plates with basket inserts (Fig. 21.8a). In fact, this technique has been successfully automated and scaled up using similar plates (Fig. 21.8b) in the Hybrimaster HS-5100 machine (made by Aloka, Japan). Please note that all recipes are given at the end in Sect. 21.4.3.

21.4.1.1 Embryo Fixation

1. Decide on which stage embryos you want to work on; usually the stages present days 4–5 are a good starting point (Niwa et al. 2000; Donoughe and Extavour 2015).
2. Collect eggs and transfer them to a glass depression slide or a small Petri dish. Add a sufficient volume of PBS to submerge the eggs.
3. Identify the anterior end (pointy) and the posterior end (rounded).
4. Hold onto the egg at the posterior end, and use a fine tungsten needle to *poke a hole in the anterior end (this* will ensure that during dissection, the pressure comes out the small hole and doesn't damage the embryo; asterisk in Fig. 21.9a).
5. Use fine dissection scissors to cut off the anterior end of the egg, *posterior to the hole* (Fig. 21.9b).

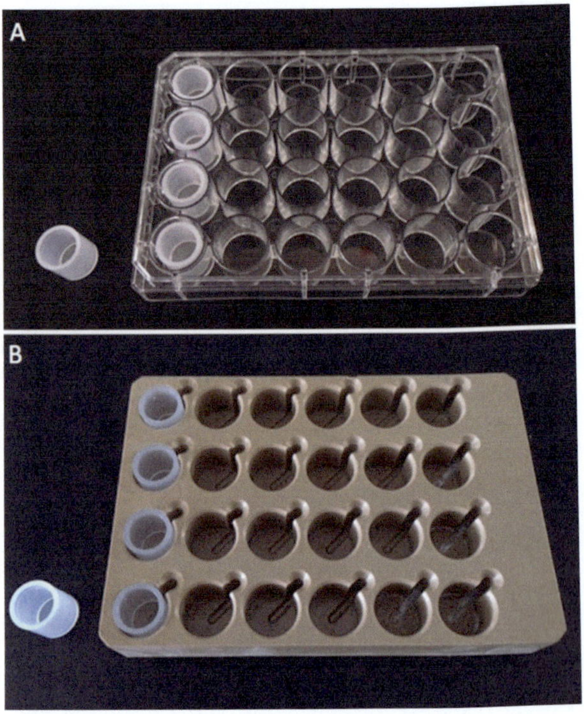

Fig. 21.8 24-well plates can be used with small "baskets" for ISH. Baskets have fine mesh at the bottom to hold tissue. (**a**) 24-well plates are commercially available. (**b**) Metal plates containing a small channel can also be used. This design makes it easy to remove and add solution from below the wells, which minimizes tissue disruption. These plates can also be used with ISH robots such as the Hybrimaster

6. Use the forceps to immobilize the posterior part of the egg, while gradually pushing the embryo out of the egg with another pair of forceps or scissors (Fig. 21.9c).
7. Use a pair of sharpened tungsten needles to separate the embryo from yolk (try to remove as much yolk as possible).
8. Transfer dissected and cleaned embryos to a sterile (RNAse free) 2 ml, round-bottom Eppendorf tube (or small glass scintillation vials can be used). Add a sufficient volume of fix solution (4 % PFA/PBS).
9. Keep at 4 °C and fix overnight (at least 10–12 h).
10. Rinse in PBST several times.
11. Dehydrate fixed embryos in the MeOH series: 25 % MeOH/PBST, 50 % MeOH/PBST, 75 % MeOH/PBST, and 100 % MeOH, 20 min each step on orbital shaker.
12. Store at −20 °C until ready to use.

21.4.1.2 Riboprobe Template Synthesis

Probe Design

Ideally the probe should be about 500 bp in length (~300 bp on the low end and no more than 1,000 bp on high end). *The probe should include a UTR if at all possible,*

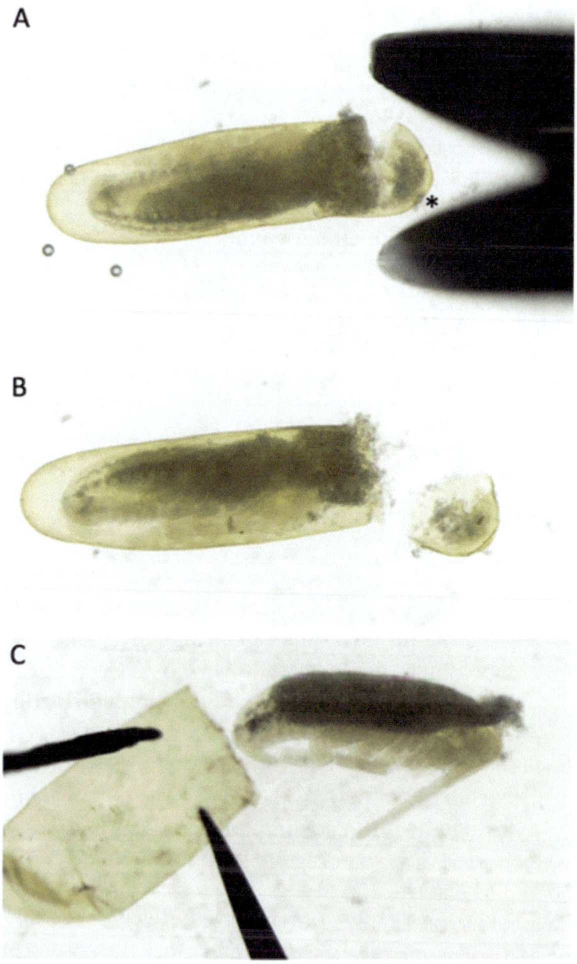

Fig. 21.9 Embryo dissection for ISH. (**a**) A hole is poked in the anterior end (*asterisk*) to relieve pressure on the embryo. Scissors can be used to make a cut posterior to this hole, (**b**) removing the anterior tip of the egg. (**c**) After immobilizing the posterior end of the egg, the embryo can be pushed out of the egg

which decreases the chances for nonspecific binding of the probe. If the probe is significantly longer than 1,000 bp, it will need to be hydrolyzed. Depending on the type of experiment, sometimes it may be desirable to have the probe and dsRNA targeted against different regions of the gene.

Template DNA Creation

1. Amplify desired sequence and ligate into a commercially available vector (such as pGEM-T Easy or pBluescript).
2. Sequence the generated clone to determine the orientation of the insert.
3. Make a riboprobe template by amplifying the insert using either M13 forward and reverse primers or gene-specific primers (GSP) with RNA polymerase binding sites added.

Primer	Sequence
M13 Fwd	CGCCAGGGTTTTCCCAGTCACGAC
M13 Rev	TCACACAGGAAACAGCTATGAC
or	
T7 + GSP sequence	5′-TAATACGACTCACTATAGGG-(GSPseq)-3′
Sp6 + GSP sequence	5′-CAAGCTATTTAGGTGACACTATAGA-(GSPseq)-3′

4. Set up PCR with a total volume of 80 μl.

	1×	4×
2.5 mM dNTP mix	2 μl	8 μl
10× Ex Taq buffer	2 μl	8 μl
DNA vector template (~2–5 ng is all you need)	1 μl	4 μl
10 μM Fw primer	1 μl	4 μl
10 μM Rv primer	1 μl	4 μl
EX-HS Taq	0.1 μl	0.4 μl
ddH_2O water	12.9 μl	51.6 μl
Total	20 μl	80 μl

Use the following PCR cycle parameters: 95 °C for 1 min, 95 °C for 30 s, 56 °C for 30 s, 72 °C for 60 s (this can change depending on size of product). Repeat this cycle 35 times. Include the last extension step at 72 °C for 4 min, and store 4 °C.

5. Check the quality of generated PCR template by running 2 μl of the reaction on a 2 % agarose gel. A single band should be observed.
6. Purify and quantify PCR product as follows. Bring volume of PCR up to 200 μl using ddH_2O water. Add 200 μl of PCI (phenol/chloroform/isopropanol) and centrifuge at RT for 5 min at maximum speed. Transfer top phase to a fresh tube. Add 2.5× the original 200 μl volume (or 500 μl) of EtOH and 0.1 volume of 3 M NaOAC (20 μl).
7. Incubate on ice for 10–20 min.
8. Mix and centrifuge at 4 °C for 15 min (at 15,000 RPM). A small pellet the size of a pinhead should be observed.
9. Rinse the pellet in 1 ml of 70 % EtOH. Vortex and centrifuge at 4 °C for 5 min (at 15,000 RPM). Remove EtOH, dry pellet, and resuspend in 11 μl TE (depending on size of pellet). Check concentration and dilute to 0.5 μg/μl with TE. Store at −20 °C.

21.4.1.3 DIG-Labeled Riboprobe Synthesis

1. Set up the sense and antisense transcription reactions as in the table below.

Component (Megascript kit)	SP6	T7
Transcription buffer	2 μl	4 μl
100 mM DTT or 0.1 M DTT	2 μl	2 μl

(continued)

Component (Megascript kit)	SP6	T7
BSA	2 μl	–
10× DIG-labeling mixture	2 μl	2 μl
Amplified (linearized) template DNA	1 μg	1 μg
RNase out (inhibitor)	1 μl	1 μl
RNA polymerase	2 μl	2 μl
ddH_2O	Up to 20 μl	Up to 20 μl
Total	20 μl	20 μl

Follow the Megascript kit directions (Ambion).

2. Mix the reagents
3. Incubate at 37 °C for 2 h (longer time is OK).
4. Spin down for few seconds.
5. Add 2 μl DNaseI
6. Incubate at 37 °C for 15 min.
7. Add 30 μl RNAse-free water and 30 μl Lithium Chloride Precipitation Solution.
8. Vortex and incubate at −20 °C for 20–30 min.
9. Centrifuge for 15 min at 4 °C (at 15,000 rpm).
10. Aspirate and discard the supernatant.
11. Add 100 μl RNase-free 80 % EtOH.
12. Centrifuge for 5 min at 4 °C (15,000 rpm).
13. Aspirate and discard the supernatant.
14. Dry pellet at RT for a 5–10 min (until it changes color from white to translucent).
15. Resuspend in TE buffer (RNase-free grade). Make sure to vortex tube to ensure the entire pellet is resuspended.
16. Check the quality of the riboprobe by running 1 μl on 1.5 % TAE agarose gel. Measure the concentration (using a NanoDrop) and adjust to 40 ng/μL.
17. Store at −20 °C. *Note*: the preferred way of storing the riboprobe is to use a Watson preservation plate: aliquot 3 μl per well, let dry about 30 min (cover loosely with foil while drying), and then seal with supplied seal. Additional information is available at:

    (a) http://www.watson.co.jp/english/products/pre-plate24.html

21.4.1.4 In Situ Hybridization

Originally, the protocol used single 2.0 mL Eppendorf tubes (containing embryos), which have subsequently been modified for the use of a 24-well plate and small baskets (Fig. 21.8) to make exchanging solutions easier. Before each use, treat these plates and baskets with a small amount of 5 % H_2O_2 for 5–10 min. In addition, a plastic container (Tupperware) that will be used for containing a 24-well plate should also be treated with 5 % H_2O_2 for 60 min in advance of each use. For each step, place the plate in a Tupperware on top of an orbital shaker. NOTE: H_2O_2 solution can be reused many times (15–20×).

Day 1

NOTE: Keep RNase-free conditions throughout day 1.
In advance:

Set up a hybridization oven and water bath to 70 °C.

- Filter 75 %, 50 %, 25 % MeOH/PBST, PBST, and KTBT with 0.22 μm Sterile Millex Filter Unit (Millipore, cat # SLGS033SS) to remove precipitations of Tween 20.
- Transfer MeOH stored samples into a small Petri dish, and view through microscope to select the appropriate stages (and to eliminate damaged embryos).
- At this point, it is best to group embryos by their approximate stage (e.g., separate days 2–3 embryos, days 4–5, days 6–7, etc.) This improves the success of future Proteinase K steps. These embryos can then be recombined after the Proteinase K step.

Rehydration step (NOTE: use only filtered reagents, see above)

100 % MeOH	RT	Transfer embryos from a stock vial
75 % MeOH	RT	10 min
50 % MeOH	RT	10 min
25 % MeOH	RT	10 min
PBST buffer	RT	5 min
PBST buffer	RT	5 min

Proteinase K step

Proteinase K solution (2 μg/ml in PBST)	RT	5 min

NOTE: The length of this step depends on the age of the embryos. For example, 5 min incubation is sufficient for early- and middle-stage embryos, but should be increased to 10 min for late and just before hatching stages. Optimization of the time may be required.

PBST buffer	RT	5 min
PBST buffer	RT	5 min

Refixation step

G&P solution	RT	20 min
PBST buffer	RT	5 min
PBST buffer	RT	5 min

If using a 24-well plate, the refixed embryos can be sorted into the appropriate baskets for antisense or sense treatment at this stage. To do this, pick up a basket and move it quickly to a dish that is filled with PBST buffer. Have baskets in the hybridization plate ready (with PBST in them), and select embryos to be placed in each well. Transfer embryos using a P1000 pipetter, but use a tip with the end cut off to make for a wider opening. It is probably a good idea to draw a "map" of your wells and each well's content.

Prehybridization step

Lay a paper towel soaked in hybridization solution in the bottom of the plastic Tupperware to keep the inside of the container moist during prehybridization and hybridization steps (this "moisturing" buffer can be reused).

Hybridization solution	70 °C	5 min
Hybridization solution	70 °C	120 min

Hybridization step

Add the RNA probe (final 0.1 ng/μl) to hybridization solution and incubate overnight at 70 °C (NOTE: the optimal time is 16 h).

NOTE 1: if using a Watson preservation plate, add 1 mL hyb solution to a tube, and pop in one filter paper circle containing stored probe from the preservation plate. Vortex and incubate @ 70 °C for 30 min to extract the probe from the filter, then add the appropriate amount of the solution to each well.

NOTE 2: In our experience, 70 °C is a good starting point for hybridization, as this temperature has worked well for a majority of genes (i.e., produced very little to no background). Of course, if weak signal is seen, the temperature may need to be reduced to 65 or even 60 °C.

Day 2

NOTE 1: RNase-free conditions are no longer necessary.

NOTE 2: Ahead of time, prepare buffers that will be used (Solution 1, Wash 1, and Wash 2) and *pre-warm to a temperature that was used for the riboprobe hybridization (in this case to 70 °C).*

Riboprobe washes

Solution1	70 °C	20 min
Wash 1	70 °C	20 min
Wash 1	70 °C	20 min
Wash 1	70 °C	20 min
Wash 2	70 °C	20 min
Wash 2	70 °C	20 min
Wash 2	70 °C	20 min
KTBT	RT	10 min
KTBT	RT	10 min
KTBT	RT	10 min

Antibody reaction
Pro-block:

1.5 % blocking solution/KTBT	RT	60 min
(1:2,500) anti-FITC or DIG-AP, Fab fragments in 1.5 % blocking solution	RT	60 min

Antibody Washes

KTBT	RT	60 min
KTBT	RT	60 min
KTBT	RT	60 min
KTBT	RT	60 min
KTBT	RT	60 min
KTBT	RT	60 min OR leave at 4 °C O/N[a]

[a]At any of these steps, depending on one's preference, the embryos can be washed overnight in KTBT at 4 °C, or you can proceed directly to the next step

Day 3

Equilibration step

NTMT (pH 9.5)	RT	15 min
NTMT (pH 9.5)	RT	15 min
NTMT – AP buffer (pH 9.5)	RT	15 min
NTMT – AP buffer (pH 9.5)	RT	15 min
AP buffer (pH 9.5)	RT	15 min
AP buffer (pH 9.5)	RT	15 min

NOTE: The pH is very important here. Check and adjust the pH of these solutions before using them

AP staining reaction

NBT/BCIP	RT or 4 °C	Time to be determined

NOTE1: The staining reaction takes place in the dark; hence wrap the containers (Eppendorf tubes or 24-well plates) in aluminum foil if exposed to direct light. You may also place the tubes or plate in a drawer
NOTE2: The embryos should be examined at regular intervals to check for the intensity of signal and to avoid nonspecific background staining. The actual staining time will be gene specific (i.e., riboprobe specific) and may range from as little as 10 min to several hours. In general, it is best to overstain the embryos, as subsequent washings will tend to "wash out" stain. If samples are to be sectioned, then overstaining is highly recommended. If necessary, the color reaction can be paused by washing in NTMT, then in TBTX overnight, and then restarted the next morning

Stopping AP reaction

PBST	RT	5 min (even less time is OK)
PBST	RT	5 min

(continued)

PBST	RT	5 min
PBST	RT	5 min
PBST	RT	5 min

NOTE: If using individual baskets, transfer them to a fresh, 24-well plate with filtered PBST already aliquoted into each well. When transferring, blot each basket bottom first on a Kimwipe to drain the excess stain, and then move into the plate well with fresh PBST

Destaining

50 % EtOH/PBST	RT	#
100 % EtOH	RT	#
50 % EtOH/PBST	RT	#
PBST	RT	5 min × 3

EtOH stops the staining reaction
#In these steps, keep examining the embryos using a stereomicroscope to avoid destaining too much. Usually, each wash takes 2–3 min

Clearing

25 % glycerol/PBST	RT	Until the embryos sink
50 % glycerol/PBST	RT	Store at 4 °C

Store embryos in 50 % glycerol/PBST in a 24-well plate without baskets (transfer with pippetor).

Mounting

Use Aqua-Poly/Mount (Polysciences, Inc., USA) as a mounting media. Place few drops of Aqua-Poly/Mount on a slide, add a few embryos, and arrange them into the desired orientation. Note that the media will harden in about 10 min, so there is a time window for mounting. Let the slide sit for 20–30 min, then add one to two drops of Aqua-Poly/Mount to the top, and cover slip.

NOTE: Because this media is water-soluble, if necessary the embryos can be repositioned by placing the slide in a dish of water, waiting for 5 min, then using a razor blade to lift off the cover slip.

21.4.1.5 ISH Optimized for Neuronal and Brain Tissue

For CNS studies, the previously described embryonic protocol (Sects. 21.4.1.1, 21.4.1.2, 21.4.1.3, and 21.4.1.4) should be followed with only a few exceptions. Most importantly, the fixation and Proteinase K steps must be modified. There are also several important dissection details that will improve the success. Details are provided below.

Larval or Adult CNS Tissue Fixation

1. Anesthetize crickets on ice.

2. For prothoracic ganglion, immobilize as usual and remove ganglion and place into 4 % PFA/PBS for 2–4 h.
   For brain, cut off head with ophthalmic scissors and fix the head on a paraffin dish with pins, being careful not to damage the brain, which is located between antennae. Cut away the cuticle between the two antennae into a square with a small scalpel blade and ophthalmic scissors to reveal the brain (Fig. 21.10a). Add 20 μl of 4 % paraformaldehyde (PFA)/PBS into the brain and wait 2 min. Carefully remove the brain with forceps from the head immersed in PBS (Fig. 21.10b). Place brain in 4 % PFA/PBS. Fix all tissue at 4 °C.
3. After about 2–4 h, move brain/ganglia into a dish of PBS and carefully remove sheath. Using "biology" fine forceps, make small scratches in the sheath and then peel it away. Flip and repeat. Ideally this comes off in somewhat rigid pieces with no damage to the underlying tissue (Fig. 21.10c).
   (a) Alternative approach: Place vial of tissue fixed in 4 %PFA/PBS in waterbath at 65 °C and shake by hand for 5 min. Cool down vial with running water, then desheath tissue in PBS dish.
   (b) In either case, by the end of your desheathing, make sure that exterior tracheal branches are removed. Trachea that protrude from the tissue can take up and trap probe and cause background staining problems.

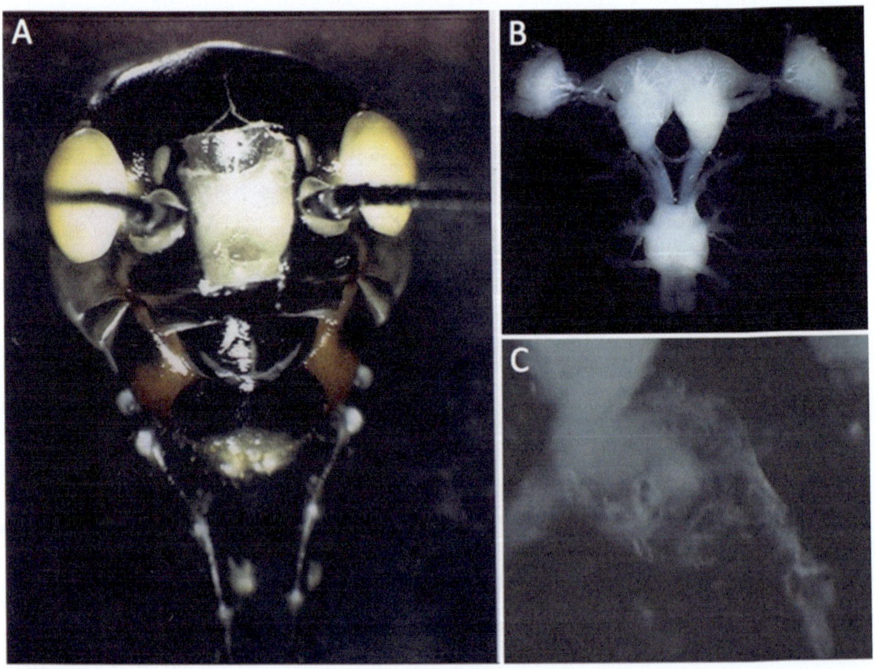

Fig. 21.10 Brain dissection and preparation for ISH. (**a**) Remove the cuticle between the antennae to expose the brain. (**b**) The dissected brain, optic lobes, and subesophageal ganglion. (**c**) After a brief fixation, remove the sheath from the brain

4. Return de-sheathed tissue to 4 % PFA/PBS. Keep the vial at 4 °C overnight to fix tissue.
5. Rinse tissue in 2–3 mL PBST five times.
6. Dehydrate the brain in the MeOH series: 25 % MeOH/PBST, 50 % MeOH/PBST, 75 % MeOH/PBST, and 100 % MeOH, each with shaking for 20 min.
7. Store tissue in 100 % MeOH at −20 °C until use.

Proteinase K Step Modifications

As already pointed out in Sect. 21.4.1.4, the neuronal tissue requires longer incubation times in Proteinase K as compared to embryos. Generally, 10 min for ganglia and 20 min for brains is appropriate. The exact time may need to be optimized. Note that because there is a 5 min time difference between the two tissue types, Proteinase K should be first added to the brain, and only after 5 min it should also be added to ganglia. This way, after 15 min, all the tissue is finished at the same time.

21.4.1.6 ISH Optimized for Nymphal Legs

The basic embryonic protocol can also be successfully applied to studies of gene expression in nymphal legs. The main modification includes removal of the cuticle during leg tissue fixation. In addition, the Proteinase K step is longer (20 min). This procedure was successfully and repeatedly used on stage 3 nymphs, but should also work on older nymphal stages as well.

Fixation of Nymphal Legs Undergoing Regeneration

This procedure encompasses three steps (leg amputation, cuticle removal, and the actual leg tissue fixation), which are described in more details below.

(a) *Leg amputation*

1. Choose a third instar cricket nymph, immediately after it has molted.
2. Anesthetize on ice for 20 min.
3. Amputate the distal region of the tibia using ophthalmological scissors (position *a* in Fig. 21.11a).

(b) *Cuticle removal*

1. Two days after amputation (this is 1 day prior to next molt; the third nymphal stage is only 3 days long), anesthetize the cricket by placing it on ice.
   Note: It is very difficult to remove cuticle at earlier times, because the epithelium is too soft. In contrast, while it is very easy to remove the cuticle

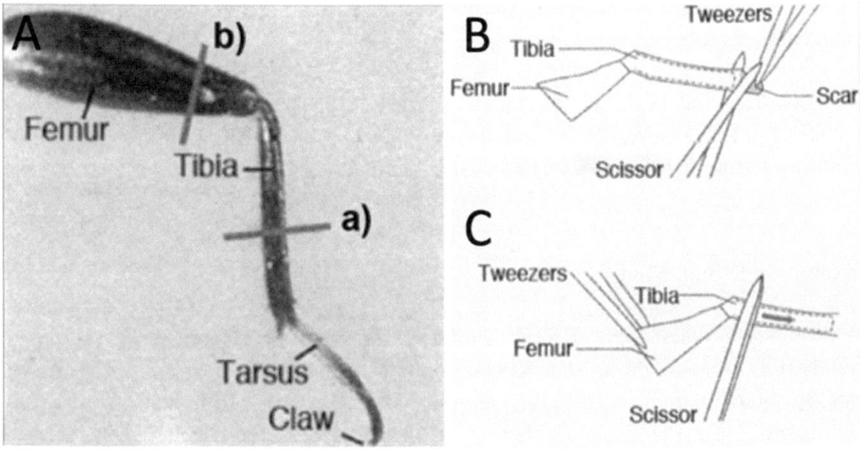

Fig. 21.11 Larval leg dissection for ISH. (**a**) The leg is first amputated at the distal tibia (position a) and 2 days later at the distal femur (position b). (**b**) The distal scar is removed so (**c**) the leg tissue can be extruded

at a later time, those samples cannot be used for ISH because the newly formed cuticle prevents the entry of the riboprobe into the cells.

2. Prepare 2 % PFA (paraformaldehyde)/PBT (0.1 % Tween 20 in PBS) and 4 % PFA/PBT.
3. Cut off the leg at distal position of femur (position *b* in Fig. 21.11a).
4. Heat-fix the cut legs in 2 % PFA/PBT for 6 min at 55 °C (use microwave).
5. Remove the scar in the amputated region by pinching with tweezers (see Fig. 21.11b).
6. Hold the femur with another pair of tweezers, then extrude the leg tissue from cuticle by gradually pushing leg with scissors (or forceps) from the proximal to distal region of the tibia (see Fig. 21.11c).

(c) *Fixation*

1. Following cuticle removal, fix regenerated legs again in 4 % PFA/PBT for 20 min at room temperature.
2. Discard PFA/PBT, then wash three times with PBT.
3. Discard PBT, then dehydrate the legs using 25 %, 50 %, and 75 % methanol/PBT and 100 % methanol sequentially.
4. Keep at −20 °C until ready to use.

Proteinase K Step

For nymphal leg tissue, this step is extended to 20 min. Note that this time is specific for the third nymphal stage. If using older stages, the Proteinase K step should be extended accordingly (in 2-min increments) until empirically determining the optimal incubation time.

21.4.1.7 ISH Optimized for Nymphal Wings

This protocol was originally developed to study wing expression patterns in the last nymphs (stage 8), the stage before they molt into adults. The same procedure can be applied to earlier nymphal stages; the critical issue is to empirically determine the duration of the Proteinase K step for each stage. Note that this protocol was first described and applied by Liu and colleagues (2016) in a study of the hemipteran insect, *Oncopeltus fasciatus*. As such, it is somewhat different in terms of reagents from the *Gryllus* embryonic protocol we described in the previous section. For this reason, a separate reagent list is provided specifically for this application (nymphal wing studies). Please note that all recipes are given at the end in Sect. 21.4.4.

Fixation of Wing Tissue

1. In DEPC treated PBT, dissect the winglets (wing pads) from eighth instar nymphs that are 4–6 days old. Transfer these wing pads to a dish with fixation buffer solution for at least 10 min.
   (a) The winglets must be at the perfect stage, i.e., the wing tissue must be thin but not too thin. Younger tissue will be too thin to be lifted off of a wing pad without falling apart. The "perfect" tissue will slide off easily. Note that if tissue is noticeably thicker and/or visible wrinkles are present on its surface, then wings are too old. The best way to learn the differences between "too young," "just right," and "too old" is to perform a set of staged winglet dissections that are set 1 day apart.
2. Dissect the wing tissue from cuticle, and place each sample in fixation buffer for 1 h.
3. Wash in PBT two times for 5 min each at RT.
4. Perform a set of methanol washes with 25 %, 50 %, and 80 % methanol/PBT (5 min each).
5. Wash again in methanol three times for 5 min each.
6. Store at −20 °C for at least 24 h before starting the ISH. Tissues can be stored indefinitely at −20 °C.

The following steps should be performed in an RNase-free environment:

Day 1
Rehydration and Proteinase K treatment

1. Wash tissues with 100 % ethanol for 5 min.
2. Gradual rehydration with 80 %, 50 %, and 25 % methanol/PBT. Each wash is for 5 min at RT.
3. Incubate the tissues in 10 μg/ml Proteinase K/PBT for 8–10 min. DO NOT SHAKE.
4. Rinse in 2 mg/ml glycine/PBT. Do two rinses for 5 min at RT.

5. Wash in PBT two times for 5 min.
6. Postfix in postfix buffer for 20 min.

Prehybridization and hybridization

1. Wash in PBT three times for 5 min each.
2. Wash in PBT: hyb buffer (1:1) for 5 min.
3. Wash in hyb buffer for 5 min.
4. Prehybridization for at least 1 h at 55 °C.
5. Boil DIG-labeled riboprobe (at least 100 ng) in 700 μL of hybridization buffer for 5 min.
6. Incubate riboprobe solution on ice for 2 min.
7. Add probe solution to the wing tissues and incubate for at least 14 h at 55 °C.

The following steps do not need to be in an RNase-free environment:

Day 2
Washing and antibody staining

1. Pre-warm wash buffer I–V in 1.5 ml Eppendorf tubes to 55 °C.
2. Wash in wash buffer I for 30 min at 55 °C.
3. Wash in wash buffer II for 30 min at 55 °C.
4. Wash in wash buffer III for 30 min at 55 °C.
5. Wash in wash buffer IV for 30 min at 55 °C.
6. Wash in wash buffer V two times for 40 min each at 55 °C.
7. Wash in MAw three times for 5 min at RT.
8. Incubate in blocking buffer for at least 1 h at RT.
9. Incubate with anti-DIG (1:2,000 in blocking buffer) at 4 °C overnight.

Day3
Color development

1. Wash in MAw buffer four times for 10 min each at RT.
2. Wash in staining buffer two times for 5 min each at RT.
3. Add 15 μl NBT-BCIP solution (Roche) into a tube with the staining buffer. Observe at regular intervals (every 5 min or so), checking for the first signs of staining. This step may range from 25 min to overnight and will be gene-dependent (riboprobe-dependent).
4. Wash in PBT four times at RT for 5 min each.
5. Fix tissues on glass slides using Aqua-Poly/Mount. Take images.

21.4.2 IAb in Whole-Mount Embryos (Embryonic IAb)

Immunohistochemistry (IAb) is a method for detecting the localization of specific proteins in the tissue of an organism. There are direct and indirect ways to detect a specific antigen with antibodies. In the direct method, primary antibodies labeled with fluorescent molecules or enzymes are used. In the indirect method, chemically labeled secondary antibodies, which specifically recognize the primary antibodies, are used. Here we describe only the latter approach because it provides several advantages, such as multiple signal detection methods and enhanced sensitivity. Traditional indirect signal amplification methods are (1) fluorescence, which uses fluorescently labeled secondary antibodies, and (2) enzymatic, which uses secondary antibodies that are linked to an enzyme molecule. Enzymes, such as horseradish peroxidase (HRP), are used to catalyze color-developing substrates for visualizing antigen localization. (3) Amplification, such as the tyramide signal amplification (TSA) method, which relies on tyramide conjugates incubated in the presence of secondary antibodies linked to HRP. In this case HRP catalyzes the binding of tyramide and an associated label to adjacent tyrosines, resulting in signal enhancement. In the protocol described below, the biotinylated tyramide is used and tyrosine-bound biotinylated tyramides are recognized by streptavidin-HRP to enhance the color-developing reaction.

Note that the success of immunostaining experiments primarily depends on the titers of the antibody used. It is essential to carefully choose an antibody and dilution appropriate for your experiments, if such information is available from previous studies. Recipes can be found in Sect. 21.4.3.

21.4.2.1 Embryo Fixation

1. Fix the samples in 4 % PFA 1 h at RT. Alternatively, this step can be performed at 4 °C O/N.
2. Wash in PBT (0.1 % Tween-20 in PBS) at RT, 10 min, three times.
   Fixed samples can be dehydrated in MeOH (as explained above) and then stored at −20 °C for prolonged periods of time.
   These samples can be used after rehydrating and washing in PBT (10 min, three times).
3. 1.5 % blocking/PBT RT for 1 h.
4. Add the primary antibody and incubate for 1 h at RT. Alternatively, this step can be performed at 4 °C O/N.
5. PBT wash, RT, 30 min, three times.

21.4.2.2 IAb with Secondary Antibodies

1. *Fluorescence* (*Cy3*, *Alexa*, *FITC*, etc.)
   1. Add Cy3, Alexa, FITC, etc., conjugated secondary antibody at RT, for 1 h or 4 °C, O/N.
   2. PBT wash, RT, 30 min, three times.
   3. *OPTIONAL*: counter staining with Hoechst or DAPI in PBT at RT, 10 min.

4. PBT wash, RT, 5 min.
5. Twenty-five percent glycerol/PBT, RT, 5 min.
6. Fifty percent glycerol/PBT, 4 °C, store and mount on slides as desired.

2. *Enzyme method* (*DAB*)

1. *HRP-conjugated* secondary antibody RT, 1 h or 4 °C, O/N.
2. PBT Wash, RT, 30 min, three times.
3. Prepare DAB solution
4. Staining reaction with DAB. When appropriate, stop the reaction with PBT.
5. *OPTIONAL*: counter staining with Hoechst or DAPI in PBT RT 10 min.
6. PBT Wash, RT, 5 min, one times
7. Twenty-five percent glycerol/PBT, RT, 5 min
8. Fifty percent glycerol/PBT, 4 °C; store and mount on slides as desired.

3. *Signal amplification* (*TSA system-DAB*)

1. *HRP-conjugated* secondary antibody RT, 1 h or 4 °C, O/N.
2. PBT wash, RT 30 min, three times.
3. Dilute biotinyl-tyramide with 1× Plus Amplification Diluent, RT, 20 min.
4. TNT wash, RT, 30 min, three times.
5. Dilute streptavidin-HRP/TNB buffer to 1/100, RT, 30 min.
6. TNT wash, RT, 30 min, three times.
7. Prepare DAB solution.
8. Staining reaction with DAB. Stop the reaction with PBT.
9. OPTIONAL: counter staining with Hoechst or DAPI in PBT at RT, 10 min.
10. PBT wash, RT, 5 min, one time.
11. Twenty-five percent glycerol/PBT, RT., 5 min.
12. Fifty percent glycerol/PBT, 4 °C; store and mount on slides as desired.

Notes:

(A) The use of fresh samples (tissues) is important. If possible, start IAb immediately after dissection and fixation.
(B) Set up the appropriate positive and negative controls. For a positive control, the use of anti-Engrailed antibody (4D9) is recommended because it results in distinct stripes in each segment of an embryo. For a negative control, perform a "no primary control," using only the secondary antibody.
(C) Primary and secondary antibodies can be reused up to three or four times.
(D) If reusing the antibodies, store them at 4 °C with an addition of NaN_3 (sodium azide to a final concentration of 0.1–0.02 %). However, avoid antibodies stored in NaN_3 if performing HRP staining, as this will inhibit HRP enzyme activity. Instead, use thimerosal to a final concentration of 0.02 %.
(E) In instances where multicolor detection of multiple antigens is required, all primary antibodies can be combined and added simultaneously. However, the detection is performed sequentially, using a specific secondary antibody for each primary antibody added one at a time.

(F) The concentration of an antibody dilution must be optimized. The optimal concentration may vary depending on the actual room temperature during incubation.
(G) Be careful in handling fluorescence-labeled antibodies because photobleaching can easily occur. After adding the secondary antibodies, samples should be kept in the dark (e.g., by wrapping tubes in aluminum foil).
(H) When handling DAB, make sure to wear gloves because it is a potential carcinogen. Used containers for DAB reaction should be washed by soaking in chlorine bleach.
(I) If using the signal amplification method, be careful when handling and storing biotinyl-tyramide. This chemical breaks down easily and should be stored in the dark at 4 °C. Alternatively, it can be dissolved in high-grade DMSO and stored at −20 °C.
(J) Thicker pieces of tissue, such as central nervous system tissue, may require much longer incubation times, such as 36–48 h at 4 °C.

21.4.3 *ISH Materials, Reagents, and Recipes*

ISH Materials		
Reagents		Solutions
Probe preparation	*Egg collection*	(see Sect. 21.4.3.1)
Primers (table)	Paper towels or sand	PBST
pGEM-T Easy		Hybridization solution
dNTPs	Petri dishes	KTBT
Taq and buffer	Mesh sieve	Blocking solution
PCI (phenol/chloroform/isoamyl alcohol; 25:24:1)	Tupperware	Staining solution
RNAse-free H_2O (ddH_2Owater)	Bleach	Solution 1
TE (10 mM Tris-HCl, 1 mM EDTA; pH8.0)	*In situ hybridization*	Wash 1 and 2
Megascript kit (Ambion)	PBS	KTBS
RNase inhibitory (Invitrogen)	PFA	NTMT
Sp6 RNA polymerase (TaKaRa)	Triton	NTMT-AP
10× transcription buffer	MeOH	AP Buffer
100 mM DTT	H_2O_2	
		Equipment
BSA	Proteinase K	PCR machine
T7 RNA polymerase (Invitrogen)	Gluteraldehyde	Microdissecting equipment
5× transcription buffer	RNA probe (see Sects. 21.4.1.2 and 21.4.1.3)	Dissecting microscope
0.1 M DTT	EtOH	0.22 μm Sterile Millex Filter Unit
10× Dig-labeling mix (Roche)	Glycerol	Orbital shaker

(continued)

ISH Materials		
Reagents		Equipment
DNase I (Roche)	Formamide	24-well plates with baskets
10 mg/ml yeast tRNA (Invitrogen)	SSC	Hybridization oven
Sodium carbonate (40 mM $NaHCO_3$, 60 mM Na_2CO_3; pH10.2)	CHAPS	
Glacial acetic acid	EDTA	
LiCl solution (Ambion)	Heparin	
100 % EtOH (RNAse free)	Yeast tRNA	
80 % EtOH (RNAse free)	RNAse-free H_2O	
Watson preservation plates	SDS	
	NaCl	
	KCl	
	Tris-HCl (pH 7.5 and 9.5)	
	Blocking reagent	
	$MgCl_2$	
	Polyvinyl alcohol	
	Dimethylformamide	
	NBT	
	BCIP	
	Aqua-Poly/Mount	

21.4.3.1 ISH Recipes

Solutions

PBST (pH7.4)

	For 1 L stock
a. PBS	1 L
b. Triton X-100	1 mL

- Stir overnight because Triton X-100 is viscous.
- May prepare and store stock.
- Filter with syringe filter before use. (Triton can form aggregates that damage tissue.)

Proteinase K

Dilute Proteinase K stock solution (20 mg/mL, kept frozen) with filtered PBST.

ProK(final conc. μg/mL):	ProK2	ProK5
a. ProK stock solution (20 mg/mL)	1 μL	1 μL
b. Filtered PBST	10 mL	5 mL

- Prepare for each use and keep on ice.

G&P (0.2 % glutaraldehyde/PBST, 4 %PFA/PBST)

	For 14 mL	For 28 mL
a. 50 % glutaraldehyde	56 μL	112 μL
b. 16 % PFA	3.5 mL	7 mL
c. Filtered PBST	10.4 mL	20.9 mL

- Prepare for each use and keep on ice.
- PFA should be less than 2 weeks old.
- Use 140 μL 20 % glutaraldehyde for 14 mL solution or 280 μL 20 % glutaraldehyde for 28 mL solution.
- Use 2.8 mL 20 % PFA for 14 mL solution or 5.6 mL 20 % PFA for 28 mL solution.

Hybridization solution

	For 10 mL	For 20 mL	For 50 mL
a. Blocking reagent	0.2 g	0.4 g	1.0 g
b. Formamide	5.0 mL	10 mL	25 mL
c. 20×SSC (pH 7.0)	2.5 mL	5.0 mL	12.5 mL
d. 10 % TritonX-100	0.1 mL	0.2 mL	0.5 mL
e. 10 % CHAPS	0.1 mL	0.2 mL	0.5 mL
f. 0.5 M EDTA	0.1 mL	0.2 mL	0.5 mL
g. Heparin (50 mg/mL)	10 μL	20 μL	50 μL
h. Yeast tRNA (10 mg/mL)	1.0 μL	2.0 μL	5.0 μL
i. ddH_2O	Up to 10 mL	Up to 20 mL	Up to 50 mL

- Dissolve blocking reagent in formamide in oven at 70 °C with vigorous shaking for 20–30 min. (A thick gel-like mixture results; solid does not fully disolve.) Then add remaining reagents. Dissolve and keep mixture at 70 °C until use.
- Do not vortex yeast tRNA solution.
- Roche blocking solution (11 096 176 001) comes in powder form.

Solution1 (pH 4.5)

	For 10 mL	For 20 mL
a. Formamide	5.0 mL	10 mL
b. 20× SSC (pH 4.5)	2.5 mL	5.0 mL
c. 10 % SDS	1.0 mL	2.0 mL
d. ddH_2O	1.5 mL	3.0 mL

- May prepare and store stock.

Wash1 (pH 7.0)

	For 500 mL
a. 20× SSC (pH 7.0)	50 mL
b. 10 % CHAPS	5.0 mL
c. ddH_2O	Up to 500 mL

- May prepare and store stock.

Wash2 (pH 7.0)

	For 500 mL
a. 20× SSC (pH 7.0)	5.0 mL
b. 10 % CHAPS	5.0 mL
c. ddH_2O	Up to 500 mL

- May prepare and store stock.

10X KTBS

	For 1 L
a. NaCl (1.5 M final concentration)	87.66 g
b. KCl (0.1 M final concentration)	7.46 g
c. 1 M Tris-HCl (pH 7.5, 0.5 final conc)	500 mL
d. ddH_2O	1000 mL

- May autoclave and store stock.

KTBT

	For 1 L
a. 10× KTBS	100 mL
b. TritonX-100	1.0 mL
c. ddH_2O	Up to 1000 mL

- Stir prepared solution overnight because Triton X-100 is viscous.
- May prepare and store stock.

1.5 % blocking solution/KTBT

	For 50 mL	For 200 mL
a. Blocking reagent	0.75 g	3.0 g
b. KTBT	50 mL	200 mL

- Dissolve at 70 °C. Store aliquots at −20 °C.
- Do not freeze-thaw multiple times. Fresh is best, but aliquots are okay.

1:2500 anti-DIG or FITC-AP, Fab fragments

	10 mL	15 mL	50 mL
a. 1.5 % blocking solution/KTBT	10 mL	15 mL	50 mL
b. Anti-DIG or FITC-AP, Fab fragments	4 μL	6 μL	20 μL

- Prepare just before use. Keep on ice after making.
- Use pre-absorbed antibody.

Preparing absorbed antibody

(a) After refixation, transfer cricket embryos with PBST buffer into a 1.5 mL tube.
(b) Store at 4 °C.
(c) Remove PBST buffer and add 1 mL 1.5 % blocking buffer.
(d) Add 4 μl antibody (in case of making total 10 ml solution) into the tube.
(e) Gently shake for 60 min on ice.
(f) Centrifuge for 20 min at 12,000 rpm in 4 °C.
(g) Use the supernatant to make the antibody solution.

NTMT (pH 9.5)

	For 10 mL	For 50 mL
a. 1 M Tris-HCl (pH 9.5)	1.0 mL	5.0 mL
b. 5 M NaCl	0.2 mL	1.0 mL
c. 10 % TritonX-100	0.1 mL	0.5 mL
d. 1 M $MgCl_2$	0.5 mL	2.5 mL
e. ddH_2O	up to 10 mL	50 mL

- Add 1 M $MgCl_2$ just before use to avoid pH change.

NTMT-AP
Mix NTMT and AP buffer in 1:1 ratio.

AP buffer

	For 10 mL	For 50 mL
a. Polyvinyl alcohol (PVA)	0.5 g	2.5 g
b. 1 M Tris-HCl (pH = 9.5)	1.0 mL	5.0 mL
c. 5 M NaCl	0.2 mL	1.0 mL
d. 10 % TritonX-100	0.1 mL	0.5 mL
e. 1 M $MgCl_2$	0.5 mL	2.5 mL
f. ddH_2O	8.2 mL	41 mL

- Dissolve PVA powder, Tris-HCl, and NaCl in ddH_2O in 90 °C waterbath for 30 min.
- Cool down before adding 10 % TritonX-100 to prevent formation of aggregates.
- Do not put on ice to avoid deposition of PVA.
- Add 1 M $MgCl_2$ just before use to avoid pH change.

NBT(4-Nitro blue tetrazolium chloride)
Prepare 75 mg/mL solution if using NBT powder:

	For 1.0 mL
a. NBT	0.075 g
b. 70 % dimethylformamide (in water)	1.0 mL

- Protect from light with foil.
- Prepare 70 % dimethylformamide (100 % dimethylformamide in ddH_2O) for each use.
- Vortex until completely dissolved. Store at −20 °C in the dark.

BCIP (X-phosphate/5-Bromo-4-chloro-3-indolyl phosphate)

	For 1.0 mL
a. BCIP	0.050 g
b. 100 % dimethylformamide	1.0 mL

- Protect from light with foil.
- Vortex until completely dissolved. Store at −20 °C in the dark.

Staining solution (NBT/BCIP)

	For 10 mL	For 20 mL	For 30 mL	For 40 mL	For 50 mL
AP buffer	10 mL	20 mL	30 mL	40 mL	50 mL
NBT (75 mg/mL)	45 μL	90 μL	135 μL	180 μL	225 μL
NBT (100 mg/mL)	34 μL	68 μL	101 μL	135 μL	169 μL
BCIP	30 μL	60 μL	90 μL	120 μL	150 μL

- Protect from light with foil.

Troubleshooting

- If no staining can be observed, try to design a probe against a different region, and repeat the ISH.
- If there is still no staining present, then try lowering the hybridization temperature to 65 °C or even lower.
- If only a faint signal is observed, double the amount of the probe. Alternatively, this may indicate that there may be a problem with riboprobe penetrance (in which case incubation time in Proteinase K should be increased).

21.4.4 Wing Tissue ISH Recipes

A few of these recipes are used in the IAb protocol (Sect. 21.4.2) as well.

10X PBS (10X = 0.1 M)
In 800 ml dH_2O, dissolve:

11.92 g	Na_2HPO_4
2.18 g	KH_2PO_4
80.00 g	NaCl
2.00	KCl

Adjust pH to 7.5 with 10 N NaOH, and then adjust to 1 L (total volume) with dH_2O.
Autoclave for 20 min.
Store at RT.

1X PBT (Treat with DEPC)
One hundred milliliter 10X PBS.
Three milliliter Tween 20.
Adjust to 1 L with dH_2O, add 1 ml DEPC, and stir overnight at RT.
Autoclave for 20 min.
Store at RT.

Calf Thymus DNA (0.01 g/ml)
Dissolve 0.4 g of DNA in 40 ml TE.
Heat to 65 °C for 20 min.
Shear DNA by pumping through an 18 gauge needle.
Store at −20 °C.

Fixation buffer:

1 ml	10X PBS
2.5 ml	37 % formaldehyde
1 ml	0.5 M EGTA (pH 7.5)
5.5 ml	DEPC H_2O
10 ml	Total volume
Store at 4 °C. Good for 1 week.	

Postfix buffer:

1.35 ml	37 % formaldehyde
8.65 ml	1X PBT (DEPC treated)
10 ml	Total
Make fresh each time.	

2 mg/ml Glycine/PBT
Dissolve 0.1 g glycine in 50 ml 1X PBT (DEPC treated).
Store at RT.

Hybridization buffer:

Content	Volume	Final conc.
Formamide (add last)	300 ml	50 %
20XSSC	150 ml	5X
Glycogen (dissolve first)	0.6 g	1 mg/ml
Tween 20	0.6 ml	0.01 %
Calf thymus DNA (0.01 g/ml)	6 ml	100 μg/ml
ddH_2O	143 ml	
Total	600 ml	

Keep at −20 °C

Wash buffer I:

25 ml	Formamide
12.5 ml	20XSSC
50 μl	Tween 20
Add water to 50 ml.	

Wash buffer II:

25 ml	Formamide
5 ml	20XSSC
50 μl	Tween 20
Add water to 50 ml.	

Wash buffer III:

12.5 ml	Formamide
5 ml	20XSSC
50 μl	Tween 20
Add water to 50 ml.	

Wash buffer IV:

5 ml	Formamide
50 μl	Tween 20
Add water to 50 ml.	

Wash buffer V:

500 μl	20X SSC
50 μl	Tween 20
Add water to 50 ml.	

MAB (Maleic Acid Buffer)

In 400 ml of dH_2O, dissolve:	
5.8 g	Maleic acid
4.38 g	NaCl
3.5 g	NaOH
Adjust pH to 7.5 with solid NaOH.	
Adjust volume to 500 ml with dH_2O	
Add 1 ml DEPC, stir overnight.	
Autoclave for 20 min.	
Store at RT.	

MAw (MAB ±0.1 % Tw)
Mix 50 ul Tween 20 into 50 ml MAB

Blocking buffer:

5 ml	10 % blocking reagent
500 ul	Normal goat serum (NGS)
20 ul	Triton X
Adjust to 10 ml using MAB	
Make fresh before use each time.	

Staining buffer:

1 ml	1 M Tris pH 9.5
200 ul	5 M NaCl
500 ul	1 M $MgCl_2$
10 ul	Tween 20
Add water to 10 ml	
Make fresh before use each time.	

References

Auer TO, Duroure K, De Cian A, Concordet JP, Del Bene F (2014) Highly efficient CRISPR/Cas9-mediated knock-in in zebrafish by homology-independent DNA repair. Genome Res (1):142–153

Awata H, Watanabe T, Hamanaka Y, Mito T, Noji S, Mizunami M (2015) Knockout crickets for the study of learning and memory: dopamine receptor Dop1 mediates aversive but not appetitive reinforcement in crickets. Sci Rep (5):15885

Cong L, Ran FA, Cox D, Lin S, Barretto R, Habib N, Hsu PD, Wu X, Jiang W, Marraffini LA, Zhang F (2013) Multiplex genome engineering using CRISPR/Cas systems. Science 3 (39):819–823

Donoughe S, Extavour C (2015) Embryonic development of the cricket *Gryllus bimaculatus*. Dev Biol (411):140–156

Hagmann M, Bruggmann R, Xue L, Georgiev O, Schaffner W, Rungger D, Spaniol P, Gerster T (1998) Homologous recombination and DNA-end joining reactions in zygotes and early embryos of zebrafish (*Danio rerio*) and *Drosophila melanogaster*. Biol Chem (379):673–681

Hwang WY, Fu Y, Reyon D, Maeder ML, Tsai SQ, Sander JD, Peterson RT, Yeh JR, Joung JK (2013) Efficient genome editing in zebrafish using a CRISPR-Cas system. Nat Biotechnol 31 (3):227–229

Jinek M, Chylinski K, Fonfara I, Hauer M, Doudna JA, Charpentier E (2012) A programmable dual-RNA-guided DNA endonuclease in adaptive bacterial immunity. Science 337:816–821

Lieber MR, Ma Y, Pannicke U, Schwarz K (2003) Mechanism and regulation of human non-homologous DNA end-joining. Nat Rev Mol Cell Biol 4:712–720

Liu J, Lemonds TR, Marden JH, Popadić A (2016) A pathway analysis of melanin patterning in a Hemimetabolous insect. Genetics 203:403–413

Miller JC, Tan S, Qiao G, Barlow KA, Wang J, Xia DF, Meng X, Paschon DE, Leung E, Hinkley SJ, Dulay GP, Hua KL, Ankoudinova I, Cost GJ, Urnov FD, Zhang HS, Holms MC, Zhang L, Gregory PD, Rebar EJ (2011) A TALE nuclease architecture for efficient genome editing. Nat Biotechnol 29:143–148

Mito T, Sarashina I, Zhang H, Iwahashi A, Okamoto H, Miyawaki K, Shinmyo Y, Ohuchi H, Noji S (2005) Non-canonical functions of *hunchback* in segment patterning of the intermediate germ cricket *Gryllus bimaculatus*. Development 32:2069–2079

Mojica FJ, Díez-Villaseñor C, García-Martínez J, Almendros C (2009) Short motif sequences determine the targets of the prokaryotic CRISPR defence system. Microbiology 155:733–740

Moscou MJ, Bogdanove AJ (2009) A simple cipher governs DNA recognition by TAL effectors. Science 326:1501

Nakamura T, Mito T, Bando T, Ohuchi H, Noji S (2008) Dissecting insect leg regeneration through RNA interference. Cell Mol Life Sci 65:64–72

Nakamura T, Yoshizaki M, Ogawa S, Okamoto H, Shinmyo Y, Bando T, Ohuchi H, Noji S, Mito T (2010) Imaging of transgenic cricket embryos reveals cell movements consistent with a syncytial patterning mechanism. Curr Biol 20:1641–1647

Niwa N, Inoue Y, Nozawa A, Saito M, Misumi Y, Ohuchi H, Yoshioka H, Noji S (2000) Correlation of diversity of leg morphology in *Gryllus bimaculatus* (cricket) with divergence in dpp expression pattern during leg development. Development 127:4373–4381

Porteus MH, Carroll D (2005) Gene targeting using zinc finger nucleases. Nat Biotechnol 23:967–973

Rémy S, Tesson L, Ménoret S, Usal C, Scharenberg AM, Anegon I (2010) Zinc-finger nucleases: a powerful tool for genetic engineering of animals. Transgenic Res 19:363–371

Sander JD, Maeder ML, Reyon D, Voytas DF, Joung JK, Dobbs D (2010) ZiFiT (Zinc Finger Targeter): an updated zinc finger engineering tool. Nucleic Acids Res 38(Web Server issue): W462–W468

Shinmyo Y, Mito T, Matsushita T, Sarashina I, Miyawaki K, Ohuchi H, Noji S (2004) *piggyBac*-mediated somatic transformation of the two-spotted cricket, *Gryllus bimaculatus*. Dev Growth Differ 46:343–349

Watanabe T, Ochiai H, Sakuma T, Horch HW, Hamaguchi N, Nakamura T, Bando T, Ohuchi H, Yamamoto T, Noji S, Mito T (2012) Non-transgenic genome modifications in a hemimetabolous insect using zinc-finger and TAL effector nucleases. Nat Commun 3:1017

Weiner A, Zauberman N, Minsky A (2009) Recombinational DNA repair in a cellular context: a search for the homology search. Nat Rev Microbiol 7:748–755

Xiao A, Cheng Z, Kong L, Zhu Z, Lin S, Gao G, Zhang B (2014) CasOT: a genome-wide Cas9/gRNA off-target searching tool. Bioinformatics [Epub ahead of print]

Index

H.W. Horch et al. (eds.), *The Cricket as a Model Organism*,
DOI 10.1007/978-4-431-56478-2

MIX
Papier aus verantwortungsvollen Quellen
Paper from responsible sources
FSC® C105338

If you have any concerns about our products,
you can contact us on
ProductSafety@springernature.com

In case Publisher is established outside the EU,
the EU authorized representative is:
Springer Nature Customer Service Center GmbH
Europaplatz 3, 69115 Heidelberg, Germany

Printed by Libri Plureos GmbH
in Hamburg, Germany